INDOOR
AIR *and*
HUMAN
HEALTH

Second Edition

INDOOR AIR and HUMAN HEALTH

Second Edition

Edited by
Richard B. Gammage
Barry A. Berven

LEWIS PUBLISHERS

Boca Raton   New York   London   Tokyo

Library of Congress Cataloging-in-Publication Data

Indoor air and human health / edited by Richard B. Gammage and Barry
   A. Berven. -- 2nd ed.
        p.   cm.
   Includes bibliographical references and index.
   ISBN 1-56670-144-9 (alk. paper)
   1. Indoor air pollution--Toxicology--Congresses.   I. Gammage,
Richard B.   II. Berven, B. A.
RA577.5.I53   1996
613'.5--dc20   95-43001
                                                                                      CIP

© 1996 by CRC Press, Inc.
Lewis Publishers is an imprint of CRC Press

No claim to original U.S. Government works
International Standard Book Number 1-56670-144-9
Library of Congress Card Number 95-43001
Printed in the United States of America 1 2 3 4 5 6 7 8 9 0
Printed on acid-free paper

Preface

The Oak Ridge National Laboratory (ORNL) Life Sciences Symposia series is conducted under the Associate Director for Life Sciences and Environmental Technologies. This series began in 1978 and provides a forum to discuss subjects of interest to the U.S. Department of Energy, the scientific community, and the public.

Following the oil embargo in 1973, there began a concerted effort by the U.S. government and its citizens to use energy resources more efficiently. Renewable resources, more fuel-efficient cars, and conservation were areas of increased attention. A by-product of conservation efforts was to make buildings more energy efficient which reduced air exchange within structures and subsequently increased the concentration of many indoor air contaminants including organics and organisms such as mold and bacteria. As a result of degrading indoor air quality in buildings around the U.S., public and scientific concern was raised to understand the consequences of potential health impacts to the occupants of these buildings. Claims of hypersensitivity, hypoallergenic, and even neurotoxicity and carcinogenic responses in people were identified as a price we paid for improving energy efficiency. Because of this conundrum, federal research sponsors requested that ORNL sponsor a Life Sciences Symposium in 1984 on Indoor Air and Human Health.

The intent of that symposium was to bring together experts who would present data on indoor pollutant levels and associated health effects in humans and animals from radon, microorganisms, passive cigarette smoke, combustion products, and organics. The symposium was a tremendous success in quality of presented material, discussions which occurred during the meeting, and the reception of the proceedings resulting from the symposium which went into three printings and was used as a college text at some universities.

Because of the continuing interest of the public, scientists, and government officials in the topic of indoor air quality, in 1994 we were asked to organize a follow-up symposium to review progress in this field over the last ten years. Thus, this book represents the proceedings of the Eleventh ORNL Life Sciences Symposium conducted during March 28-31, 1994, in Knoxville, TN, on Indoor Air and Human Health Revisited.

The specific focus of this symposium was to have some of the world's best experts present the current state-of-the-art research and offer their opinions concerning sensory and sensitivity effects, allergy and respiratory disease, neurotoxicity, and cancer health effects associated with exposure to various contaminants typical of the contemporary indoor air environment. The reader will find an impressive state of knowledge of indoor air conditions and responses by the occupants who live in these structures. The presenters also strove to identify much of the work yet to be performed to better understand causality and associated risks of people from exposure to low levels of indoor air pollutants. The difficulty in understanding the effects of a single pollutant at low levels in a population is significant. The difficulty in understanding the effects of multiple contaminants in different mixtures is monumental. We hope the reader finds information in this book useful and the issues discussed stimulating. We look forward to revisiting indoor air and human health again in ten years.

We gratefully acknowledge the efforts of many others who have contributed to the success of this symposium. They include: William Cain, Yale University; Hal Levin, Santa Cruz, CA; Jim Otten, Martin Marietta Energy Systems; David Reichle, Oak Ridge National Laboratory; Cecile Rose, National Jewish Center for Immunology and Respiratory Medicine at the University of Colorado Health Sciences Center; Susan Rose, U.S. Department of Energy; Jonathan Samet, Johns Hopkins University; Kevin Teichman, U.S. Environmental Protection Agency; and Hugh Tilson, Health Effects Research Laboratory at Research Triangle Park, NC. The generous support of the following sponsors is appreciated: U.S. Environmental Protection Agency, U.S. Department of Energy, and Martin Marietta Energy Systems Hazardous Waste Remedial Action Program.

We are especially grateful to Carolyn Householder and Debbie Dickerson for many hours invested in the success of the symposium and this publication. We also acknowledge the contributions of Jimmie Wade at ORNL, who contributed to the smooth functioning of the symposium. For the management of the symposium and the efficient arrangements made with the conference hotel, we thank Joy S. Lee of the ORNL Conference Office. We also thank the authors for their presentations and manuscripts that appear in this book.

Barry A. Berven
Richard B. Gammage

The Editors

Richard B. Gammage, Ph.D., is leader of the Measurement Systems Research Group in the Health Sciences Research Division at the Oak Ridge National Laboratory (ORNL). He and his research group have conducted studies of indoor air quality since 1980. He was co-chairman of the 7th ORNL Life Sciences symposium on Indoor Air and Human Health in 1984. The 11th ORNL Life Sciences symposium is a revisitation of this same subject a decade later. He is a founding member, past chair, and current member of the Indoor Environmental Quality Committee of the American Industrial Hygiene Committee. He has organized several international workshops and symposia on IAQ, lately

focusing on problems of human response to low-level, complex mixtures of volatile organic compounds. His research group is also engaged in developing *in situ* and field-screening monitoring devices and techniques for both chemical and radiological contaminants. His open literature papers, book chapters, and conference proceedings number over 200.

Dr. Gammage received his honors B.S. (1960) and Ph.D. (1964) in chemistry from Exeter University, England, and a D.Sc. for distinguished research in 1984.

Barry A. Berven, Ph.D., is Director of the Health Sciences Research Division at Oak Ridge National Laboratory. As director, he has overall responsibility for an integrated laboratory and field research program concerned with human health and safety impacts of energy technologies. This work includes both basic and applied studies and is augmented with information management and health risk assessments. He directs a research organization of over 450 staff involved in basic physics, chemistry, laser photonics, advanced instrumentation, biomedical technologies, radiation dosimetry, radiopharmaceutical development, environmental assessments, and health risk analysis.

Dr. Berven has published approximately 22 journal, abstract, or proceedings articles, published over 100 environmental reports, co-edited one book chapter, and co-edited one book.

Dr. Berven received his Ph.D. in Environmental Health Physics from Colorado State University, Fort Collins, Colorado. He has an M.S. in Radiology and Radiation Biology, and a B.S. in Biological Sciences from Colorado State University. He is a member of the Health Physics Society and Sigma Xi. He is also the Environmental Effects co-editor for Nuclear Safety.

Acknowledgment

The authors would especially like to acknowledge the speakers and session chairs for their contributions to this publication. It takes an unusual effort for these people to take time out of their very busy schedules to prepare for not only outstanding presentations, but also for publications as well. What is especially unusual about this symposium is that all presenters have completed their manuscripts and submitted them for publication. We believe this is a tangible demonstration of the interest and commitment of the presenters to indoor air issues.

Contributors

Michael H. Abraham
Chemistry Department
University College London
London, England

Juan C. Bartolomei
Department of Neurosurgery and
  Neurobiology
Yale University School of Medicine
New Haven, Connecticut

Rebecca Bascom
Environmental and Airway Disease
  Research Facility
University of Maryland School of
  Medicine
Baltimore, Maryland

Vernon A. Benignus
Health Effects Research Laboratory
U.S. Environmental Protection
  Agency
Research Triangle Park, North
  Carolina

Barry A. Berven
Health Sciences Research Division
Oak Ridge National Laboratory
Oak Ridge, Tennessee

Harriet A. Burge
Department of Environmental Health
Harvard School of Public Health
Boston, Massachusetts

William S. Cain
Chemosensory Perception Laboratory
Department of Surgery
  (Otolaryngology)
University of California, San Diego
La Jolla, California

Jacqueline Carter
Health Effects Research Laboratory
U.S. Environmental Protection
  Agency
Research Triangle Park, North
  Carolina

Lisa Cazares
Center for Environmental Medicine
  and Lung Biology
University of North Carolina at
  Chapel Hill
Chapel Hill, North Carolina

J. Enrique Cometto-Muñiz
Chemosensory Perception Laboratory
Department of Surgery
  (Otolaryngology)
University of California, San Diego
La Jolla, California

Lisa A. Dailey
Center for Environmental Medicine
  and Lung Biology
University of North Carolina at
  Chapel Hill
Chapel Hill, North Carolina

Robert B. Devlin
Health Effects Research Laboratory
U.S. Environmental Protection
  Agency
Research Triangle Park, North
  Carolina

Douglas W. Dockery
Department of Environmental Health
Harvard School of Public Health
Boston, Massachusetts

Robert S. Dyer
National Health and Environmental
  Effects Research Laboratory
U.S. Environmental Protection
  Agency
Research Triangle Park, North
  Carolina

Richard B. Gammage
Health and Safety Research Division
Oak Ridge National Laboratory
Oak Ridge, Tennessee

Charles A. Greer
Department of Neurosurgery and
  Neurobiology
Yale University School of Medicine
New Haven, Connecticut

Jana Kesavanathan
Environmental Health Engineering
The Johns Hopkins School of Hygiene
  and Public Health
Baltimore, Maryland

Susan C. Knasko
Monell Chemical Senses Center
Philadelphia, Pennsylvania

Gerd Kobal
Department of Pharmacology and
  Toxicology
University of Erlangen
Nuremburg, Germany

George P. Kubica
Tuberculosis and Airflow Dynamics
  Consultant
Dunwoody, Georgia

Donald T. Lysle
Department of Psychology
University of North Carolina at
  Chapel Hill
Chapel Hill, North Carolina

Donald K. Milton
Department of Environmental Health
Harvard School of Public Health
Boston, Massachusetts

M. Christopher Newland
Department of Psychology
Auburn University
Auburn, Alabama

David B. Peden
Center for Environmental Medicine
  and Lung Biology
University of North Carolina at
  Chapel Hill
Chapel Hill, North Carolina

Thomas A. E. Platts-Mills
Asthma and Allergic Diseases Center
University of Virginia Medical Center
Charlottesville, Virginia

Bruce S. Rabin
Clinical Immunopathology
University of Pittsburgh Medical
  Center
Pittsburgh, Pennsylvania

Joseph V. Rodricks
ENVIRON International Corporation
Arlington, Virginia

Cecile Rose
National Jewish Center for
  Immunology and Respiratory
  Medicine
Denver, Colorado

Jonathan M. Samet
The Johns Hopkins School of Hygiene
  and Public Health
Baltimore, Maryland

Woodrow A. Setzer
Health Effects Research Laboratory
U. S. Environmental Protection
  Agency
Research Triangle Park, North
  Carolina

Ken Sexton
Environmental and Occupational
  Health
School of Public Health
University of Minnesota
Minneapolis, Minnesota

Rashid A. Shaikh
Health Effects Institute
Cambridge, Massachusetts

Steven K. Shapiro
Department of Psychology
Auburn University
Auburn, Alabama

Andrew J. Streifel
Department of Environmental Health
  and Safety
School of Public Health
University of Minnesota
Minneapolis, Minnesota

David L. Swift
Environmental Health Engineering
The Johns Hopkins School of Hygiene
  and Public Health
Baltimore, MD

Hugh A. Tilson
Health Effects Research Laboratory
U.S. Environmental Protection Agency
Research Triangle Park, North Carolina

Table of Contents

Effects of Indoor Air Quality on Human Health: Setting Strategic Research Directions and Priorities*

Ken Sexton and Robert S. Dyer

ABSTRACT

Exposure to indoor air pollution is a potentially serious public health problem in a wide variety of nonindustrial settings, for example, residences, offices, schools, and vehicles. Research, along with assessment, management, and communication of risks, is an integral part of decision making about protecting public health indoors. The role of research is twofold: to improve the quantity and quality of available scientific and technical information and to improve our capabilities to interpret the public health significance of the information on hand. Setting strategic research directions and priorities is an important aspect of informed decision making that requires future-oriented choices about goals, purposes, and activities. One area that should be a high priority for indoor air research is the general topic of adverse health effects, with four groups of effects deserving particular attention: sensory and sensitivity; allergy and respiratory; neurotoxic; and cancer. Three cross-cutting research themes, relevant to achieving better understanding of indoor air-related health effects, are (1) the need for research to develop markers of health-related events, (2) the need for research to elucidate the health effects of indoor air mixtures, and (3) the need for multidisciplinary research approaches to investigate and resolve indoor air problems.

* This manuscript has been reviewed and approved for publication by the National Health and Environmental Effects Research Laboratory, U.S. Environmental Protection Agency. The views expressed are those of the authors and not of the Agency. Mention of trade names or commercial products does not constitute endorsement or recommendation for use.

INTRODUCTION

In the U.S., it has become increasingly apparent since the early 1970s that air quality inside nonindustrial settings, such as residences, office buildings, and public modes of transportation, can, under certain conditions, be harmful to people's health (NRC, 1981a,b, 1984, 1985, 1986; Spengler and Sexton, 1983; Samet and Spengler, 1991; Gold, 1992; Sexton et al., 1993). Breathing contaminated indoor air is now recognized as an important, and in some cases dominant, exposure pathway for many air pollutants, including environmental tobacco smoke, aeroallergens, radon, microorganisms, organic vapors, particles and fibers (NRC, 1981a,b, 1985, 1986, 1991, 1994). In response, efforts to understand (e.g., research, testing) and address (e.g., product labeling, ventilation standards, rules and regulations) indoor air problems are now routinely carried out by federal, state, and local governments (Sexton, 1986; Samet and Spengler, 1991; EPA, 1993a; *EPA Journal,* 1993).

In 1985, publication of *Indoor Air and Human Health* (Gammage and Kaye, 1985) marked one of the first attempts to have a knowledgeable, multidisciplinary group of scientists look at the health implications of indoor air pollution. That book, based on proceedings of the Seventh Oak Ridge National Laboratory (ORNL) Life Sciences Symposium, summarized available data on indoor air levels and associated health effects for five classes of pollutants: radon, microorganisms, passive tobacco smoke; combustion products; and organics. Ten years later, the proceedings of the Eleventh ORNL Life Sciences Symposium serve as the basis for this volume, titled *Indoor Air and Human Health, Second Edition* (Gammage and Berven, 1996). In contrast to the earlier book, which followed the traditional approach of discussing health effects related to particular classes of airborne agents, *Indoor Air and Human Health, Second Edition* is organized around specific health outcomes that are known or suspected to be linked with indoor air exposures: sensory and sensitivity effects; allergy and respiratory effects; neurotoxic effects; and cancer. This organizational structure directs attention to the kinds of adverse health consequences that are of most concern indoors, thereby encouraging us to examine indoor air quality from a broad-based, health-oriented perspective instead of the more conventional agent- or source-specific approaches.

Our purpose in this introductory chapter is to highlight the ongoing need for research in support of informed and credible policy decisions about safeguarding public health indoors and to emphasize the importance of a rational approach to setting strategic research directions and priorities. The discussion is divided into three major sections: (1) a summary of the scientific basis for concern about the healthfulness of indoor air quality; (2) a review of the role of research in public policy decisions about unhealthful indoor environments, including a proposed conceptual approach for analyzing the research-related dimensions of complex indoor air quality issues; and (3) a survey of the need for and approaches to establishing strategic directions and priorities for indoor air research, with special emphasis on research opportunities related to sensory and sensitivity effects, allergy and respiratory effects, neurotoxic effects, and cancer.

CONCERNS ABOUT THE HEALTHFULNESS
OF INDOOR AIR QUALITY

There are five major factors driving concerns about the healthfulness of indoor air quality: people's time-activity patterns and the contribution of indoor air pollution to total exposures; presence of sources indoors; measured indoor concentrations; health-related complaints associated with indoor environments; and formal assessments of indoor air-related health risks (Sexton, 1993). First, we now know that most people in the U.S. spend 90% or more of their time indoors, primarily at home and at work. Groups potentially more susceptible to the effects of air pollution, like infants, the infirm, and the elderly, are inside virtually all the time. Because most of us spend so much time inside, indoor air concentrations, even if they are uniformly lower than outdoor levels, make a significant contribution to our time-weighted, average exposures to most air pollutants. The evidence is compelling that realistic exposure estimates depend on a thorough understanding of pollutant concentrations in both indoor and outdoor settings, as well as an appreciation of the time that people spend there. (NRC, 1981a; Spengler and Sexton, 1983; Sexton and Ryan, 1988; Samet and Spengler, 1991; Sexton et al., 1993).

Second, modern indoor environments contain a complex array of potential sources of air pollution, including synthetic building materials, consumer products, and furnishings. Airborne emissions also occur because of the people, pets, and plants that inhabit these spaces. Moreover, efforts to lower energy costs by reducing ventilation rates have increased the likelihood that pollutants generated indoors will accumulate (NRC, 1981a; Spengler and Sexton, 1983; Samet and Spengler, 1991; Gold, 1992).

Third, monitoring studies inside buildings and vehicles have consistently found that concentrations of many air pollutants tend to be higher indoors than out. Indoor air has been shown to be a complex mixture of chemical, biological, and physical agents, only a small fraction of which has been characterized adequately. This complexity is illustrated by the fact that more than 4000 different compounds have been identified in tobacco smoke alone (NRC, 1981a; 1986, 1991; Spengler and Sexton, 1983; Samet and Spengler, 1991).

Fourth among the factors driving indoor air concerns is the fact that complaints about inadequate indoor air quality and associated discomfort and dysfunction are a burgeoning problem in our society. It is now commonplace to hear reports of illness outbreaks among building occupants, particularly office workers, with no secondary spread of the illness to others outside the building with whom affected individuals come into contact. The U.S. Environmental Protection Agency (EPA) classifies these reports into two general categories. Building-related illness refers to episodes when symptoms of diagnosable illness are identified and can be attributed directly to airborne contaminants in the building. In contrast, sick-building syndrome refers to situations in which the building occupants experience acute health and comfort effects that appear to be linked to time spent in the building, but no specific illness or cause can be identified (EPA, 1988; Mendell, 1993; Samet and Spengler, 1991).

Different, but potentially related matters are the so-called chemical sensitivity syndromes, which may be caused wholly, only partially, or not at all, by chemicals. Broadly defined, "multiple chemical sensitivity" (MCS) is postulated to be the development of responsiveness, including manifestation of often disabling symptoms, to extremely low concentrations of chemicals following sensitization. A controversial and emotional topic, the concept of MCS as a distinct entity caused by exposure to chemicals has been challenged by the medical and scientific communities, and there appears to be consensus that substantially more research is needed before MCS (and similar ill-defined, poorly understood syndromes, such as chronic fatigue syndrome, Persian Gulf War syndrome, chemical hypersensitivity syndrome) should be considered as a clinical diagnosis (e.g., Mitchell and Price, 1994). Nevertheless, many sufferers of MCS continue to believe that their condition is either caused or exacerbated by indoor air pollution (NRC, 1992).

And fifth, exposures to many indoor air pollutants are known or suspected to occur at levels sufficient to cause increased rates of discomfort, dysfunction, disease, or death. Scientific evidence suggests that indoor air exposures may be linked with higher incidence of respiratory disease, allergy, mucous membrane irritation, and lung cancer, as well as adverse effects on the nervous, cardiovascular, and reproductive systems. The scientific community has expressed its concern by consistently ranking indoor air pollution at or near the top of environmental health risks faced by the U.S. population (NRC, 1981a,b, 1986; Spengler and Sexton, 1983; McCann et al., 1988; EPA, 1990; Stolwijk, 1990; Samet and Spengler, 1991; Gold, 1992; Berglund et al., 1992).

RESEARCH IN SUPPORT OF INFORMED
AND CREDIBLE DECISIONS

The specter of potential public health risks from breathing contaminated indoor air presents elected officials and government regulators with a dilemma. They must decide whether the problem is serious enough to warrant intervention, and, if so, what preventive or remedial actions will be most cost effective and politically acceptable. In the course of answering these public policy questions, government decision makers must confront several pervasive and interrelated problems: the need to make hard choices about allocation of scarce environmental protection resources; an absence of explicit statutory direction and/or authority to address indoor air pollution problems; the need to balance government's responsibility for protecting the public health against its responsibility to protect individual privacy (e.g., inside private residences); public perceptions that indoor air pollution is less serious than more traditional sources of pollution, like toxic waste dumps and industrial smoke stacks; and scarcity of scientific evidence about past, present, and future health risks from indoor air exposures. The following discussion examines this last issue, focusing on the need for research to improve the scientific basis for informed and credible decisions.

The Role of Research in Decision Making

Decision making about indoor air quality can be thought of as involving four distinct elements (see Figure 1): risk assessment to determine the actual human health risks from indoor exposures; risk management to decide whether risks are unacceptable and, if so, what should be done about them; risk communication to explain the scientific and policy justification for risk assessment and risk management decisions; and research to provide the scientific knowledge and understanding necessary to answer important questions generated in the course of assessing, managing, and communicating risks.

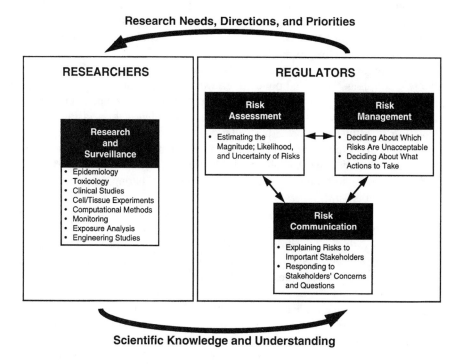

Figure 1   The feedback loop between researchers and regulators.

Risk assessment is primarily concerned with using science and, where necessary, science policy judgments to estimate, either qualitatively or quantitatively, the magnitude, likelihood, and uncertainty of human health risks that are caused by indoor air pollution. Examples of assessment questions requiring scientific input include: What are typical and high-end indoor exposures, and how do these exposures relate to indoor and outdoor sources? What are the most serious health risks associated with indoor air pollution? What are the cumulative risks of exposures to complex mixtures of airborne contaminants in indoor settings? How do the risks of exposures to indoor air pollution compare to risks related to other environments (e.g., outdoor air, industrial workplace air) and to other environmental media (e.g., food, beverages, soil, dust, surfaces, consumer products)?

Risk management uses science-based risk assessment, in combination with consideration of relevant economic, social, legal, political, and engineering issues, to answer questions about whether risks from unhealthful indoor air quality are serious enough to require intervention and, if so, what forms of intervention will be most effective and efficient in preventing or reducing those risks deemed to be unacceptable. This requires answers to a variety of pertinent questions, such as: What are the relative costs and benefits of government intervention to protect public health in indoor nonindustrial environments? What are the tradeoffs between protecting public health and protecting individuals' right to privacy? How can indoor pollutant sources be eliminated or modified to address indoor air quality problems? What are the most cost-effective ways to design, construct, operate, and maintain buildings to optimize indoor air quality and energy efficiency? When steps are taken to prevent or reduce health risks from indoor air pollution, how effective, expensive, and durable are they?

Risk communication engages stakeholders and regulators in a dialog about the scientific (e.g., facts) and policy (e.g., values) justification for risk estimates and for risk-related actions or inactions. Effective risk communication requires credible answers to questions like: How much confidence do decision makers have in risk estimates? Does a particular risk-related action mean that affected individuals and groups are subsequently "safe" from the adverse health effects of indoor air pollution? Are observed health effects (e.g., cancer cases, birth defects) or reported health complaints (e.g., headaches, burning eyes) caused by indoor air pollution?

Because of the many scientific and technical questions that are raised by the need to assess, manage, and communicate risks, research is an integral part of the regulatory decision making process. The role of research is twofold: to improve the quantity and quality of available scientific and technical information and to improve our capabilities to interpret the public health significance of the information on hand. Research (and surveillance) in support of informed decision making should form a feedback loop with the information needs of policy makers and regulators. The feedback loop requires that information, including both facts and values, flow in two directions. First, the information needs identified as part of the assessment, management, and communication of health risks should drive the direction and priorities of supporting research. Second, the knowledge and understanding generated by research activities should directly improve the scientific basis for decisions (Sexton, 1995).

To be successful, the feedback loop portrayed in Figure 1 requires timely and effective communication between regulators and researchers. Still, the complicated, multidisciplinary nature of indoor air quality issues is often a serious impediment to effective interactions and information transfer between researchers and regulators, as well as between researchers with different backgrounds and training. It is helpful, therefore, to simplify complex indoor air quality issues by breaking them into smaller, more manageable pieces that can be used to describe and elucidate a wide range of indoor air problems.

Conceptual Analytical Approach

A conceptual approach for analyzing the major research-related dimensions of indoor air quality is illustrated in Table 1 (Sexton and Teichman, 1993). It disaggregates indoor air quality issues into eight common domains (i.e., indoor environments, sources, agents, exposures, health effects, people, complicating factors, solutions), which together provide a useful method for describing the important characteristics of a particular indoor air issue or for identifying the crucial differences that distinguish one indoor air issue from another. This approach also provides a mechanism for promoting constructive dialog among regulators, researchers, and stakeholders by encouraging consensus about key concepts, principles, and terms.

Table 1   Conceptual Domains for Analyzing Indoor Air Quality Issues

Indoor Environments	Sources	Agents	Solutions
• Residential	• Occupants	• Criteria pollutants	• Source removal
• Occupational	• Outdoor Air	• Pesticides	• Source modification/ substitution
- industrial	• Soil	• Organic vapors	• Structural barriers
- nonindustrial	• Water	• Polycyclic organic matter	• Pressure control
	• Combustion appliances	• Metals	
• Commercial	• Tobacco products	• Fibers	• Ventilation
• Public	• Consumer products	• Aeroallergens	• Air cleaning
• Transportation	• Building materials	• Microorganisms	• Time and use adjustment
	• Maintenance products	• Radiation	• Encapsulation
	• Furnishings	• Mixtures	• Education
	• HVAC systems		

Health effects	Exposures	People	Complicating factors
• Cancer	• Magnitude	• General population	• Odor
• Noncancer	• Duration	• At-risk subgroups	• Thermal comfort
- illness/injury	• Frequency	• Highly exposed individuals	• Lighting
- signs	• Route		• Noise
- symptoms	• Distribution		• Ergonomics
	• Time Frame		• Stress
	• Geography		• Psychosocial factors

In addition to the five traditional domains (i.e., sources, agents, exposures, health effects, solutions) that one would expect to use when describing an indoor air issue, the conceptual approach also includes three additional domains that have important ramifications for research and policy decisions:

1. Indoor environments — within which type(s) of indoor space (e.g., residence, office, convention hall, restaurant) are problems occurring?

2. People — who is the focus of concern, the general population (e.g., all U.S. residents), potentially at-risk subgroups (e.g., mobile home residents, asthmatics, smokers, infants), or the highest exposed persons in the population?

3. Complicating factors — to what extent do variables like odor, noise, temperature, and stress need to be taken into account?

This greatly simplified analytical approach serves to remind us that indoor air quality is a multidimensional issue requiring multidisciplinary approaches to identify, evaluate, and resolve health-related problems. There are many different types of indoor environments, containing a diverse array of potential sources that are capable of contributing to increased levels of potentially harmful biological, chemical, and physical agents. Indoor exposures to these agents can occur via a plethora of pathways and may vary substantially over time and space. Establishing causal links between these exposures and related adverse health effects, in the general population and in potentially at-risk groups, is complicated by the confounding effects of certain indoor factors, such as odors, temperature, lighting, noise, ergonomics, stress, and psychosocial variables. Although in most cases it is difficult, if not impossible, to establish unequivocally whether indoor exposures cause observed health outcomes, there is solid evidence that exposures to many indoor toxicants can, under the right circumstances, contribute to increased rates of dysfunction (e.g., headache), disease (e.g., Legionnaires disease), and death (e.g., mortality from lung cancer). Ultimately, there is a relatively wide range of possible mitigation options available to address indoor air quality problems, however, selection of appropriate solutions depends directly on an appreciation and understanding of the other seven domains.

One valuable attribute of this analytical approach is that it allows us to look across several different indoor air quality issues, comparing and contrasting them on the basis of eight common domains. This approach has been used previously (Teichman, 1994) to analyze similarities and differences for six indoor air issues: airline cabin smoking regulations; environmental tobacco smoke in the workplace; residential radon; nonresidential problem buildings, the American Society of Heating, Refrigerating, and Air-conditioning Engineers (ASHRAE) Standard 62-1989 (ventilation rate procedure); and the Commission of European Communities (CEC) ventilation guidelines. The results, which are summarized in Table 2, demonstrate that this is a simple, effective, easy-to-use method for comparing indoor air quality issues that are both diverse and complex.

SETTING STRATEGIC RESEARCH DIRECTIONS AND PRIORITIES

Even if we use the approach described above to analyze indoor air issues, we are still left with the problem of what research will be most effective in improving the scientific basis for informed decision making. All too often, the state-of-the-science is insufficient to answer the most important risk-related questions with a reasonable degree of certainty. This invariably leads to controversy over "whether regulators have the facts right," and raises concerns that, in the absence of a firm

Table 2 Comparing Indoor Air Quality Issues on the Basis of Eight Common Domains

Indoor air domains	Airline cabins	Environmental tobacco smoke (workplace)	Indoor air quality issues			
			Radon	Problem buildings	ASHRAE 62-1989	CEC ventilation guidelines
Environments	Transportation	Occupational Institutional Commercial Public	Residential	Occupational Institutional Commercial Public	All	Residential Occupational Institutional Commercial Public
Sources	Tobacco products	Tobacco products	Soil Water	Uncertain, but potentially many different types	Occupants Outdoor air	Occupants Outdoor air Tobacco products Building materials HVAC systems
Agents	CO Particles Organic vapors POM mixtures	CO Particles Organic vapors POM mixtures	Radiation	Uncertain, but potentially many different types	Criteria pollutants	Potentially all
Exposures	Concentration Duration	Concentration Duration	Concentration Duration Geography	All aspects of exposure	Concentration Duration	All aspects of exposure
Effects	Cancer Noncancer -illness -signs -symptoms	Cancer Noncancer -illness -signs -symptoms	Cancer	Noncancer -illness -signs -symptoms	Noncancer -illness -signs -symptoms	Cancer Noncancer -illness -signs -symptoms
People	General population	General population	General population At-risk subgroups	General population At-risk subgroups	General population	General population
Complicating factors	Odor	Odor	Energy	Uncertain, but potentially many different types	Odor	Odor
Solutions	Source removal	Source removal Ventilation Air cleaning	Ventilation Structural barriers Pressure control	Uncertain, but potentially many different types	Ventilation Air cleaning	Ventilation Air cleaning

Adapted from Teichman, K.Y., *Indoor Air*, 4, 202-211, 1994.

factual basis, decisions may be more easily driven by political agendas, media pressures, special interests, and bureaucratic inertia (Sexton, 1995). An obvious solution is to invest adequate resources in targeted research to reduce the most critical scientific uncertainties that currently limit our ability to assess, manage, and communicate risks. Budget limitations force us to confront questions about which research directions and priorities will make the best use of available resources. One way to set research priorities is through a formalized process called strategic planning.

Strategic Planning

Strategic planning is future-oriented decision making about organizational goals, purposes, and activities, and it necessarily involves choices about what will and will not be done to achieve objectives within resource constraints. Simply stated, strategic planning is the formal expression of a clear, concise vision of what an organization is working toward and an explicit plan for how it intends to get there. Strategic planning is used routinely by public and private organizations as a management tool for deciding where they want to be in 5 to 10 years and for choosing the means by which they intend to get there. Linking the strategic planning process to the budget formulation process ensures that strategic decisions will drive the distribution of human and financial resources, thereby making strategic objectives an integral part of everyday management decisions.

Ideally, strategic planning should be an important factor in funding decisions about indoor air research, particularly in federal agencies, such as the Department of Energy and the EPA, that are a primary source of indoor air-related research funds. Strategic planning is a way to institutionalize the idea that the primary goal of indoor air research is to improve scientific knowledge and understanding in ways that will contribute directly to more informed, more credible, and more cost-effective decisions about safeguarding public health. The planning process itself helps to highlight the fact that attaining this goal requires a clear understanding of the scientific evidence that is driving concerns about health-related problems associated with indoor environments; the critical questions confronting decision makers and the role of science in answering them; the complex, multidimensional nature of indoor air quality issues; and the suitability of alternative research approaches, methods, and techniques for producing the necessary scientific data in a timely and cost-effective manner.

For funding agencies, the strategic planning challenge is to build on this understanding to determine directions and set priorities for a multidisciplinary research program that will, within resource constraints and subject to political and regulatory realities, improve the scientific and technical basis for the identification, assessment, management, and communication of health risks.

EPA's Strategic Plan for Indoor Air Research

An example of a strategic plan is provided by EPA's Indoor Air Research Plan for fiscal years 1993 to 1997 (Sexton and Teichman, 1993). The EPA, Office of Research and Development, identified three major goals for its indoor air research program (fiscal year 1993 budget approximately $9 million): (1) improving scientific understanding of the key determinants that underlie indoor air pollution health risks and of the effectiveness of risk reduction strategies; (2) providing critical scientific information to EPA program offices (e.g., Office of Air and Radiation, Office of Pesticides, Pollution Prevention, and Toxic Substances) and to EPA regional offices in support of developing, implementing, and evaluating risk management options (e.g., prudent exposure avoidance, public information programs); and (3) promoting private sector involvement in identifying, understanding, and addressing important indoor air pollution problems (e.g., industry testing based on standardized methods and guidelines, voluntary consensus standards). In addition to these three goals, the EPA also identified other key factors to be considered in setting research priorities, including: potential importance for protecting public health (i.e., relative risk); potential for the EPA research program to make a substantial contribution to improving scientific understanding; probability of success (i.e., scientific tractability); potential to complement related research efforts in EPA, in other federal and state agencies, and in the private sector; relevance to EPA program offices and regions; and opportunities to leverage resources from other agencies and the private sector.

In accordance with stated goals, and taking account of other key factors, as noted above, the plan states that

"...the primary focus and direction of EPA's indoor air research program over the next 5 years will be to understand the relative contributions of organic vapors, particulate matter, and microorganisms to observed symptoms (e.g., headache, eye irritation) and signs (e.g., antibody formation, skin rash) associated with air pollution exposures inside residences and office buildings. A major part of this effort will be devoted to understanding the role of confounding factors (e.g., odor, thermal comfort, noise, stress) in causing the observed responses. If an association is found between indoor air pollution (either individual agents or mixtures) and signs and symptoms, further research will be conducted both to examine the possibility that repeated or prolonged exposures may lead to damage or disease and to develop risk management strategies to reduce these exposures."

To implement the plan and make progress toward achieving its strategic goals, the EPA has divided its indoor air research program into five, multidisciplinary research areas:

1. Source characterization — What is the nature of emitted material, and how does it contribute to human exposures?
2. Exposure assessment — How many people are exposed to what levels of pollution?
3. Health effects assessment — What is the quantitative relationship between exposure/dose and adverse health outcomes in people?
4. Risk assessment — What is the likelihood, magnitude, and uncertainty of health risks associated with indoor air pollution?
5. Solutions — What are the most cost-effective strategies to prevent or control indoor pollution?

The strategic plan includes a section on each research area that summarizes purpose and goals, scientific approach, research projects and time lines, and anticipated results.

Research Approaches

Within the strategic planning process there is a need to make decisions about which research approaches are likely to be most effective and efficient in attaining the desired objectives. This question can be thought of as having two parts: (1) what is the concern about single agents (e.g., benzene, chlorpyrifos, asbestos, dust mites, noise) or mixtures of agents (e.g., environmental tobacco smoke, volatile organic compounds, coincidental mixtures that occur at the time and place of environmental sampling); and (2) in which of the EPA's five research areas should efforts be concentrated (i.e., sources, exposure, effects, assessment, solutions). Obviously, answers to these and related questions depend on a variety of parameters, such as the needs of decision makers, the state-of-the-science, and the level of information already available.

Traditionally, indoor air pollution research has been rather narrowly focused on individual chemicals (e.g., lead, asbestos, nitrogen dioxide), or sometimes other agents (e.g., Legionella), on single sources (e.g., unvented combustion appliances, carpets, tobacco smoking) or source categories (e.g., building materials, consumer products), and on specific types of indoor environments (e.g., residences, office buildings). Moreover, much of the research emphasis has been on source characterization and solutions, with relatively fewer resources devoted to assessment of exposure, health effects, and risk.

While there is ample evidence that previous research efforts have been successful in helping us to better understand and address such traditional indoor air problems as formaldehyde, carbon monoxide, nitrogen dioxide, asbestos, radon, lead, and environmental tobacco smoke, it is not clear that continuing with this agent-by-agent, source-by-source approach will remain equally productive in the future. As we turn our attention to other, less straightforward indoor air issues, like health-related complaints of building occupants, cumulative health effects of exposures to multiple agents, health risks to susceptible groups and individuals, and comparative health risks of indoor vs. outdoor exposures, it is becoming obvious that a more balanced research approach is needed. One aspect of a more balanced research portfolio is an increased emphasis on research to better understand the nature and extent of health outcomes that are known or suspected to be linked with indoor exposures.

Health Effects of Indoor Air Pollution

Exposures to airborne toxicants indoors can potentially lead to a variety of adverse health outcomes, depending on variables like inherent toxicity of the agent or agents in question, exposure magnitude, duration, frequency and route, and intra-individual and interindividual variability in biological susceptibility to the effects of exposures. Generally speaking, indoor air pollutants have the potential, given the right circumstances, to cause a wide diversity of negative health effects, including dysfunction, disability, disease, and even death, in at-risk (e.g., more exposed, more susceptible) groups and individuals (Samet and Spengler, 1991; Berglund et al., 1992). Two general schemes for evaluating the severity of health-related impacts from exposures are illustrated in Table 3. The first part of Table 3 is an example of a generic approach for classifying health effects into three broad categories by severity: catastrophic; serious; and adverse (EPA, 1993b). The second part of Table 3 provides an example of a scheme, proposed by Berglund and her colleagues (1992), for evaluating the severity of health-related impacts from indoor air pollution for a particular community. Although conceptually straightforward, in reality it is difficult, if not impossible, to apply either approach rigorously and unambiguously to specific indoor agents or to particular communities because of a scarcity of information about exposures, doses, and related health effects.

Table 3   Examples of Conceptual Schemes for Classifying the Severity of Health Effects

Classifying Health Effects

Catastrophic	Serious	Adverse
Death	Organ dysfunction	Loss in body weight
Shortened life span	Nervous system dysfunction	Hyperplasia
Severe disability	Developmental dysfunction	Hypertrophy/atrophy
Mental retardation	Behavior dysfunction	Enzyme changes
Hereditary disorder	Behavioral	Reversible organ dysfunction

From EPA, 1993b.

Assessing Priority of Indoor Air Health Risks for Research and/or Remedial Action

Size of affected population	Severity of effect	
	High[a]	Low[b]
Large[c]	Highest Priority	Medium Priority
Small[d]	Medium Priority	Lowest Priority

[a] Examples include cancer and serious noncancer diseases.
[b] Examples include mild disease, discomfort, annoyance, and reduced productivity.
[c] More than 10% of the population affected.
[d] Less than 10% of the population affected.
Adapted from Berglund et al., 1992.

One way to think about the diversity of possible human health effects from exposures to indoor air is to divide effects into four broad categories (Figure 2): cancer (e.g., of the lung or stomach, leukemia, Hodgkin's disease); noncancer illness or injury (e.g., clinically well-defined disease endpoints that can be either reversible or nonreversible); noncancer-related signs (e.g., objective, measurable evidence of exposure-related effects, without a well-defined etiology, that are usually reversible), and noncancer-related symptoms (e.g., subjective, difficult to measure and verify exposure-related effects, without a well-defined etiology, that are generally reversible). Under the right conditions, indoor air pollutants, singly or in combination, have the potential to cause any or all of these four types of health effects.

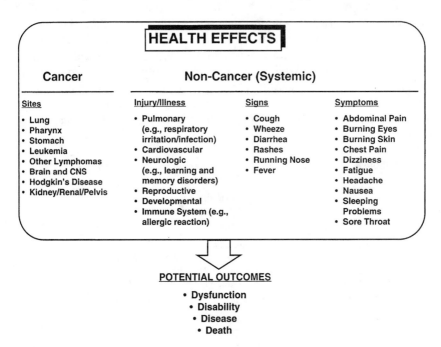

Figure 2   Potential health effects and related outcomes from exposures to indoor air pollution.

Berglund et al. (1992) have identified eight groups of effects that they consider to be especially important in relation to indoor air pollution (see Table 4): effects on the respiratory system; allergy and other effects on the immune system; effects on reproduction; effects on skin and mucous membranes; sensory effects and other effects on the nervous system; effects on the cardiovascular system; systemic effects on the liver, kidney and gastrointestinal system; and cancer. They point out that relationships between certain groups of effects and exposures to indoor air pollution are well established; citing as examples, respiratory disease in children, allergy to house dust mites, and mucous membrane irritation to formaldehyde. Moreover, they remind us that many indoor air pollutants are known or suspected to cause sensory irritation and stimulation, and may, therefore, give rise to some or all of the symptoms associated with sick building syndrome and building-related illness. In this regard, they emphasize the potential for complex mixtures of indoor air pollutants, such as occur routinely in most indoor settings, to invoke subtle effects on the central and peripheral nervous system and potentially lead to changes in behavior and performance.

Table 4   Groups of Health Effects that are Thought to be Important in Relation to Indoor Air Pollution

- Effects on the respiratory system
- Allergy and other effects on the immune system
- Cancer
- Effects on reproduction
- Effects on skin and mucous membranes
- Sensory effects and other effects on the nervous system
- Effects of the cardiovascular system
- Systemic effects on the liver, kidney, and gastrointestinal system

Adapted from Berglund et al., *Indoor Air*, 2, 2-25, 1992.

The well-documented association between an increased risk of developing lung cancer and exposure to radon and environmental tobacco smoke is also cited by Berglund and her colleagues, although they believe that the number of people affected is much lower than the number of people contracting respiratory disease or allergies, or experiencing irritative effects due to indoor air pollution. The effects of indoor air pollution on reproduction, cardiovascular disease, and on other systems and organs are described by the authors as having not been well documented, at least in part, because of a scarcity of relevant research.

Indoor Air and Human Health Revisited

Four groups of health effects, considered to be important in relation to indoor air pollution, were chosen as the focus for the Eleventh ORNL Life Sciences Symposium, titled "Indoor Air and Human Health Revisited": sensory and sensitivity effects; allergy and respiratory effects; neurotoxic effects; and cancer. The goal was to summarize the state-of-the-science for each topic area and to identify research issues, needs, and opportunities.

The emphasis within the broad category of sensory and sensitivity effects was on exploring our current understanding of olfaction and irritation in response to chemicals and chemical mixtures. Among the topics considered were the neurobiology of smell, physicochemical determinants and functional properties of the senses of irritation and smell, theoretical and practical models to predict the potency of irritants and odorants, electrophysiological indices of human chemosensory functioning, and behavioral effects of odors on people. The fundamental research question posed was "do odors and irritants cause adverse human health effects?" Although there has been considerable improvement in our knowledge and understanding about these and related issues, we still confront difficult questions about how to solve current indoor air quality problems and how to prevent future problems. One of the keys to making further progress is the development of standardized research protocols.

The issues of allergy and respiratory effects were examined from the perspective of four different kinds of causative agents; infectious agents (e.g., mycobacteria); upper airway irritants (e.g., ozone); bioaerosols (e.g., fungi, endotoxins); and allergens (e.g., dust mites). The fundamental research question posed was: What are the critical factors in the complex interplay between the nature of the airborne agent(s), host factors, and the circumstances of exposure, which cause allergy and respiratory effects? Because respiratory infection, irritation, and hypersensitivity diseases are likely to be more widespread and more complex than is currently recognized, a concentrated, multidisciplinary effort will be required to resolve important research questions.

The fundamental research questions posed in regard to neurotoxic effects from exposures to indoor air pollutants were what do we measure and how do we measure it. While it is well established that people can be exposed indoors to known neurotoxicants, such as volatile organic solvents, carbon monoxide, and pesticides, there is a dearth of information to document associated neurotoxic effects, or the lack thereof. This is in spite of the fact that suitable measurement methods and techniques are available to investigate many facets of this issue. Discussion focused on available techniques to measure distractibility and how such techniques might be used to study effects of indoor air pollution, and on the physiological consequences of stress, including the possibility that learning or conditioning might contribute to development of stress-induced alterations in immune function.

Three indoor air pollutants associated with increased respiratory cancer, environmental tobacco smoke, asbestos, and radon were the focus of discussions about the cancer risk from indoor air pollution. Although there is clear and convincing epidemiologic evidence that exposures experienced by active smokers (tobacco smoke), asbestos workers (asbestos), and underground miners (radon) cause an increased risk of developing cancer, questions remain about the cancer risks associated with exposures at levels typically found in indoor, nonindustrial environments (e.g., residences). The fundamental research questions posed were how do we reduce the uncertainties in extrapolating from higher occupational exposures to lower residential exposures, and how do we reduce the imprecision of direct risk estimates from epidemiologic studies.

In looking across the groups of health effects described above, three broad themes emerged: (1) the need for more research on markers; (2) the need for more research on mixtures; and (3) the fact that indoor air pollution is a multifactorial problem requiring multidisciplinary research approaches.

Research on Markers

It is obvious that the healthfulness of indoor air quality depends on complicated and poorly understood interactions among a plethora of everchanging variables, few of which are well characterized. Even if we had perfect understanding of this complex reality, it would still be infeasible, impractical, or unaffordable to measure all important variables for all important situations and circumstances. Consequently, there is an ongoing need for research to develop easy-to-measure "markers" and easy-to-interpret surrogates, indicators, and/or estimators for important determinants of indoor air-related health effects. For example, markers are needed for airborne concentrations of important classes of indoor agents (e.g., volatile organic chemicals, pesticides, polycyclic aromatic hydrocarbons, fungi, endotoxins), for contributions of important sources (e.g., tobacco smoke) and source categories (e.g., furnishings, consumer products, pets, and plants) to observed airborne concentrations, for exposure to important indoor air pollutants (e.g., cotinine in urine as a marker of tobacco smoke exposure), for the dose resulting from exposure (e.g., lead in blood as a marker of absorbed dose), and for related health effects (e.g., altered gene expression as a marker of liver cancer).

Research on Mixtures

It is a fact that indoor air is composed of a complex mixture of biological (e.g., bacteria, fungi, viruses), chemical (e.g., volatile organic compounds, heavy metals), and physical (e.g., noise, light, heat) agents. Indoor air mixtures, in combination with psychosocial dynamics of human interactions in indoor settings, are important determinants of the healthfulness of indoor air quality. It is likely that in many, if not most instances, the cumulative impact of multiple exposures to multiple toxicants, occurring either simultaneously or sequentially, is a primary factor in causing indoor air-related health effects. There is a pervasive need, therefore, for research to elucidate the nature, magnitude, and extent of adverse health consequences (e.g., are they additive, antagonistic, or synergistic?) resulting from exposures to real-world mixtures of indoor air pollutants. Because research on this topic has not been a high priority in the past, efforts should focus on developing methods to study mixtures, and using these methods to measure health effects of mixtures. The goals of this research should be to develop principles which govern the nature of biological responses to mixtures, and construct models to predict or estimate health effects of mixtures.

Multifactorial Problems/Multidisciplinary Approaches

As we have learned more about the complexity of indoor air quality and its complicated interrelationships with human health, a basic truth has gradually emerged: indoor air pollution is a multifactorial problem requiring multidisciplinary research approaches to answer important scientific questions related to assessing, managing, and communicating health risks. It is apparent that no single agent, source, setting, or pathway is responsible for the broad array of health effects that are known or suspected to be caused by indoor air exposures. A corollary is that there is no single research approach or discipline that can adequately address the whole spectrum of indoor air-related health issues. Consequently, multidisciplinary approaches are needed for framing important research questions, identifying and prioritizing research opportunities, generating and testing relevant hypotheses, and resolving indoor air problems cost effectively.

SUMMARY AND CONCLUSIONS

There is reason for legitimate concern about the healthfulness of indoor air quality, including the amount of time people spend indoors and the associated implications for indoor contributions to total exposures, the presence indoors of significant sources of biological, chemical, and physical agents that are known human toxicants, the fact that airborne concentrations of many toxicants are higher indoors than out, the increasing number of health-related complaints related to indoor air quality, and the relative magnitude and likelihood of health risks ascribed to indoor air pollutants. The role of research in decision making about safeguarding public health indoors is to provide the necessary scientific and technical basis for assessing, managing, and communicating risks. A useful conceptual approach for analyzing the research-related dimensions of indoor air quality issues is to divide them into eight common domains: indoor environments; sources; agents; exposures; health effects; people; complicating factors; and solutions.

Because indoor air pollution is an important public health issue, and in light of budgetary limitations, there is a critical need to set research priorities and directions so as to make the best use of available resources. One way to do this is through the process of strategic planning: future-oriented decision making about organizational goals, purposes, and activities, including choices about what will and will not be done to achieve stated objectives. If the overall goal is to inform decisions about the healthfulness of indoor air quality, then a high priority must be given to research on the nature of health effects related to indoor air pollution. Four groups of health effects particularly deserve more research attention: sensory and sensitivity effects; allergy and respiratory effects; neurotoxic effects; and cancer. Cross-cutting research themes that are important for improving our understanding of each group of effects are: the need for more research on markers of health-related events (e.g., exposure, dose, effects); the need for more research on health effects of indoor air mixtures (e.g., neurotoxic effects of air inside office buildings); and the need for multidisciplinary research approaches (e.g., toxicology, epidemiology, chemistry, engineering, mathematical modeling) to understand and resolve indoor air-related health problems.

In summary, strategic planning provides a way to institutionalize the idea that the primary goal of indoor air research is to improve scientific knowledge and understanding in support of more informed, more credible, and more cost-effective decisions about safeguarding public health. The planning process itself highlights the fact that attaining this goal requires a clear understanding of (1) the scientific evidence that is driving concerns about health-related problems associated with indoor air quality, (2) the critical questions confronting decision makers and the role of science in answering them, (3) the complex, multidimensional nature of indoor air quality issues, and (4) the suitability of alternative research approaches, methods, and techniques for producing the necessary scientific information in a timely and cost-effective manner.

REFERENCES

Berglund, B., Brunekreef, B., Knoppel, H., Lindvall, T., Maroni, M., Møhave, L., and Skov, P., Effects of indoor air pollution on human health, *Indoor Air,* 2, 2-25, 1992.
Environmental Protection Agency (EPA), The Inside Story: A Guide to Indoor Air Quality, EPA/400/1-88/004, Washington, D.C., 1988.

Environmental Protection Agency, Science Advisory Board (A-101), Reducing Risk: Setting Priorities and Strategies for Environmental Protection, SAB-EC-90-021, Washington, D.C., 1990.

Environmental Protection Agency, Program Description: Office of Radiation and Indoor Air, EPA 402-K-93-002, Washington, D.C., 1993a.

Environmental Protection Agency, A Guidebook to Comparing Risks and Setting Environmental Priorities, EPA 230-8-93-003, Washington, D.C., 1993b.

Indoor air, *EPA J.* 19 (No. 4), 6-43, EPA 175-N-93-027, 1993.

Gammage, R.B. and Berven, B.A., *Indoor Air and Human Health Revisited,* Lewis Publishers, Chelsea, MI, 1995.

Gammage, R.B. and Kaye, S.V., *Indoor Air and Human Health,* Lewis Publishers, Chelsea, MI, 1985.

Gold, D.R., Indoor air pollution, in *Clinics in Chest Medicine,* Epler, E.D., Ed., *Occup. Lung Dis.,* 13, 215-229, 1992.

McCann, J., Horn, L., Girman, J., and Nero, A.V., Potential Risks from Exposure to Organic Compounds in Indoor Air, LBL-22473, Lawrence Berkeley Laboratory, Berkeley, CA, 1988.

Mendell, M.J., Non-specific symptoms of office workers: a review and summary of the epidemiologic literature, *Indoor Air,* 3, 227-236, 1993.

Mitchell, F.L. and Price, P. (Eds.), Proceedings of the conference on low-level exposure to chemicals and neurobiologic sensitivity. *Toxicol. Indust. Health,* 10, 253-666, 1994.

National Research Council, *Indoor Pollutants,* National Academy Press, Washington, D.C., 1981a.

National Research Council, *Formaldehyde and Other Aldehydes,* National Academy Press, Washington, D.C., 1981b.

National Research Council, *Asbestiform Fibers: Nonoccupational Health Risks,* National Academy Press, Washington, D.C., 1984.

National Research Council, *Epidemiology and Air Pollution,* National Academy Press, Washington, D.C., 1985.

National Research Council, *Environmental Tobacco Smoke: Measuring and Assessing Health Effects,* National Academy Press, Washington, D.C., 1986.

National Research Council, *Human Exposure Assessment for Airborne Pollutants: Advances and Opportunities,* National Academy Press, Washington, D.C., 1991.

National Research Council, *Multiple Chemical Sensitivities,* National Academy Press, Washington, D.C., 1992.

National Research Council, *Science and Judgment in Risk Assessment,* National Academy Press, Washington, D.C., 1994.

Samet, J.M. and Spengler, J.D., *Indoor Air Pollution: A Health Perspective,* The Johns Hopkins University Press, Baltimore, MD, 1991.

Sexton, K., Indoor air quality: an overview of policy and regulatory issues, *Sci. Technol. Human Values,* 11, 53-67, 1986.

Sexton, K., An inside look at air pollution, *EPA J.,* 19, 9-12, 1993.

Sexton, K., Science and policy in regulatory decision making: getting the facts right about hazardous air pollutants, *Environ. Health Perspect.,* 103 (Supplement 6), 213-222, 1995.

Sexton, K. and Ryan, P.B., Human exposure to air pollution: methods, measurements, and models, in *Air Pollution, the Automobile, and Public Health,* Watson, A., Bates, R.R., and Kennedy, D., Eds., National Academy Press, Washington, D.C., 1988.

Sexton, K. and Teichman, K.Y., Indoor Air Research Plan: FY 1993-FY 1997, Environmental Protection Agency, Office of Research and Development, Washington, D.C., February, 1993.

Sexton, K., Gong, H., Bailar, J.C., Ford, J.G., Gold, D.R., Lambert, W.E., and Utell, M.J., Air pollution health risks: do class and race matter?, *Toxicol. Indust. Health,* 9, 843-878, 1993.

Spengler, J.D. and Sexton, K., Indoor air pollution: a public health perspective, *Science,* 221, 9-17, 1983.

Stolwijk, J.A.J., Assessment of population exposure and carcinogenic risk posed by volatile organic compounds in indoor air, *Risk Analysis,* 10, 49-57, 1990.

Teichman, K.Y., Indoor air quality: exploring the policy options to reduce human exposures, *Indoor Air,* 4, 202-211, 1994.

Part I
Sensory Sensitivity

Overview: Odors and Irritation in Indoor Pollution

William S. Cain

INTRODUCTION

When air smells bad or even just unusual, or when it irritates the nose or eyes, people commonly feel threatened. Observations about odor and irritation therefore often serve as reason to launch an investigation into the quality of air in a building. If remedial actions eliminate the odors or irritation, occupants will generally feel the threat lift.

Because these chemosensory cues of odor and irritation have such leverage over people's impression of air quality, it seems important to know how the cues are sensed. What properties endow a molecule with ability to stimulate odor at all and with ability to arouse a particular odor quality? Are the receptors and neural apparatus for the mediation of odors and irritants well characterized? Can the same vapor stimulate both odor and irritation? Are there predictive models for the potency of odors and irritants? Are there objective ways to show whether a person has experienced odor or irritation? How does odor influence behavior?

NEUROBIOLOGY AND PSYCHOPHYSICS OF OLFACTION

The neurobiology of chemoreception is known with a level of detail inconceivable just a few years ago. Consideration of the neural structure and functioning of the sense of smell reveals it as enormously complex, but with ultimately decipherable workings. The neural complexities of the olfactory system within the central nervous system derive in part from duality in its role. On the one hand, it transmits information about whatever organic chemicals happen to be airborne and, on the other hand, it plays a role in regulation of bodily functions, such as alimentation. A role in regulation requires stimulation to be interpreted in terms of the body's needs. In this role, smell endows chemicals with "good" and "bad" properties. One odor can be inviting and another repulsive. Those hedonically between the extremes will be viewed with suspicion unless known to be safe. We may note in passing that olfaction protects more intelligently against ingestion than about inhalation of harmful materials; the relative pleasantness of an odor will guide well toward what may be eaten or not, and this dimension will predominate. Bad smelling air accordingly does not necessarily mean "bad to breathe" and good smelling air does not necessarily mean "good to breathe." The irritation sense probably functions more accurately than olfaction regarding whether air is bad to breathe.

In Chapter 1 Dr. Charles Greer of the Yale University School of Medicine focuses on the workings of the peripheral portions of the olfactory nervous system, particularly the olfactory epithelium where impinging molecules of odorant trigger transduction from their own energy to the neurochemical activity of the nervous system, and the olfactory bulb, where the organization of olfactory information begins (Figure 1). The olfactory system, and perhaps more especially the olfactory epithelium, poses the last great chemical mystery of the body. Ironically, neurobiologists know more about the chemistry of the brain, e.g., the types and the actions of neurotransmitters, than they do about peripheral olfaction. The olfactory mystery lies both in how transduction occurs at the epithelium and in the subsequent processing of the transduced information. Through Greer, we learn about anatomical and neurochemical details and complexities of the circuitry.

Long before any scientist could have expected to know the workings of brain chemistry, we perceiving organisms knew about the chemical detectors inside our noses. Some great scientists, such as Tyndall and Ramsay in the 19th century and Linus Pauling in the 20th, sought to give insights into the workings of smell as a radiation detector or a stereochemical detector when almost nothing was known about mechanism. The rationale behind their musings was undoubtedly that one great insight might set the field on a firm path for others to extend.

Pauling was correct when he urged consideration of stereochemistry. The importance of the shape and size of molecules in endowing them with distinctive qualities is not in dispute, but the story is more complex. For smaller molecules (less than about 100 Da), functional group seems to play more of a role than shape and size. Hence, among these smaller molecules amines smell fishy urinous, esters smell fruity, ethers smell ethereal, etc. For molecules above about 200 Da, the particular atoms that make up a molecule have much less importance than does its stereochemistry, as represented in what has been termed its oriented profile as it interacts with the protein molecule that forms the receptor.

One might expect that the search to specify quantitative structure-activity relationships in olfaction would be intense. It is not, at either of the levels where it could occur: in the search for determinants of potency or in the search for determinants of quality. The field even lacks a good assay to determine odor quality at either the neurophysiological level or the psychophysical level. What metric should we devise to measure the discriminability of odor qualities? The task is easier psychophysically. One could measure, for example, how much of one odorant he needs to put into another in order to produce a just discriminable change in odor quality. Research has begun on this, but to consider whether to go with one's measurements is at this stage a bit like peering at the sea. There

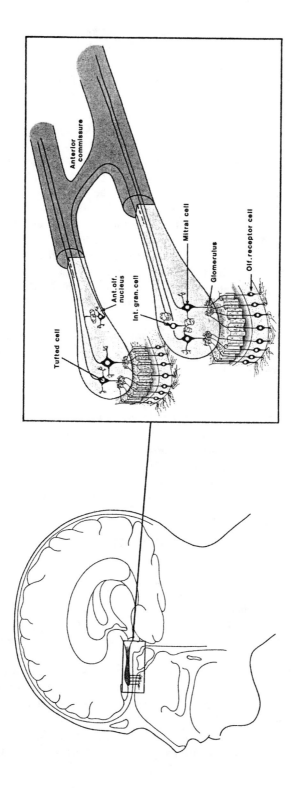

Figure 1   Diagram illustrating the location of the olfactory bulbs at the inferior surface of the brain and their connections via the anterior commissure. (Adapted from Ottoson, D., *Physiology of the Nervous System*, Oxford University Press, New York, 1983.)

are about 500,000 odorants and 25 million pairs to consider. Clearly one starts with particular hypotheses about structural determinants of quality, but we cannot expect to crack the code any time soon.

The neurobiologist and psychophysicist generally look at olfaction through lenses of very different "focal lengths." The neurobiologist can look at small details of the system but commonly at the expense of the "big" picture. The neurobiologist may be able to say whether a particular cell, by its neural response, can discriminate between two molecules, but he may not be able to tell what the whole organism can do. Will the neurobiologist be able to distill order into the pursuit by new techniques that unearth mechanism? Most likely, but one of the long hoped for outcomes has not materialized, viz., the existence of a small group of receptor types that might correspond to categories of odors.

Upon inspection, categories in smell seem rather arbitrary and reflective of behavioral needs and goals, such as food selection, rather than reflective of peripheral chemosensory physiology per se. Recent findings that perhaps a thousand genes express the receptors for olfaction seem consistent with the conclusion that each odoriferous molecule interacts with various receptors. The story therefore seems the converse of "many odorants, few receptor types" in that each molecule apparently paints a rather elaborate pattern on the receptor mosaic. More than likely, such stimulus-receptor interaction essentially rules out the existence of an inherent classification scheme with just a few categories in the most peripheral message. Those who investigate the olfactory bulb and more central structures, nevertheless, see signs of organization of the message into a simpler scheme.

The complexity of the interaction of molecules with receptors does not prevent progress on how the olfactory system processes information once transduction has occurred. Dr. Greer elucidates the wiring in the olfactory bulb as the first way station for olfactory neural information on the way to neocortical targets. Understanding processing more central than the bulb will rest upon the understanding achieved in Dr. Greer's investigations and those similar.

SENSORY IRRITATION

Complaints about indoor air pollution probably begin more frequently with symptoms of irritation than with symptoms of odor. The irritation sense, also known as the common chemical sense, is very primitive and, compared to olfaction, simple. This does not mean, however, that its workings have been well elucidated. Chemosensory scientists generally treated irritation as somewhat of a nuisance incidental to smell and taste. Lately, however, the market for spicy foods has sparked interest in the molecular determinants of the pungency in such products as chili pepper, curries, etc. The active ingredients, capsicum compounds, in chilies and other peppers have been found to have major pharmacological properties when administered systemically. The compounds can, for example, deplete the body of particular mediators of irritation and thereby render the organism insensitive to chemical irritation. Appropriate application topically can accomplish similar effects and actually reduce clinically significant burning pain. Curiously, then, but hardly for the first time, an ingredient valued as a food has turned out to have importance pharmacologically and medically.

Toxicologists have long shown some interest in irritation, for irritation can be seen as a toxic reaction to a chemical. Sensory irritation forms the basis for threshold limit values (TLVs) for about half the chemicals regulated in the industrial workplace. In many cases, the irritation results from chemical reactions between airborne molecules and tissue. Some understanding exists of structure-activity relations for what we might call such "reactive" irritants, but only at a rather rudimentary level. These irritants may stimulate sensory neurons directly or may stimulate them indirectly through byproducts of tissue damage. If, for example, a corrosive material dissolves the wall of an epithelial cell, the damaged cell will secrete chemicals into its local environment. These endogenous substances will stimulate nerve fibers and an ensuing axon reflex that leads to remote secretion of another endogenous chemical will spread the site of perceived irritation to a larger area.

Understanding the basis for sensory responses to nonreactive irritants has unfortunately lagged. Almost every chemical can cause irritation at high concentration. We all know that such materials as rubbing alcohol and most solvents can trigger some transient, reversible irritation of the eyes or nose.

Reports that indoor pollution may have caused symptoms of irritation led to profiling of the air in many buildings with the general outcome that, aside from formaldehyde, environmental tobacco smoke, and perhaps ozone, most places had no obviously reactive irritants at detectable levels. In the absence of these agents, it must have been nonreactive irritants acting alone or in concert with other agents, such as viable particles, that led to the irritation. This has motivated concern with quantitative structure-activity relationships for nonreactive irritants.

As Drs. Cometto-Muñiz and Cain note in Chapter 2, the common chemical modality should prove easier to understand than olfaction because of the relative absence of qualitative information in the sensory message. Although irritation may admit to some qualitative variation, such as "burn" vs. "stinging," the variety is much smaller than for smell, where the focus is more on how the vast qualitative variation of odoriferous chemicals is encoded than on almost anything else.

Ironically, the absence of qualitative variation in sensory irritation adds to the problems it causes in indoor air. As noted above, virtually every airborne organic compound can potentially cause irritation. The vapors of rubbing alcohol have an odor when dilute, but can burn the nose when concentrated. For rubbing alcohol, the gap in concentration between where it first causes odor and where it first causes irritation is about two orders of magnitude. For a few substances, this gap is smaller, but for most it is larger. Nevertheless, the absolute concentration at which irritation might occur is often lower for those chemicals with a larger gap. Octanol, a chemical neighbor of rubbing alcohol, has a gap between its odor and its irritation of about four orders of magnitude, but it has both a lower odor threshold and a lower irritation threshold than rubbing alcohol. When more than one organic substance is present in the atmosphere, as is virtually always the case, irritation that comes from that substance may not be distinguishable from the irritation that comes from another. Hence, it becomes more difficult to trace the source. In the real world, where potential sources of irritation may lie in the scores or hundreds, tracing the sources might be virtually impossible.

Psychophysical measurement of sensitivity to odors and irritants will ultimately point to the physicochemical or molecular determinants of potency and thereby allow prediction. Measurements made in recent years on subjects without an olfactory sense have yielded information on pure irritation sensitivity to various series of chemicals. As Cometto-Muñiz and Cain point out in Chapter 2 and as Dr. Michael Abraham does in Chapter 3, data from such anosmic subjects and data on irritant-induced respiratory depression in rats provide complementary pictures of physicochemical determinants of irritation and have permitted a beginning for modeling of potency. This in turn has led to the conclusion that long-invoked thermodynamic principles for biological efficacy have been misapplied, but that appropriate principles of the energy of solvation can replace them. Abraham shows that principles important for the modeling of anesthetic gases and other biological phenomena hold for irritation as well. Modeling of the potency of odorants unfortunately lags behind that of irritants, another manifestation of the complexity of olfaction.

OBJECTIVE INDICES OF CHEMOSENSORY FUNCTIONING

If I complained of irritation of the nose and eyes in your house and you felt no such thing, would you believe me? It might depend on my general relationship with you. If you suspected me of an ulterior motive, if I commonly complained about insignificant things, or if my complaints seemed likely to cost you money in remediation, you might decide not to believe me. You might ask for evidence, but I might not be able to give you any. In many cases where occupants have complained about indoor pollution, there has existed no objective way to verify the complaints. Just as interest in indoor pollution has necessitated concern with nonreactive irritants, interest in pollution has necessitated concern with validation of symptoms. Symptoms that would largely go unquestioned in a physician's office raise serious issues of credibility in the commercial office.

Research on objective measures of symptoms can take many avenues from measurement of reflexes involved in breathing, to staining the eye to look for scoring of an irritated cornea, to assaying markers of inflammation in fluid from an irritated nose. In the chemosensory domain, evoked electrical potentials obtained from the scalp or from mucosa of the human subject can complement psychophysical measurements as objective indices of functioning and with a degree of isolation of systems not readily obtainable psychophysically. Research on chemosensory evoked potentials has been relatively slow to develop, but is now being applied to evaluation of subjects not readily evaluated psychophysically and to topics, such as hemispheric locus of processing, not directly open to psychophysical observation.

Symptom validation is a byproduct of the research on evoked potentials that Dr. Kobal explicates in Chapter 4. Because such potentials require precise time-locking between stimulus and response, sophisticated control of the chemical stimulus is an important requirement of the work and ultimately a limiting factor for its potential widespread use. Kobal pioneered the recording of the electro-olfacto-gram (EOG) from the olfactory mucosa and its cortically measured analog, the olfactory event-related potential. His work on the peripheral potential has con-firmed, for example, that little of the phenomenon of olfactory adaptation occurs at the mucosa. Insofar as adaptation has its locus in the central nervous system, then it may come under the control of psychological variables, such as expectation and pleasantness.

Kobal and associates have shown by a series of experiments that have included desensitizing injections of capsiacin that a potential measurable from the respi-ratory mucosa, the negative mucosal potential (NMP), provides an objective measure of nasal irritation. As an objective index of sensory irritation, it has value both for validation of symptoms and for clinical assessments of the magnitude of pain. The peripheral potentials also play a key role for interpretation of more central potentials, chemosensory event-related potentials (CSERPs), which are highly integrated responses of the brain. Kobal has shown how to separate an irritant-stimulated CSERP from an olfactory-stimulated CSERP. He has studied the neural generators of the potentials in various ways, including the application of magnetoencephalography to map the topographic distribution of the responses. The application of technology to cases where the brain has been damaged offers ways to peer into the central nervous system of humans. He illustrates this with studies of persons with temporal lobe epilepsy.

ODORS AND BEHAVIOR

Understanding the molecular basis of odor quality or the neurobiology of olfaction will in principle give us ways to manage our olfactory environment better and thereby to avoid certain problems with undesirable stimuli. A key to control, we should note, will be to understand complex as well as simple stimuli. In the chemosensory domain, this means understanding how the perception of simple stimuli can be extended to the perception of mixtures. There is presently little predictability regarding how more than a simple mixture will be perceived in its intensity, its quality, and its pleasantness. For this reason, studies of venti-lation requirements to control body odors, cooking odors, and the like, need to depend on natural stimuli (e.g., actual people who occupy a chamber and thereby create the odor of occupancy) and what we might think of as "bottom-line" judgments of intensity and acceptability. The odor of occupancy is too complex by far, relative to existing knowledge, for us to replace human judgments with instruments, though we might be able to calibrate instruments against the bottom-line judgments. This means, however, that the human observer will remain an important part of indoor air quality research.

The level of complexity needed to understand real-world mixtures pales in comparison to the level needed to understand how odors influence behavior. The topic has seen only spotty inquiry, though now Dr. Knasko has made it her specialty and has uncovered various relationships. In Chapter 5 Knasko reviews the entire field as well as her own work. She reports that studies of the influence of odors on task performance have generally proven negative. One probably should not, therefore, pump an odor into an office with the expectation that productivity will increase. Where studies have found statistically significant effects, they have generally shown decreases in performance.

Odors formulated to influence our feelings, if not necessarily our behavior, surround us so thoroughly that we take them for granted. The scent of a laundry product can convince us that our clothes are clean, the scent of shampooed hair can make us feel invigorated, the scent of another person's perfume may excite us. Not all odors are unwanted, even though we may not always have a conscious sense of their presence. Odors are often assimilated into an event rather than noted separately. They may accordingly "sneak up" and color perception of the environment. As Knasko and associates showed, even odor placebos can influence feelings. She sees odors as part of a feedback loop; odors can influence how we feel and how we feel can influence how we interpret odors.

Studies of ambient odor pollution have sometimes revealed a rather dramatic progression of reactions to an odor that is out of the control of people. What may start as an annoyance may grow over time to a source of adverse physiological reactions and may eventually become the focus of people's existence with attributions of dire health consequences from exposure to the odor. The feedback loop, including a stress response, may cause exacerbation of existing symptoms as well as the emergence of new symptoms. This is not to say that exposure to odoriferous pollutants causes health effects only through psychological factors. Every odorant is, after all, a chemical. These various matters deserve more attention within the context of indoor pollution.

CHAPTER 1

The Neurobiology of Olfaction

Charles A. Greer and Juan C. Bartolomei

INTRODUCTION

The olfactory system is specialized for detecting and processing odorants. Together with gustation, olfaction is a unique sensory system in that the appropriate stimuli are molecules that bind to specific receptor sites. Because the olfactory system offers a direct route to the central nervous system, there has been much interest in how environmental conditions may influence odor processing. As will be discussed below, there is evidence that as toxins increase in the environment there is a corresponding increase in the death of olfactory receptor cells. Moreover, a variety of compounds, including heavy metals, have been shown to be transported intracellularly from the olfactory epithelium in the nose into the central nervous system where they may accumulate in cells throughout the olfactory pathway. Thus, analyses of cellular organization and olfactory function are useful indices of environmental conditions and the mechanisms via which toxins or pollutants may exert their effects.

Given the importance of the olfactory system in studying the interface between the environment and central nervous system structure/function, the purpose of this chapter is to review the basic cellular, synaptic, and functional organization of the vertebrate olfactory system. When appropriate, particular attention will be given to issues of organization that bear upon the influence of the environment on olfactory system function.

ANATOMY OF THE OLFACTORY SYSTEM

Gross Anatomy

The olfactory system in most vertebrates can be divided into three gross anatomical compartments: (1) the olfactory epithelium; (2) the olfactory bulb;

and (3) piriform cortex (Figure 1) (Greer, 1991). The relative size of the olfactory structures appears to correlate closely with the degree to which the species is dependent on the utilization of odor signals. For example, macrosmatic mammals such as rodents and dogs depend on the olfactory system for behaviors that are critical for the survival of the species including alimentation and mating. Appropriately, the olfactory systems of these species compromise a large proportion of their brain. Conversely, the olfactory systems of microsmatic mammals, such as primates who depend heavily on sight and hearing, occupy a smaller proportion of their brain.

In order to gain an overall understanding of odor processing, the pathway followed by odor signals may be briefly summarized as follows. Odor molecules within the nasal cavity interact with the olfactory receptor cells located in the olfactory epithelium. The axons of the receptor cells pierce the cribriform plate and establish synaptic contact with projection and intrinsic neurons located within the olfactory bulb. The axons of the projection neurons exit via the lateral olfactory tract and terminate in higher cortical regions including the piriform cortex (Greer, 1991). From the cortical regions further tertiary connections are made that integrate and modify the response of the animal to olfactory information (McLean and Shipley, 1992). In this review we will focus on the organization of the first two centers of processing, the olfactory epithelium and olfactory bulb. Readers are directed to several excellent reports for further discussion of cortical representation and processing of odor information (Haberly, 1983, 1990; Haberly and Feig, 1983).

Olfactory Epithelium

The olfactory epithelium is distributed along the surfaces of complex turbinates within the nasal cavity. Embryologically, the epithelium is derived from the olfactory placode on the neural tube. Three principal populations of cells are found within the epithelium (Morrison and Costanzo, 1992). The first, the olfactory receptor cells, are characterized by a small cell body, approximately 8 μm in diameter, and a long thin apical dendrite that extends to the surface of the epithelium. At the surface the dendrite enlarges into a knob-like structure from which emanate 6 to 8 cilia that distribute across the epithelial surface. The cell bodies of the receptor cells are generally located in the deep two thirds of the epithelium. The small, 0.2-μm, unmyelinated axons exit the epithelium through the basal lamina and bundle to form the olfactory nerve that penetrates the cribriform plate. One of the most intriguing aspects of the olfactory epithelium is its capacity for renewal (Graziadei and Monti-Graziadei, 1979; Mackay-Sim and Kittel, 1991). The olfactory receptor cells are the only neural-derived sensory cells in direct contact with the brain and the environment (Shepherd and Greer, 1990). The receptor cells have a limited lifespan of 4 to 12 weeks, dependent in part on environmental conditions (Schwob, 1992; Hinds et al., 1984; Hinds and McNelly, 1981). As receptor cells die, new cells are generated from the stem cell population at the base of the epithelium (Caggiano et al., 1994). These new cells differentiate and innervate the appropriate region of the olfactory bulb.

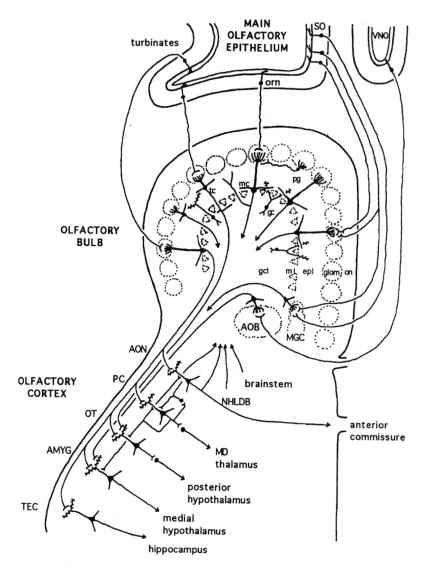

Figure 1   An overview of the olfactory pathway. The olfactory bulb receives input from the receptor neurons in the olfactory epithelium and projects to the olfactory cortex. The diagram indicates some essential aspects of the projection patterns between the regions as well as the main neural elements within the olfactory bulb. Note that the olfactory epithelium is arranged in overlapping populations of receptor neurons which project to individual glomeruli. Some of the central connections to limbic brain structures are also indicated. AMYG, amygdala; AOB, accessory olfactory bulb; AON, anterior olfactory nucleus; MD, mediodorsal; MGC, modified glomerular complex; NHLDB, nucleus of the horizontal limb of the diagonal band; OT, olfactory tubercle; PC, piriform cortex; TEC, temporal entorhinal cortex; VNO, vomeronasal organ; epl, external plexiform layer; gc, granule cell; gcl, granule cell layer; mc, mitral cell; ml, mitral cell layer; on, olfactory nerve layer; orn, olfactory receptor neuron; pg, periglomerular cell; so, septal organ; tc, tufted cell. (Adapted from Shepherd, G.M. and C.A. Greer (1990) *The Synaptic Organization of the Brain*. New York: Oxford University Press, pp. 133-169 and Greer, C.A. (1991) *Smell and Taste in Health and Disease*. New York: Raven Press, pp. 65-81. With permission.)

The second population of cells are the nonsensory sustentacular or supporting cells. They are elongated and extend from the basal lamina to the epithelial surface where they are covered with microvilli. The nuclei of the supporting cells are found in a single layer in the upper one-third of the epithelium. Beyond their role as supporting cells, reports have speculated that the sustentacular cells may also contribute to the removal of odorants and toxins within the epithelium (Chen et al., 1992; Getchell et al., 1984).

The third population is the stem or basal cell. They are found adjacent to the basal lamina of the olfactory epithelium and are the progenitors of receptor cells (Caggiano et al., 1994). Destruction of the existing receptor cell population, by cutting the olfactory nerve, for example, will increase mitotic activity among the basal cells until the epithelium is restored.

Olfactory Bulb

In most species the olfactory bulb appears as a protrusion beyond the rostral pole of frontal cortex. In man and other primates, the olfactory bulb is found in the olfactory groove beneath the frontal-orbital gyri of frontal cortex. The olfactory bulb is organized in concentric laminae that facilitate understanding the cellular and synaptic organization of the initial sites of odor processing within the central nervous system.

Laminar Organization of the Olfactory Bulb

Olfactory Nerve Layer

In mammals the outermost lamina of the olfactory bulb is the olfactory nerve layer which contains olfactory receptor cell axons and glial cells (Shepherd and Greer, 1990; McLean and Shipley, 1992). After piercing through the cribriform plate these axons are partially wrapped by specialized astrocytes called "ensheathing cells" (Doucette, 1993). The olfactory receptor cell axons are thus bundled into discrete fascicles that form a superficial stratified layer of varying thickness around the main olfactory bulb (Halász, 1990). Numerous studies have shown a degree of topological organization of the projections from the olfactory epithelium onto the bulb (Costanzo and O'Connell, 1978; Jastreboff et al., 1984; Mackay-Sim and Nathan, 1984; Astic et al., 1987; Duncan et al., 1990; Greer, 1991; Schoenfeld, 1994). However, the topographical point-to-point specificity characteristic of the visual or somatosensory systems is not apparent in the olfactory system.

Glomerular Layer

Immediately deep to the olfactory nerve layer lies the glomerular layer where the olfactory receptor cell axons terminate (Halász and Greer, 1993) and establish axodendritic synapses with processes arising from intrinsic neurons (periglomerular cells) and projection neurons (mitral and tufted cells); (Pinching and Powell, 1971b; Shepherd and Greer, 1990). The glomerular layer is the first tier in olfactory coding processing and is considered by many to be the most distinctive anatomical and physiological unit of the olfactory bulb (Andres, 1965; Shepherd, 1972; Halász, 1990; McLean and Shipley, 1992). Individual glomeruli are cell-free spherical structures that are sharply delineated by a surrounding shell composed of the cell bodies of periglomerular and short axon intrinsic neurons and deeply, the cell bodies of superficial tufted cells. Depending on the species, the glomeruli range from 30 to 200 μm in diameter (Allison and Warwick, 1949; Baier and Korsching, 1994). It is estimated that the number of glomeruli is approximately 2000 in rabbits (Allison and Warwick, 1949) and mice (White, 1972); in rats it has been estimated at 3000 (Meisami and Safari, 1981). There is a high convergence of approximately 25,000 olfactory axons into a single glomerulus (Halász and Greer, 1994).

Glomeruli contain, in addition to olfactory receptor cell axon terminals, the arborized dendritic processes of projection neurons, tufted and mitral cells as will be discussed below, and the dendritic processes of periglomerular cells whose cell bodies delimit the glomeruli. The periglomerular cells are a population of intrinsic neurons whose intraglomerular dendritic processes receive afferent input from receptor cell axons as well as establishing local circuit interactions with the dendritic processes of the projection neurons (Pinching and Powell, 1971a–c; White, 1972). In addition to these synaptic components the glomeruli receive limited centrifugal input from several cortical regions such as locus coeruleus, raphe nuclei, and the diagonal band of Broca (Mori, 1987; Shepherd and Greer, 1990).

External Plexiform Layer

Immediately deep to the glomerular layer is the external plexiform layer. This layer consists mostly of basal dendrites from projection neurons (mitral/tufted cells) and granule cell dendrites which together form an intricate network of local circuits. In addition, several centrifugal axonal projections from cortical centers terminate within the external plexiform layer (Halász, 1990; Mori, 1987). The external plexiform layer also includes the cell bodies of tufted cells which can be divided into superficial, middle, and deep according to their relative depth in the external plexiform layer (Orona et al., 1984).

Mitral Cell Layer

Deep to the external plexiform layer lies the mitral cell layer. It is only one to two cells thick and is often considered a monolayer. The mitral cell layer is composed predominately of mitral cell bodies with only occasional granule cell bodies (Greer and Shepherd, 1982; Shepherd and Greer, 1990).

Granule Cell Layer

The granule cell layer is the deepest neuronal layer of the olfactory bulb and is composed largely of granule cell bodies. These intrinsic neurons are arranged in tightly packed row-like clusters. Interspersed between the cell bodies, the myelinated axons from the tufted/mitral cells gather as they form the lateral olfactory tract. Also included within the granule cell layer are the centrifugal axons en route to more superficial layers of the olfactory bulb as noted above (Halász, 1990; Shepherd and Greer, 1990; McLean and Shipley, 1992).

Cellular Organization of the Olfactory Bulb

Within the principal layers of the olfactory bulb there are five distinct neuronal classes that are generally representative of neurons found in other areas of the brain. Three are considered to be interneurons responsible for local synaptic circuits, defined as those whose connections are confined to a region of the brain and are primarily responsible for information processing. This is in contrast to long distance or projection circuits which transfer information between regions of the brain (Rakic, 1976). The three intrinsic neurons are short axon cells, periglomerular cells, and granule cells. The two projection neurons, mitral and tufted cells, send their axons via the lateral olfactory tract to higher olfactory centers. The morphological features of these cells were previously described by early anatomists including Cajal (1911). Subsequent studies have elaborated many subtypes among these general populations of neurons (Schneider and Macrides, 1978).

Short Axon Cells

Short axon neurons are local interneurons with medium-sized bodies approximately 8 to 12 μm in diameter. Golgi studies reveal that the dendritic process of these cells is entirely periglomerular and does not receive contact from the olfactory nerve (Shepherd and Greer, 1990; McLean and Shipley, 1992). Although short axon cells exhibit significant heterogeneity in their laminar distribution and morphological characteristics, they are generally not well characterized nor well understood and will not be discussed further.

Periglomerular Cells

Periglomerular cells are small neurons, 5 to 8 μm in diameter, densely distributed around glomeruli. Pinching and Powell (1971a,b) determined that the dendrites of periglomerular cells penetrated more than one but preferentially one glomerulus, usually extending 50 to 100 μm. These dendrites appeared not to fill the entire glomerulus, but are organized in a highly well-defined compartmental fashion within the glomerulus. At the electron microscopic level these intrinsic neurons are seen establishing synaptic contact with presumed mitral/tufted cells and receiving synaptic input from the olfactory nerve and mitral/tufted cell processes. The axons of periglomerular cells are restricted to intraglomerular regions extending laterally to about 800 μm (Pinching and Powell, 1971a,b; White, 1972). The periglomerular cells are not differentiated into subtypes based on their morphology. However, there are distinct subpopulations based on the neurotransmitters they utilize (Gall et al., 1987; Baker, 1988; Bartolomei and Greer, 1993).

Granule Cells

As mentioned previously, granule cells are found predominately in the deepest layer of the olfactory bulb, the granule cell layer. Granule cells are often compared to the amacrine cell of the retina based on their morphology and presumed physiologic role in local circuits. Granule cells are 6 to 8 μm and are anaxonic neurons (Price and Powell, 1970a,b; Schneider and Macrides, 1978; Greer, 1987; Woolf and Greer, 1994). Another interesting characteristic of these cells is the high density of spines along their dendritic arborization (Greer, 1987, Woolf et al., 1991a,b). The dendritic processes of granule cells are divided into apical segments, which extend radially from the somata into the external plexiform layer, and a smaller basal component that remains confined to deeper parts of the granule cell layer. The apical dendrites establish synaptic contact with mitral and tufted cells and thus provide a mechanism for modulating signal output (Shepherd and Greer, 1990; Woolf et al., 1991a,b). Based on the differential distribution of their apical dendrites and location of their somata within the granule cell layer, the granule cells have been divided into three subtypes (Orona et al., 1983; Mori et al., 1983; Greer, 1987).

Superficial granule cell bodies are closer to the mitral cell layer and extend their apical dendrites through the external plexiform layer to arborize at the most superficial sublaminae adjacent to the glomerular layer. Deeper granule cells, on the other hand, are found in the deeper aspects of the granule cell layer and restrict their dendritic arbors to the deepest sublaminae of the external plexiform layer. Intermediate granule cells have dendritic processes that are not rigidly confined to specific sublaminae of the external plexiform layer. These specific patterns of selective dendritic arborization raise the notion of potential parallel circuits for odor processing information with the local circuit organization of the olfactory bulb (Greer, 1991).

Mitral Cells

Named after their resemblance to a bishop's miter, the mitral cells are principal output neurons of the olfactory bulb. Located in the mitral cell layer, their cell bodies range from 15 to 35 μm in diameter. An apical dendrite arises from the somata and innervates a single glomerulus. Within the glomerulus the dendritic processes arborize extensively and establish synaptic contact with the olfactory nerve and intrinsic neurons. The dendrites of approximately 25 mitral cells enter each glomerulus (Cajal, 1911; Allison and Warwick, 1949). A single mitral cell also extends 2 to 9 secondary dendrites which can branch several times within the external plexiform layer. These secondary dendrites tend to run parallel to the surface of the bulb, extend up to 2 mm in the external plexiform layer, and establish synaptic contacts with the dendritic spines of granule cells (Rall et al., 1966; Macrides and Schneider, 1982; Orona et al., 1983; Greer and Halász, 1987; Nickell and Shipley, 1992).

The axons of mitral cells arise from the deep side of the somata and run into the granule cell layer where they branch into collaterals. Contrary to Cajal's

(1911) observation, these axonal collaterals do not escape the granule cell layer into the external plexiform layer. The main axons of the mitral cells coalesce to form the lateral olfactory tract and subsequently terminate throughout olfactory cortex (Ojima et al., 1984; Orona et al., 1984; Haberly and Price, 1977).

Tufted Cells

Tufted cells are also a primary projection neuron of the olfactory bulb. Initially considered by Cajal (1911) as dislocated mitral cells, it is well known today that tufted cells have a distinct morphological pattern (Macrides and Schneider, 1982; Mori et al., 1983; Kishi et al., 1984; Orona, 1984) and genetic origin (Greer and Shepherd, 1982). The tufted cell somata range from 15 to 20 μm in diameter and are distributed throughout the external plexiform layer. The diameter of the cell body differs depending on its location within the external plexiform layer; the larger cell bodies are found deep within the external plexiform layer, close to the mitral cell layer, while smaller cell bodies are found more superficially. Based on the size and differential dendritic arborization, tufted cells are divided into deep, middle, and superficial subpopulations (Macrides and Schneider, 1982; Mori et al., 1983; Kishi, 1983; Orona et al., 1984).

Deep tufted cells are scattered sparsely in the deeper one third of the external plexiform layer. The dendritic distribution of these cells is similar to that of mitral cells (Macrides and Schneider, 1982). Middle tufted cells have a medium size somata and are found in the superficial two thirds of the external plexiform layer. They typically extend a single apical dendrite that arborizes into specific domains in a single glomerulus. The primary dendrites of middle tufted cells tend to project in a tangential manner throughout the external plexiform layer. The secondary dendrites are largely restricted to the superficial part of the external plexiform layer (Macrides and Schneider, 1982). The axons of these cells often have extensive bifurcations in the internal plexiform layer (Mori, 1987; Greer, 1991). The axons of middle and deep tufted cells, that are not part of the intrabulbar associational system, coalesce and join the axons of mitral cells to form the lateral olfactory tract. The projections from the tufted and mitral cell differ in their axon collateral patterns and projections to higher olfactory centers (Scott, 1981; Haberly and Price, 1977). The extent of the olfactory cortical areas that receives tufted cell projections roughly correlates with their laminar position in the external plexiform layer. Thus, deep tufted cells project extensively to virtually all areas of the olfactory cortex, as do the mitral cells. In contrast, middle tufted cell projections are limited to the more rostrally located olfactory centers (Haberly and Price, 1977; Skeen and Hall, 1977; Scott, 1981).

The superficial tufted cells are distributed relatively densely in the periglomerular region between the glomerular and external plexiform layers. Many of them have dendritic processes that extend into a single glomerulus, while a few innervate two glomeruli (Pinching and Powell, 1971a; Schneider and Scott, 1983). The axons of the superficial tufted cells are mostly restricted to the intra-olfactory bulb associational system and do not contribute to cortical projections (Macrides et al., 1985).

Synaptic Organization

The basic organization of the olfactory bulb has been used as a model system for understanding intrinsic neuronal elements and their corresponding synaptic organization throughout the central nervous system (Shepherd and Greer, 1990). Synaptic organization in the olfactory bulb can be divided into two levels: the glomerular layer and the external plexiform layer. Each of these will be discussed in turn.

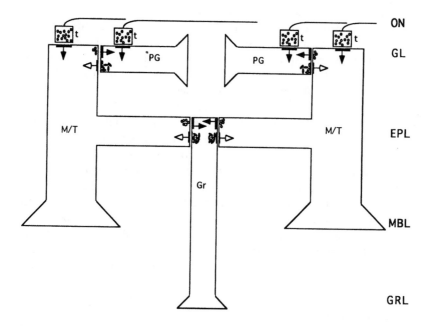

Figure 2   Synaptic organization of the olfactory bulb. In the glomerular layer (GL) the terminals of olfactory receptor cell axons (t) are seen making axodendritic Gray Type I synpases onto mitral/tufted (M/T) and periglomerular (PG) cells. Also in the glomerular the dendrites of M/T cells make Gray Type I dendrodendritic synapses onto PG cells which, in turn, make reciprocal Gray Type II synapses onto opposed M/T dendrites. In the external plexiform layer (EPL) M/T secondary dendrites make Gray Type I dendrodendritic synapses onto granule cell (Gr) dendrites which, in turn, make reciprocal Gray Type II dendrodendritic synapses onto apposed M/T dendrites. The polarity of the synapses is indicated with arrows; solid arrows are excitatory, open arrows are inhibitory. Abbreviations: ON, olfactory nerve layer; GL, glomerular layer; EPL, external plexiform layer; MBL, mitral body layer; GRL, granule cell layer; PG, periglomerular cell; M/T, mitral/tufted cell; Gr, granule cell; t, olfactory receptor cell axon terminal. See text for further discussion.

Synaptic Organization of the Glomerulus

The synaptic organization of the glomerular layer can be divided into intra-glomerular and interglomerular connections (Figure 2). Prior to entering the glomerular layer the receptor cell axons are organized into well-defined fascicles.

When the axons penetrate the neuropil of the glomerulus they appear to be organized into well-defined subglomerular domains. Within the glomerulus the olfactory receptor cell terminals, presynaptic specializations, are characteristically electron dense with many small spherical vesicles filling the axoplasm. The receptor cell axons establish axodendritic synaptic contacts with mitral/tufted and periglomerular cell dendritic processes. These synapses are categorized as Gray Type I synapses which are characterized by numerous round vesicles in the presynaptic specialization and an apposing postsynaptic process that is distinctive for the thick asymmetrical membrane specialization. It is not known if these axons establish a homogenous distribution of synaptic connections; that is, if an individual axon contacts only mitral/tufted or periglomerular cells, or rather distribute uniformly to all processes. No evidence has been found suggesting that receptor cell axons receive conventional synaptic input from any source. The axodendritic Gray Type I synapses made by the olfactory receptor cell axons constitute the primary afferent input to the olfactory bulb.

The remaining intraglomerular synaptic connections contribute to local circuit processing of incoming odor information. These circuits include dendrodendritic synapses between mitral/tufted cell dendrites and periglomerular cell dendrites. This circuit may be summarized as a reciprocal synaptic pair in which the mitral/tufted cell establishes a Gray Type I synapse onto a periglomerular cell, while the periglomerular cell makes a Gray Type II synapse onto the mitral/tufted cell process. Type II synapses are characterized by numerous pleomorphic vesicles in the presynaptic process that is juxtaposed to a postsynaptic process with a symmetrical membrane specialization (Hirata, 1964; Pinching and Powell, 1971b; Getchell and Shepherd, 1975). Within the glomerulus limited centrifugal fibers, such as those from the raphe nucleus are also seen to establish synaptic contacts (Mori, 1987). The complex synaptic arrangement described above may serve as a signal modulator and processor for incoming odor information, as will be discussed further below.

Interglomerular synaptic organization consists largely of axons from periglomerular cells traveling to other glomeruli and establishing extraglomerular Gray Type II synapses with apical dendrites of mitral/tufted cells and other periglomerular cell bodies (Pinching and Powell, 1971a,b). Also present are Type I synapses between superficial tufted cell axons and tufted cell dendrites as well as presumed centrifugal axons synapsing onto periglomerular cell bodies (Pinching and Powell, 1971a,b).

Synaptic Organization of the External Plexiform Layer

The next tier containing local microcircuits is the external plexiform layer. In this layer the dendritic processes of granule and mitral/tufted cells establish synaptic contact in a manner similar to that seen in the glomerulus. At the most superficial layer the dendritic spines of superficial granule cells establish contact with the dendritic processes of tufted cells. Deeper, the dendritic spines of deep granule cells make synaptic contact with the secondary dendrites of mitral cells

(Greer and Halász, 1987). At the level of the electron microscope the dendritic processes of mitral and tufted cells appear as large, electron lucent, tangentially oriented through the external plexiform layer with a typical distribution of microtubules and cytoplasmic organelles. These electron lucent processes are seen establishing Gray Type I, asymmetrical synapses onto granule cell dendritic spines (gemmules); (Rall et al., 1966; Rall and Shepherd, 1968; Jackowski et al., 1978; Greer and Halász, 1987). In turn, the relatively electron dense granule cell spines establish Gray Type II, symmetrical synapses onto the dendritic processes of mitral/tufted cells. The reciprocal synapses in the external plexiform layer are found in a dynamic equilibrium of 1:1 (Woolf et al., 1991a). This observation is supported by developmental studies where the asymmetrical synapse appeared first followed by the reciprocal symmetrical synapse (Hinds and Hinds, 1976a,b) and in studies of mutant mice where denervated granule cell spines formed new reciprocal synapses with tufted cell dendrites (Greer and Halász, 1987). It is interesting to note that the pattern of organization of granule cell spines is not random. Using serial reconstructions, Woolf et al. (1991a) demonstrated that the spines organize in a columnar fashion along the mitral/tufted cell dendrites.

In addition to the dendrodendritic local circuit connections, within the external plexiform layer there are also axodendritic connections from short cell axons and centrifugal projections (Mori, 1987). The central projections arise from the horizontal limb of the diagonal band, the nucleus raphe, piriform cortex, and anterior olfactory nucleus. The axodendritic connections are believed to be involved in inhibition of mitral/tufted cells through the excitation of granule cells (Shepherd and Greer, 1990).

Synaptic Organization of the Mitral-Granule Cell Layers

The synaptic organization of the mitral-granule cell layers consists largely of axon collaterals of mitral/tufted cells terminating on the basal dendrites of granule cells. These synaptic terminations are believed to be excitatory in nature which in turn inhibits the mitral/tufted cells thus further modulating the odor signal output (Price and Powell, 1970b,c). Within the granule cell layer there are also numerous terminals of centrifugal axons from areas such as the horizontal limb of the diagonal band and anterior olfactory nucleus.

NEUROCHEMISTRY

The olfactory system is rich in a variety of neurotransmitters, neuromodulators, and neuroactive proteins (Macrides and Davis, 1983; Halász and Shepherd, 1983). Many colocalize in single cells and contribute to the establishment of subpopulations of neurons, particularly among the periglomerular cells (Trombley and Shepherd, 1993). The following account is intended as a summary of the neurochemistry of the olfactory system and focuses on the predominate peptides associated with each of the populations of cells.

Olfactory Receptor Cells

Recent evidence suggests that glutamate is the most likely neurotransmitter to be utilized by olfactory receptor cells. Sassoè-Pagnetto et al. (1993) showed clearly the localization of glutamate immunoreactivity in receptor cell axon terminals while Berkowicz et al. (1994) have demonstrated the physiological actions of glutamate on postsynaptic targets. Two additional major constituents of the olfactory receptor cell are olfactory marker protein and carnosine (Margolis,1972; Margolis and Rochel, 1982), though neither have yet been ascribed a clear functional role.

In recent years a revolution in understanding the organization of the olfactory epithelium has taken place with the recognition that a super gene family is controlling the expression of up to 1000 different odor receptors within the epithelium (Buck and Axel, 1991; Buck, 1992; Ngai et al., 1993a,b; Raming et al., 1993; Shepherd, 1993a,b). Recent evidence has shown that these receptors are not homogeneously distributed but rather, are found in restricted spatial domains that are organized longitudinally within the nasal cavity along the dorsal-ventral axis (Ressler et al., 1993). While it is not yet known if single cells express one or more receptors, there is currently a great deal of excitement in understanding how the receptor domains may map topographically onto the olfactory bulb (see below). Evidence from the study of receptor cell specific glycoproteins suggests that the olfactory bulb may exhibit mosaicism of odor/function specific domains (Mori et al., 1985; Schwob and Gottlieb, 1986; Schwarting and Crandall, 1991).

Periglomerular Cells

Many of the periglomerular cells are dopaminergic (Halász et al., 1981; Davis and Macrides, 1983) or contain gamma aminobutyrate acid (GABA); (Ribak et al., 1977). In hamsters approximately 70% of dopaminergic periglomerular cells colocalize GABA, and 45% of GABAergic cells colocalize dopamine (Kosaka et al., 1985). In rats there also appears to be evidence of colocalization of both GABA and dopamine. Approximately 69% of periglomerular cells colocalized GABA and dopamine, 27% were reactive to GABA alone, and the majority of dopaminergic periglomerular cells colocalize GABA (Gall et al., 1987). Interestingly, Bartolomei and Greer (1993) have shown that the synaptic input to tyrosine hydroxylase, as a marker of dopamine, immunoreactive vs. GABA immunoreactive periglomerular cells is different. The former receive a higher density of olfactory receptor cell axon synapses while the latter receive a higher density of dendrodendritic synapses from mitral/tufted cells. Substance P has also been reported to be present in GABAergic and dopaminergic periglomerular cells (Kream et al., 1984; Davis and Kream, 1991). Thus, subpopulations of periglomerular cells may contain an inhibitory neurotransmitter, GABA, a catecholamine, dopamine, and a potential excitatory neurotransmitter, substance P. In addition, there are reports of periglomerular cells also containing met-enkephalin (Davis et al., 1982), vasoactive intestinal peptide (VIP); (Gall et al., 1986; Sanides-Kohlrausch and Wahle, 1990), NADPH-diaphorase (Davis, 1991), cholecystokinin

(CCK; Seroogy et al., 1985), thyrotropin-releasing hormone (Tsuruo et al., 1988), and protein kinase C (Saito et al., 1988).

Granule Cells

Most of the granule cells contain the inhibitory neurotransmitter GABA (Nicoll, 1971; Kosaka et al., 1987; Ribak et al., 1977; Halász et al., 1979) which inhibits the mitral and tufted cells (Shepherd, 1972; Jahr and Nicoll, 1982). Met-enkephalin has also been localized in granule cells, (Bogan et al., 1982) although the physiological importance of this neuromodulator in odor processing is not known.

Mitral Cells

The most likely neurotransmitter used by mitral cells is the excitatory amino acid glutamate. This is supported by extensive literature demonstrating both localization as well as physiological effects of glutamate on postsynaptic processes (Halász and Shepherd, 1983; Anderson et al., 1986; Fuller and Price, 1988; Berkowicz et al., 1994).

Tufted Cells

A variety of compounds localize to tufted cells. While there is a general consensus that the neurotransmitter is also likely to be an excitatory amino acid, there are studies that reveal the presence of dopamine, especially in superficial tufted cells (Halász and Shepherd, 1983; Gall et al., 1987). Also, substance P (Burd et al., 1982), VIP (Gall et al., 1986), and CCK (Seroogy, 1985) have been found in subpopulations of tufted cells. The functional significance of these neuromodulators remains to be established.

FUNCTIONAL ORGANIZATION OF ODOR PROCESSING

Following transduction by the olfactory receptor cells (Firestein 1991; Firestein et al., 1991; Shepherd, 1991) odor signals from the olfactory epithelium are carried via the receptor cell axons to the olfactory bulb glomeruli. Within the glomeruli the receptor axons establish excitatory axodendritic contact with principal (mitral/tufted) cell dendrites and intrinsic (periglomerular) cell dendrites (vide supra). Following the initial interaction between the sensory afferents and the primary/intrinsic neuronal elements, further odor processing takes place within the glomerulus and in the external plexiform layer (Shepherd and Greer, 1990).

At the level of the epithelium it seems evident that subpopulations of receptor cells respond to different odorants. Several laboratories have shown that the magnitude of response across the surface of the epithelium exhibits an odor-specific topography (Kauer and Moulton, 1974; Mackay-Sim et al., 1982). It is generally believed that these response patterns reflect the differential distribution of molecular odor receptors on individual cells within the epithelium (Ressler et al., 1993).

At the level of the olfactory bulb, several lines of evidence suggest that the glomerulus is a functional unit for odor processing. Anatomical tract-tracing studies employing horseradish peroxidase (HRP) lavaged into the olfactory cavity and taken up and transported to the olfactory bulb where it distributes among glomeruli (Stewart and Pedersen, 1987). Decreasing the extent of the lavage leads to partially heterogeneous labeling in glomeruli that correlates with the number of receptor cells exposed to the tracer. This concept of glomerular organization first gained support from surface EEG recordings where glomeruli were shown to have distinct specificities in their responses to different odors (Leveateau and MacLeod, 1966). This pioneering effort was followed by studies using carbon labeled 2-deoxyglucose in which autoradiograms showed unequivocally that odors elicited unique patterns of activity across the glomerular surface (Greer et al., 1981, 1982; Lancet et al., 1982; Benson et al., 1985). Moreover, recent high resolution studies employing early intermediate genes, such as cFOS, have shown distinct patterns of expression among and within glomeruli following stimulation with different odors (Guthrie et al., 1993). If a glomerulus has a defined functional specificity, then it follows that the receptor cell axons innervating that glomerulus also share a similar functional specificity. Consequently, it is reasonable to suggest that the formation of glomerular-specific fascicles of axons *(vide supra)* also reflects the convergence of receptor cell axons that may share a similar specificity in their molecular odor receptor repertoire. The nature of the glomerular specificity of odor representation also suggests that within the olfactory bulb there is a horizontal or columnar functional organization similar to those seen in other areas of the brain such as the cerebellum (Shepherd, 1972; Shepherd and Greer, 1990).

Although a regional specificity of odor representation within the external plexiform layer of the olfactory bulb has not been forthcoming, we can, nevertheless, speculate on the nature of local circuit processing occurring within this layer. As noted above, the principle microcircuit is the reciprocal dendrodendritic synapse between mitral/tufted cells and granule cells. The mitral-to-granule synapse is excitatory and the reciprocal granule-to-mitral is inhibitory (Mori and Takagi, 1978; Shepherd and Greer, 1990). As previously noted the secondary dendrites of mitral/tufted cells expand through long distances along the olfactory bulb. Therefore, the overall function of the intricate synaptic connectivity of the external plexiform layer is to provide an inhibitory mechanism for mitral/tufted cells which confines lateralization of odor information thus controlling the output signal (Shepherd and Greer, 1990; Woolf et al., 1991a,b; Woolf and Greer, 1994). Lateral inhibition may also serve as a mechanism to enhance differences between active and inactive sites in the bulb (Nickell and Shipley, 1992).

SUMMARY

The olfactory system is unique among sensory systems in that the primary receptors in contact with the central nervous system are directly exposed to the environment and capable of transporting volatile and nonvolatile substances into the brain. In addition, a variety of studies show the potential effects of environmental toxins or abnormal odor exposure on the ability of the olfactory system to detect and process odorants (Laing et al., 1985; Pinching and Doving, 1974; Doving and Pinching, 1973; Rehn et al., 1988; Matulionis, 1974; Smith and Duncan, 1992). The preceding review has summarized the cellular and synaptic organization of the olfactory system with the intent of providing a basic foundation for understanding where such environmental effects may be manifested.

ACKNOWLEDGMENTS

This work supported in part by NIH DC00210 and NS10174.

REFERENCES

Allison, A.C. and R.T.T. Warwick (1949) Quantitative observations on the olfactory system of the rabbit. *J. Anat.* 72:186-197.

Anderson, K.J., D.T. Monaghan, C.B. Cangro, M.A. Namboodiri, J.H. Neale, and C.W. Cotman (1986) Localization of N-acetylaspartylglutamate-like immunoreactivity in selected areas of the rat brain. *Neurosci. Lett.* 72:14-20.

Andres, K.H. (1965) Der feinbau des bulbus olfactorius der ratte unter besonderer berucksichtigung der synaptischen verbindungen. *Z. Zellforsch.* 65:530-561.

Andres, K.H. (1970) Anatomy and ultrastructure of the olfactory bulb in fish, amphibia, reptiles, birds and mammals. In G.E.W. Wolstenholme and J. Knight (eds): Taste and Smell in Vertebrates. London: J. & A. Churchill, pp. 177-196.

Astic, L., D. Saucier, and A. Holley (1987) Topographical relationships between olfactory receptor cells and glomerular foci in the rat olfactory bulb. *Brain Res.* 424:144-152.

Baier, H. and S. Korsching (1994) Olfactory glomeruli in the zebrafish form an invariant pattern and are identifiable across animals. *J. Neurosci.* 14:219-230.

Baker, H. (1988) Neurotransmitter plasticity in the juxtaglomerular cells of the olfactory bulb. In F.L. Margolis and T.V. Getchell (eds): Molecular Neurobiology of the Olfactory System. New York: Plenum Press, pp. 185-216.

Bartolomei, J.C. and C.A. Greer (1993) Synaptic organization of immunocytochemically identified GABA and TH processes in rat olfactory bulb glomeruli. *Soc. Neurosci. Abstr.* 19:125.

Benson, T.E., G.D. Burd, C.A. Greer, D.M. Landis, and G.M. Shepherd (1985) High-resolution 2-deoxyglucose autoradiography in quick-frozen slabs of neonatal rat olfactory bulb. *Brain Res.* 339:67-78.

Berkowicz, D.A., P.Q. Trombley, and G.M. Shepherd (1994) Evidence for glutamate as the olfactory receptor cell neurotransmitter. *J. Neurophysiol.* 71:2557-2561.

Bogan, N., N. Brecha, C. Gall, and H.J. Karten (1982) Distribution of enkephalin-like immunoreactivity in the rat main olfactory bulb. *J. Neurosci.* 7:895-906.

Buck, L. and R. Axel (1991) A novel multigene family may encode odorant receptors: a molecular basis for odor recognition. *Cell* 65:175-187.

Buck, L.B. (1992) The olfactory multigene family. *Curr. Opinion Neurobiol.* 2:282-288.

Burd, G.D., B.J. Davis, and F. Macrides (1982) Ultrastructural identification of substance P immunoreactive neurons in the main olfactory bulb of the hamster. *J. Neurosci.* 7:2697-2704.

Caggiano, M., J.S. Kauer, and D.D. Hunter (1994) Globose basal cells are neuronal progenitors in the olfactory epithelium: a lineage analysis using a replication-incompetent retrovirus. *Neuron* 13:339-352.

Cajal, S. Ramon y (1911) Histologie du système nerveux de l'homme et des vertebres. Paris: Maloine.

Chen, Y., M.L. Getchell, X. Ding, and T.V. Getchell (1992) Immunolocalization of two cytochrome P450 isozymes in rat nasal chemosensory tissue. *NeuroReport* 3:749-752.

Clancy, A.N., T.A. Schoenfeld, W.B. Forbes, and F. Macrides (1994) The spatial organization of the peripheral olfactory system of the hamster. Part II: Receptor surfaces and odorant passageways within the nasal cavity. *Brain Res. Bull.* 34:211-241.

Costanzo, R.M. and R.J. O'Connell (1978) Spatially organized projections of hamster olfactory nerves. *Brain Res.* 139:327-332.

Davis, B.J., G.D. Burd, and F. Macrides (1982) Localization of methionine-enkephalin, substance P, and somatostatin immunoreactivities in the main olfactory bulb of the hamster. *J. Comp. Neurol.* 204:377-383.

Davis, B.J. and F. Macrides (1983) Tyrosine hydroxylase immunoreactive neurons and fibers in the olfactory system of the hamster. *J. Comp. Neurol.* 214:427-440.

Davis, B.J. (1991) NADPH-diaphorase activity in the olfactory system of the hamster and rat. *J. Comp. Neurol.* 314:493-511.

Davis, B.J. and R.M. Kream (1991) Substance P immunoreactivity in the superficial laminae of the hamster olfactory bulb. *NeuroReport* 2:739-742.

Doucette, R. (1993) Glial cells in the nerve fiber layer of the main olfactory bulb of embryonic and adult mammals. *Microscopy Res. Tech.* 24:113-130.

Doving, K.B. and A.J. Pinching (1973) Selective degeneration of neurones in the olfactory bulb following prolonged odour exposure. *Brain Res.* 52:115-129.

Duncan, H.J., W.T. Nickell, M.T. Shipley, and R.C. Gesteland (1990) Organization of projections from olfactory epithelium to olfactory bulb in the frog, *Rana pipiens*. *J. Comp. Neurol.* 299:299-311.

Firestein, S. (1991) A noseful of odor receptors. *Trends Neurosci.* 7:270-272.

Firestein, S., F. Zufall, and G.M. Shepherd (1991) Single odor-sensitive channels in olfactory receptor neurons are also gated by cyclic nucleotides. *J. Neurosci.* 11:3565-3572.

Fuller, T.A. and J.L. Price (1988) Putative glutamatergic and/or aspartatergic cells in the main and accessory olfactory bulbs of the rat. *J. Comp. Neurol.* 276:209-218.

Gall, C., K.B. Seroogy, and N. Brecha (1986) Distribution of VIP- and NPY-like immunoreactivities in rat main olfactory bulb. *Brain Res.* 389:389-394.

Gall, C.M., S.H.C. Hendry, K.B. Seroogy, E.G. Jones, and J.W. Haycock (1987) Evidence for co-existence of GABA and dopamine in neurons of the rat olfactory bulb. *J. Comp. Neurol.* 266:307-318.

Getchell, M.L., J.A. Rafols, and T.V. Getchell (1984) Histological and histochemical studies of the secretory components of the salamander olfactory mucosa: effects of isoproterenol and olfactory nerve section. *Anatom. Rec.* 208:553-565.

Getchell, T.V. and G.M. Shepherd (1975) Synaptic actions on mitral and tufted cells elicited by olfactory nerve volleys in the rabbit. *J. Physiol.* 251:497-522.

Getchell, T.V., F.L. Margolis, and M.L. Getchell (1984) Perireceptor and receptor events in vertebrate olfaction. *Prog. Neurobiol.* 23:317-345.

Graziadei, P.P.C. and G.A. Monti-Graziadei (1979) Neurogenesis and neuron regeneration in the olfactory system of mammals. I. Morphological aspects of differentiation and structural organization of the olfactory sensory neurons. *J. Neurocytol.* 8:1-18.

Greer, C.A., W.B. Stewart, J.S. Kauer, and G.M. Shepherd (1981) Topographical and laminar localization of 2-deoxyglucose uptake in rat olfactory bulb induced by electrical stimulation of olfactory nerves. *Brain Res.* 217:279-293.

Greer, C.A. and G.M. Shepherd (1982) Mitral cell degeneration and sensory function in the neurological mutant mouse Purkinje cell degeneration (PCD). *Brain Res.* 235:156-161.

Greer, C.A., W.B. Stewart, M.H. Teicher, and G.M. Shepherd (1982) Functional development of the olfactory bulb and a unique glomerular complex in the neonatal rat. *J. Neurosci.* 2:1744-1759.

Greer, C.A. (1987) Golgi analyses of dendritic organization among denervated olfactory bulb granule cells. *J. Comp. Neurol.* 257:442-452.

Greer, C.A. and N. Halász (1987) Plasticity of dendrodendritic microcircuits following mitral cell loss in the olfactory bulb of the murine mutant Purkinje cell degeneration. *J. Comp. Neurol.* 256:284-298.

Greer, C.A. (1991) Structural organization of the olfactory system. In et al. T.V.Getchell (ed): Smell and Taste in Health and Disease. New York: Raven Press, pp. 65-81.

Guthrie, K.M., A.J. Anderson, M. Leon, and C. Gall (1993) Odor-induced increases in c-*fos* mRNA expression reveal an anatomical "unit" for odor processing in olfactory bulb. *Proc. Natl. Acad. Sci. U.S.A.* 90:3329-3333.

Haberly, L.B. and J.L. Price (1977) The axonal projection patterns of the mitral and tufted cells of the olfactory bulb in the rat. *Brain Res.* 129:152-157.

Haberly, L.B. (1983) Structure of the piriform cortex of the opossum. I. Description of neuron types with Golgi methods. *J. Comp. Neurol.* 213:163-187.

Haberly, L.B. and S.L. Feig (1983) Structure of the piriform cortex of the opossum. II. Fine structure of cell bodies and neuropil. *J. Comp. Neurol.* 216:69-88.

Haberly, L.B. (1990) Olfactory cortex. In G.M. Shepherd (ed): Synaptic Organization of the Brain. New York: Oxford University Press, pp. 317-345.

Halász, N., A. Ljungdahl, and T. Hokfelt (1979) Transmitter histochemistry of the rat olfactory bulb. III. Autoradiographic localization of [³H]GABA. *Brain Res.* 167:221-240.

Halász, N., O. Johansson, T. Hokfelt, A. Ljungdahl, and M. Goldstein (1981a) Immuno-histochemical identification of two types of dopamine neuron in the rat olfactory bulb as seen by serial sectioning. *J. Neurocytol.* 10:251-259.

Halász, N., D.M. Parry, N.M. Blackett, A. Ljungdahl, and T. Hokfelt (1981b) [³H]g-Aminobutyrate autoradiography of the rat olfactory bulb: hypothetical grain analysis of the distribution of silver grains. *J. Neurosci.* 6:473-479.

Halász, N. and G. Shepherd (1983) Neurochemistry of the vertebrate olfactory bulb. *Neuroscience* 10:579-619.

Halász, N. (1990) Morphological basis of information processing in the olfactory bulb. In D. Schild (ed): Chemosensory Information Processing. Berlin: Springer-Verlag, pp. 175-190.

Halász, N. and C.A. Greer (1993) Terminal arborizations of olfactory nerve fibers in the glomeruli of the olfactory bulb. *J. Comp. Neurol.* 337:307-316.

Hinds, J.W. and P.L. Hinds (1976a) Synapse formation in the mouse olfactory bulb: quantitative studies. *J. Comp. Neurol.* 169:15-40.

Hinds, J.W. and P.L. Hinds (1976b) Synapse formation in the mouse olfactory bulb. II. Morphogenesis. *J. Comp. Neurol.* 169:41-61.

Hinds, J.W. and N.A. McNelly (1981) Aging in the rat olfactory system: correlation of changes in the olfactory epithelium and olfactory bulb. *J. Comp. Neurol.* 203:441-453.

Hinds, J.W., P.L. Hinds, and N.A. McNelly (1984) An autoradiographic study of the mouse olfactory epithelium: evidence for long-lived receptors. *Anatom. Rec.* 210:375-383.

Hirata, Y. (1964) Some observations on the fine structure of the synapses in the olfactory bulb of the mouse, with particular reference to the atypical synaptic configurations. *Arch. Histol. Jpn.* 24:302-317.

Jackowski, A., J.G. Parnavelas, and A.R. Lieberman (1978) The reciprocal synapse in the external plexiform layer of the mammalian olfactory bulb. *Brain Res.* 159:17-28.

Jahr, C.E. and R.A. Nicoll (1982) An intracellular analysis of dendrodendritic inhibition in the turtle *in vitro* olfactory bulb. *J. Physiol.* 326:213-234.

Jastreboff, P.J., P.E. Pedersen, C.A. Greer, W.B. Stewart, J.S. Kauer, T.E. Benson, and G.M. Shepherd (1984) Specific olfactory receptor populations projecting to identified glomeruli in the rat olfactory bulb. *Proc. Natl. Acad. Sci. U.S.A.* 81:5250-5254.

Kauer, J.S. and D.G. Moulton (1974) Responses of olfactory bulb neurons to odour stimulation of small nasal areas in the salamander. *J. Physiol.* 243:717-737.

Kishi, K., K. Mori, and H. Ojima (1984) Distribution of local axon collaterals of mitral, displaced mitral, and tufted cells in the rabbit olfactory bulb. *J. Comp. Neurol.* 225:511-526.

Kishi, K. (1987) Golgi studies on the development of granule cells of the rat olfactory bulb with reference to migration in the subependymal layer. *J. Comp. Neurol.* 258:112-124.

Kosaka, T., Y. Hataguchi, K. Hama, I. Nagatsu, and J.Y. Wu (1985) Coexistence of immunoreactivies for glutamate decarboxylase and tyrosine hydroxylase in some neurons in the periglomerular region of the rat main olfactory bulb: possible coexistence of gamma-aminobutyric acid (GABA) and dopamine. *Brain Res.* 343:166-171.

Kosaka, T., K. Kosaka, C.W. Heizmann, I. Nagatsu, J.Y. Wu, N. Yanaihara, and K. Hama (1987) An aspect of the organization of the GABAergic system in the rat main olfactory bulb: laminar distribution of immunohistochemically defined subpopulations of GABAergic neurons. *Brain Res.* 411:373-378.

Kream, R.M., B.J. Davis, T. Kawano, F.L. Margolis, and F. Macrides (1984) Substance P and catecholaminergic expression in neurons of the hamster main olfactory bulb. *J. Comp. Neurol.* 222:140-154.

Kream, R.M., A.N. Clancy, M.S.A. Kumar, T.A. Schoenfeld, and F. Macrides (1987) Substance P and luteinizing hormone-releasing hormone levels in the brain of the male golden hamster are both altered by castration and testosterone replacement. *Neuroendocrinology* 46:297-305.

Laing, D.G., H. Panhuber, E.A. Pittman, M.E. Willcox, and G.K. Eagleson (1985) Prolonged exposure to an odor or deodorized air alters the size of mitral cells in the olfactory bulb. *Brain Res.* 336:81-87.

Lancet, D., C.A. Greer, J.S. Kauer, and G.M. Shepherd (1982) Mapping of odor-related neuronal activity in the olfactory bulb by high-resolution 2-deoxyglucose autoradiography. *Proc. Natl. Acad. Sci. U.S.A.* 79:670-674.

Leveateau, J. and P. MacLeod (1966) Olfactory discrimination in the rabbit olfactory glomerulus. *Science* 153:175-176.

Mackay-Sim, A., P. Shaman, and D.G. Moulton (1982) Topographic coding of olfactory quality: odorant-specific patterns of epithelila responsivity in the salamander. *J. Neurophysiol.* 48:584-596.

Mackay-Sim, A. and M.H. Nathan (1984) The projection from the olfactory epithelium to the olfactory bulb in the salamander, *Ambystoma tigrinum. Anat. Embryol.* 170:93-97.

Mackay-Sim, A. and P. Kittel (1991) Cell dynamics in the adult mouse olfactory epithelium: a quantitative autoradiogaphic study. *J. Neurosci.* 11:979-984.

Macrides, F. and S.P. Schneider (1982) Laminar organization of mitral and tufted cells in the main olfactory bulb of the adult hamster. *J. Comp. Neurol.* 208:419-430.

Macrides, F. and B.J. Davis (1983) The olfactory bulb. In P.C. Emson (ed): Chemical Neuroanatomy. New York: Raven Press, pp. 391-426.

Macrides, F., T.A. Schoenfeld, J.E. Marchand, and A.N. Clancy (1985) Evidence for morphologically, neurochemically and functionally heterogeneous classes of mitral and tufted cells in the olfactory bulb. *Chem. Senses* 10:175-202.

Margolis, F.L. (1972) A brain protein unique to the olfactory bulb. *Proc. Natl. Acad. Sci. U.S.A.* 69:1221-1224.

Margolis, F.L. and S. Rochel (1982) Carnosine release from olfactory bulb synaptosomes is calcium-dependent and depolarization-stimulated. *J. Neurochem.* 38:1505-1514.

Matulionis, D.H. (1974) Ultrastructure of olfactory epithelia in mice after smoke exposure. *Ann. Otol. Rhinol. Laryngol.* 83:192-201.

McLean, J.H. and M.T. Shipley (1992) Neuroanatomical substrates of olfaction. In M.J. Serby and K.L. Chobor (eds): Science of Olfaction. New York: Springer-Verlag, pp. 126-171.

Meisami, E. and L. Safari (1981) A quantiative study of the effects of early unilateral olfactory deprivation on the number and distribution of mitral and tufted cells and glomeruli in the rat olfactory system. *Brain Res.* 221:81-107.

Mori, K. and S.F. Takagi (1978) An intracellular study of dendrodendritic inhibitory synapses on mitral cells in the rabbit olfactory bulb. *J. Physiol.* 279:569-588.

Mori, K., K. Kishi, and H. Ojima (1983) Distribution of dendrites of mitral, displaced mitral, tufted, and granule cells in the rabbit olfactory bulb. *J. Comp. Neurol.* 219:339-355.

Mori, K., S.C. Fujita, K. Imamura, and K. Obata (1985) Immunohistochemical study of subclasses of olfactory nerve fibers and their projections to the olfactory bulb in the rabbit. *J. Comp. Neurol.* 242:214-229.

Mori, K. (1987) Membrane and synaptic properties of identified neurons in the olfactory bulb. *Prog. Neurobiol.* 29:275-320.

Morrison, E.E. and R.M. Costanzo (1992) Morphology of olfactory epithelium in humans and other vertebrates. *Microscopy Res. Tech.* 23:49-61.

Ngai, J., A. Chess, M.M. Dowling, N. Necles, E.R. Macagno, and R. Axel (1993a) Coding in olfactory information: topography of odorant receptor expression in the catfish olfactory epithelium. *Cell* 72:667-680.

Ngai, J., M.M. Dowling, L. Buck, R. Axel, and A. Chess (1993b) The family of genes encoding odorant receptors in the channel catfish. *Cell* 72:657-666.

Nickell, W.T. and M.T. Shipley (1992) Neurophysiology of the olfactory bulb. In M.M. Serby and K.L. Chobor (eds): Science of Olfaction. New York: Spring-Verlag, pp. 172-212.

Nicoll, R.A. (1971) Pharmacological evidence for GABA as the transmitter in granule cell inhibiton in the olfactory bulb. *Brain Res.* 35:137-149.

Ojima, H., K. Mori, and K. Kishi (1984) The trajectory of mitral cell axons in the rabbit olfactory cortex revealed by intracellular HRP injection. *J. Comp. Neurol.* 230:77-87.

Orona, E., J.W. Scott, and E.C. Rainer (1983) Different granule cell populations innervate superficial and deep regions of the external plexiform layer in rat olfactory bulb. *J. Comp. Neurol.* 217:227-237.

Orona, E., E.C. Rainer, and J.W. Scott (1984) Dendritic and axonal organization of mitral and tufted cells in the rat olfactory bulb. *J. Comp. Neurol.* 226:346-356.

Pinching, A.J. and T.P.S. Powell (1971a) The neuropil of the glomeruli of the olfactory bulb. *J. Cell Sci.* 9:347-377.

Pinching, A.J. and T.P.S. Powell (1971b) The neuron types of the glomerular layer of the olfactory bulb. *J. Cell Sci.* 9:305-345.

Pinching, A.J. and T.P.S. Powell (1971c) The neuropil of the periglomerular region of the olfactory bulb. *J. Cell Sci.* 9:379-409.

Pinching, A.J. and K.B. Doving (1974) Selective degeneration in the rat olfactory bulb following exposure to different odours. *Brain Res.* 82:195-204.

Price, J.L. and T.P.S. Powell (1970a) The morphology of the granule cells of the olfactory bulb. *J. Cell Sci.* 7:91-123.

Price, J.L. and T.P.S. Powell (1970b) An electron-microscopic study of the termination of the afferent fibres to the olfactory bulb from the cerebral hemisphere. *J. Cell Sci.* 7:157-187.

Price, J.L. and T.P.S. Powell (1970c) The mitral and short axon cells of the olfactory bulb. *J. Cell Sci.* 7:631-651.

Rakic, P. (1976) Local circuit neurons. *Neurosci. Res. Progr. Bull.* 13:291-446.

Rall, W., G.M. Shepherd, T.S. Reese, and M.W. Brightman (1966) Dendrodendritic synaptic pathway for inhibition in the olfactory bulb. *Exp. Neurol.* 14:44-56.

Rall, W. and G.M. Shepherd (1968) Theoretical reconstruction of field potentials and dendrodendritic synaptic interactions in olfactory bulb. *J. Neurophysiol.* 31:884-915.

Raming, K., J. Krieger, J. Strotmann, I. Boekhoff, S. Kubick, C. Baumstark, and H. Breer (1993) Cloning and expression of odorant receptors. *Nature* 361:353-356.

Rehn, B., H. Panhuber, D.G. Laing, and W. Breipohl (1988) Spine density on olfactory granule cell dendrites is reduced in rats reared in a restricted olfactory environment. *Dev. Brain Res.* 40:143-147.

Ressler, K.J., S.L. Sullivan, and L.B. Buck (1993) A zonal organization of odorant receptor gene expression in the olfactory epithelium. *Cell* 73:597-609.

Ribak, C.E., J.E. Vaughn, K. Saito, R. Barber, and E. Roberts (1977) Glutamate decarboxylase localization in neurons in the olfactory bulb. *Brain Res.* 126:1-18.

Saito, N., N. Kikkawa, Y. Nishizuka, and C. Tanaka (1988) Distribution of protein kinase C-like immunoreactive neurons in the brain. *J. Neurosci.* 8:369-382.

Sanides-Kohlrausch, C. and P. Wahle (1990) VIP-and PHI-immunoreactivity in olfactory centers of the adult cat. *J. Comp. Neurol.* 294:325-339.

Sassoè-Pognetto, M., D. Cantino, P. Panzanelli, L.V. di Cantogno, M. Giustetto, F.L. Margolis, S. de Biasi, and A. Fasolo (1993) Presynaptic colocalization of carnosine and glutamate in olfactory neurones. *NeuroReport* 5:7-10.

Schneider, S.P. and F. Macrides (1978) Laminar distributions of interneurons in the main olfactory bulb of the adult hamster. *Brain Res. Bull.* 3:73-82.

Schneider, S.P. and J.W. Scott (1983) Orthodromic response properties of rat olfactory bulb mitral and tufted cells correlate with their projection patterns. *J. Neurophysiol.* 50:358-378.

Schoenfeld, T.A., A.N. Clancy, and W.B. Forbes (1994) The spatial organization of the peripheral olfactory system of the hamster. Part I: Receptor neuron projections to the main olfactory bulb. *Brain Res. Bull.* 34:183-210.

Schwarting, G.A. and J.E. Crandall (1991) Subsets of olfactory and vomeronasal sensory epithelial cells and axons revealed by monoclonal antibodies to carbohydrate antigens. *Brain Res.* 547:239-248.

Schwob, J.E. and D.I. Gottlieb (1986) The primary olfactory projection has two chemically distinct zones. *J. Neurosci.* 6:3393-3404.

Schwob, J.E. (1992) The biochemistry of olfactory neurons: stages of differentiation and neuronal subsets. In M.J. Serby and K.L. Chobor (eds): Science of Olfaction. New York: Springer-Verlag, pp. 80-125.

Scott, J.W. (1981) Electrophysiological identification of mitral and tufted cells and distributions of their axons in olfactory system of the rat. *J. Neurophysiol.* 46:918-931.

Scott, J.W., D.P. Wellis, M.J. Riggott, and N. Buonviso (1993) Functional organization of the main olfactory bulb. *Microscopy Res. Tech.* 24:142-156.

Seroogy, K.B., N. Brecha, and C. Gall (1985) Distribution of cholecystokinin-like immunoreactivity in the rat main olfactory bulb. *J. Comp. Neurol.* 239:373-383.

Shepherd, G.M. (1972) Synaptic organization of the mammalian olfactory bulb. *Physiol. Rev.* 52:864-917.

Shepherd, G.M. and C.A. Greer (1990) Olfactory bulb. In G.M. Shepherd (ed): The Synaptic Organization of the *Brain.* New York: Oxford University Press, pp. 133-169.

Shepherd, G.M. (1991) Sensory transduction: entering the mainstream of membrane signaling. *Cell* 67:845-851.

Shepherd, G.M. (1993a) Principles of specificity and redundancy underlying the organization of the olfactory system. *Microscopy Res. Tech.* 24:106-112.

Shepherd, G.M. (1993b) Current issues in the molecular biology of olfaction. *Chem. Senses* 18:191-198.

Skeen, L.C. and W.C. Hall (1977) Efferent projections of the main and accessory olfactory bulb in the tree shrew *(Tupaia glis). J. Comp. Neurol.* 172:1-36.

Smith, D.V. and H.J. Duncan (1992) Primary olfactory disorders: anosmia, hyposmia, and dysosmia. In M.J. Serby and K.L. Chobor (eds): Science of Olfaction. New York: Springer-Verlag, pp. 439-465.

Stewart, W.B. and P.E. Pedersen (1987) The spatial organization of olfactory nerve projections. *Brain Res.* 411:248-258.

Trombley, P.Q. and G.M. Shepherd (1993) Synaptic transmission and modulation in the olfactory bulb. *Curr. Opinion Neurobiol.* 3:540-547.

Tsuruo, Y., T. Hokfelt, and T. Visser (1988) Thyrotropin-releasing hormone (TRH)-immunoreactive neuron populations in the rat olfactory bulb. *Brain Res.* 447:183-187.

White, E.L. (1972) Synaptic organization in the olfactory glomerulus of the mouse. *Brain Res.* 37:69-80.

Woolf, T.B., G.M. Shepherd, and C.A. Greer (1991a) Serial reconstructions of granule cell spines in the mammalian olfactory bulb. *Synapse* 7:181-192.

Woolf, T.B., G.M. Shepherd, and C.A. Greer (1991b) Local information processing in dendritic trees: subsets of spines in granule cells of the mammalian olfactory bulb. *J. Neurosci.* 11:1837-1854.

Woolf, T.B. and C.A. Greer (1994) Local communication within dendritic spines: models of second messenger diffusion in granule cell spines of the mammalian olfactory bulb. *Synapse* 17:247-267.

CHAPTER 2

Physicochemical Determinants and Functional Properties of the Senses of Irritation and Smell

J. Enrique Cometto-Muñiz and William S. Cain

ABSTRACT

Airborne volatile organic compounds (VOCs) are prime suspects in indoor air related illnesses (e.g., sick building syndrome). Frequent symptoms in those cases involve sensory responses of odor, nasal pungency (irritation), and eye irritation. We separated the trigeminal (pungent) from the olfactory (odor) response of the nose to a common set of substances by measuring nasal detection thresholds in subjects lacking a functional sense of smell (i.e., anosmics) and in matched subjects with normal olfaction (i.e., normosmics). Eye irritation thresholds for the same compounds were also measured. Stimuli comprised homologous series of alcohols, acetates, ketones, and alkylbenzenes. Physicochemical properties change orderly and systematically in such series allowing us to relate those properties with the observed sensory responses. In all series and for the three sensory modalities, thresholds decreased with carbon chain length. For nasal pungency — but not for odor — thresholds *across* chemical series bore a uniform linear relationship with simple physicochemical properties (e.g., saturated vapor concentration at room temperature). Eye irritation thresholds fell remarkably close to those of nasal pungency. Mixtures of VOCs can reach odor, pungency, and eye irritation thresholds when all components are below their individual thresholds.

INTRODUCTION

Sensory irritation of eyes, nose, and throat is a frequent symptom in occupants of spaces with indoor air quality problems.[1] In most cases, the responsible agent(s)

1-56670-144-9/96/$0.00+$.50

elude specification, but VOCs emitted from indoor materials and present at relatively low levels are prime suspects.[2-4]

Detection of airborne chemicals by humans relies on two sensory channels: olfaction[5] and the so-called common chemical sense (CCS) or chemical irritation sense.[6] Odor sensations are carried by the olfactory nerve (cranial nerve I). They originate in specialized structures: the olfactory neurons located in the olfactory epithelium that covers a relatively small patch in the upper back portion of the nasal cavity. These bipolar neurons send a dendrite to the epithelium surface and an axon that joins other axons from neighboring cells to form the olfactory nerve. The nerve penetrates through perforations in the ethmoid bone (cribriform plate) and makes the first synapse of the pathway in olfactory bulb structures called glomeruli. The dendrite on the other end of each neuron reaches the epithelium surface and ends in an olfactory knob from which a number of cilia protrude and are immersed in the mucus bathing the surface. These cilia are most likely the sites of olfactory transduction.[7,8] Chapter 1 describes in detail the neurobiology of smell.

Unlike smell, the CCS lacks specific receptor structures — CCS reception relies on free nerve endings — and is widely distributed since there is general chemical sensitivity in all body mucosae (conjunctival, nasal, oral, upper and lower respiratory tract, genital, and anal). Of special relevance to sensory irritation generated by indoor air are two face mucosae: conjunctival and nasal. Common chemical sensations from the eyes and the nose are mediated by free nerve endings from the trigeminal nerve (cranial nerve V). We like to group together under the term *nasal pungency* an array of responses elicited in the nose by CCS stimulation. These pungent sensations include: stinging, prickliness, tingling, irritation, burning, piquancy, and freshness, among others. For simplicity, we will refer to CCS responses in the conjunctiva as *eye irritation,* although most of the terms just mentioned can be applied to the eye and thus, we can also talk of an ocular pungency.

SENSORY RESPONSES MEASURED
AND COMPOUNDS TESTED

It would be convenient to be able to separate the odor from the pungent response of the nose to airborne chemicals. Unfortunately, almost all chemicals evoke responses from both sensory systems. In general, at relatively low concentrations only odor is apparent, but as concentration increases pungency joins in. In this way, nasal pungency thresholds would have to be measured, in most cases, against a strong odorous background. This makes such thresholds heavily dependent on the individual criterion of when to call an odor sensation barely pungent.

In order to overcome the problem, we measured nasal pungency thresholds in persons lacking a functional sense of smell (i.e., *anosmics*) for whom odor did not interfere. *Odor* thresholds, on the other hand, were measured on subjects with normal olfaction (i.e., *normosmics*) matched to the anosmics by age, gender, and smoking status (variables known to affect chemosensory sensitivity).[9,10]

The approach was complemented by a thoughtful selection of the chemical stimuli. Our goal was to establish relationships and trends between chemical structure and potency of the compound as odorant or irritant, using substances relevant to indoor air. For these reasons we selected homologous series of relatively nonreactive chemicals. Physicochemical properties in such series change orderly and systematically, allowing us to track down how these gradual changes reflect themselves in the sensory responses. The choice of relatively nonreactive chemicals was based on the fact that, in the great majority of cases of indoor air quality complaints, chemical analysis fails to reveal the presence of any overt irritant (e.g., formaldehyde) of the type that will react readily with mucosal tissue. Given these considerations, the substances tested included: n-alcohols,[11] acetates,[12] branched alcohols and acetates, ketones,[13] and alkylbenzenes.[14] The compounds were delivered in vapor phase from squeeze bottles adapted to stimulate one nostril[15] or one eye.[12] Concentration in the headspace of the bottles was measured by gas chromatography.

NASAL PUNGENCY AND ODOR THRESHOLDS

Figure 1 shows the average pungency and odor thresholds, measured in anosmics and normosmics, respectively, obtained for normal homologous alcohols from methanol to 1-octanol. Also shown are the thresholds for some secondary alcohols: 2-propanol, 2-butanol (or sec-butyl alcohol), and 4-heptanol; and one tertiary alcohol: 2-methyl-2-propanol (or tert-butyl alcohol).

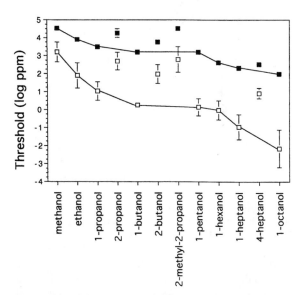

Figure 1 Average nasal pungency thresholds measured in anosmics (filled symbols) and odor thresholds measured in normosmics (empty symbols) for homologous n-alcohols, and secondary and tertiary alcohols. Only n-members of the series are joined by a line. Each point in the anosmic group represents the mean of at least 36 thresholds. Each point in the normosmic group represents the mean of 48 thresholds. Bars indicate standard deviation.

Within the *n*-alcohols, both nasal thresholds tend to decline with carbon chain length albeit not at the same rate; a feature that will be seen repeatedly in all series tested. Odor thresholds do not reach a plateau (as will be seen with the following chemical series), but ability to elicit pungency starts to fade with 1-octanol where the anosmics failed to reach threshold criterion in 25% of instances. Interestingly, switching the alcohols' chemical functional group (HO–) from a primary to a secondary carbon (e.g., 1-heptanol to 4-heptanol) or even to a tertiary carbon (e.g., 1-butanol vs. *tert*-butyl alcohol) raises both odor and pungency thresholds.

The clear concentration difference between pungency and odor thresholds does not result from an artifact of averaging, as the individual thresholds from each anosmic and normosmic demonstrate (Figure 2). The groups show no overlap of threshold values. Note that: (1) the spread of individual thresholds among anosmics is much smaller than among normosmics (something that will be repeated in all series) and (2) there seems to be a general factor of odor sensitivity: the normosmic most sensitive (i.e., having the lowest threshold) to one particular *n*-alcohol tends to be the most sensitive to the other homologous compounds (the same holds for the least sensitive normosmic).

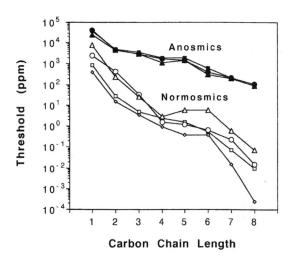

Figure 2   Individual thresholds measured in anosmics and normosmics for *n*-alcohols from methanol to 1-octanol (carbon chain length 1 to 8, respectively). Each point represents the median of 12 thresholds measured in one subject.

Figure 3 depicts pungency and odor thresholds for homologous acetates ranging from methyl to dodecyl acetate and for secondary and tertiary butyl acetate. Similarly to the outcome with the alcohols, both thresholds decline as the series progresses, there is no overlap between anosmics and normosmics (see SD bars), and odor threshold variability is higher than that for pungency. In addition, odor thresholds tend to plateau after butyl acetate, while pungency thresholds tend to

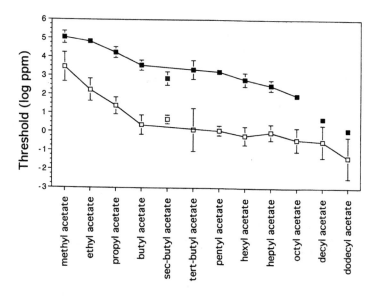

Figure 3   Average nasal pungency thresholds (filled symbols), and average odor thresholds (empty symbols) as a function of carbon chain length for homologous *n*-acetates, and secondary and tertiary butyl acetate. Nasal pungency threshold values for decyl and dodecyl acetate are not connected to the rest by lines because they represent the result from only one anosmic; the other three anosmics did not reliably detect these two stimuli. Bars indicate standard deviation.

fade after heptyl/octyl acetate since only one of the four anosmics reliably detected pungency from decyl and dodecyl acetate. The branched butyl acetates (*sec* and *tert*) did not show a systematic increase or decrease in their thresholds compared to the unbranched member.

Thresholds for the ketone series are shown in Figure 4. As seen with unbranched alcohols and acetates, pungency and odor thresholds for the ketones decline with carbon chain length, and in this case plateau after 2-heptanone.

Figure 5 presents the results for the alkylbenzenes. Many features observed in the previous chemical families are repeated here (e.g., decline of both thresholds with carbon chain length, greater variability in odor thresholds, no overlap between anosmics and normosmics) but, in the alkylbenzenes, both the fading of pungency and the plateau of odor thresholds occur relatively early in the series (members above propyl benzene).

In order to get a general sense of how our odor thresholds compare with those reported in the literature, we looked at a recent compilation, standardization, and averaging of human olfactory thresholds.[16] There is a high correlation ($r = 0.90$) between our odor thresholds and the average values from the compilation for the 25 substances common to both sets (Figure 6). From the lowest to the highest value, our thresholds cover a larger span, approximately six orders of magnitude rather than their four orders, and, thereby, offer better resolution among compounds. Averaging across studies undoubtedly accounts for much of

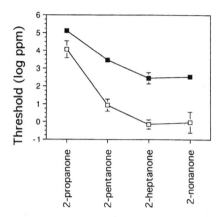

Figure 4   Nasal pungency thresholds and odor thresholds for homologous ketones. Each point represents a geometric mean across subjects. Bars indicate standard deviations.

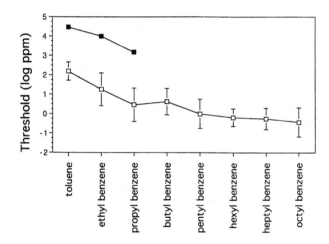

Figure 5   Nasal pungency thresholds and odor thresholds for alkylbenzenes. Each point in the pungency or odor function represents the average of 48 thresholds measured in four subjects. (The standard deviation on pungency thresholds is small enough to be covered by the symbol.)

the apparent constriction in range in the compiled data. Our thresholds, which represent the point of 100% detection, understandably often fall above those in the compilation, which presumably represent principally the points of 50% or 75% detection.

When both nasal thresholds (pungency and odor) for the total 42 VOCs studied so far are plotted as a function of a simple physicochemical property, e.g., saturated vapor concentration at room temperature (23°C), a picture such as that

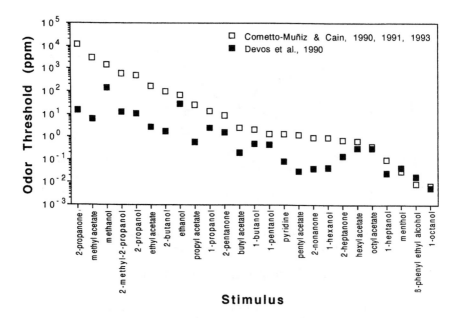

Figure 6   Comparison between our odor thresholds,[11-13] (from *Arch. Environ. Health,* 48, 313, 1993, with permission) sorted in descending order, and those compiled, standardized, and averaged by Devos et al.[16]

shown in Figure 7 emerges. Pungency thresholds taken as a whole — irrespective of molecular size or chemical functional group — exhibit a linear relationship with saturated vapor (r = 0.97), having a slope of 1.02. Pungency for individual homologous series conforms to the general picture (slopes: 0.90 for the alcohols, 1.07 for the acetates, 1.06 for the ketones, 1.17 for the alkylbenzenes, and 0.95 for miscellaneous substances). The linear correlation with slope close to 1.00 suggests that when a certain uniform percentage of vapor saturation is achieved (approximately 32%), nasal pungency would occur in the anosmics if it were to be evoked at all.

On the other hand, odor thresholds as a whole depicted more substance-to-substance scatter than pungency thresholds. They generally failed to show a linear relationship with saturated vapor, displaying systematic deviations, or, even if the odor thresholds of an individual series approximated a linear relationship, its slope departed from unity. The odor thresholds for no *individual* homologous series exhibited as strong a correlation with vapor saturation as did the pungency thresholds for *all* series *and* miscellaneous compounds grouped together. For odor, the best correlation occurred with the ketones, where r = 0.95, and the worst occurred with the acetates, where r = 0.87. The slopes of the relationships for odor thresholds commonly departed from unity: the alcohols had a slope of 1.62; the acetates, 0.83; the ketones; 1.64; the alkylbenzenes, 0.68; and the miscellaneous chemicals, 1.44.

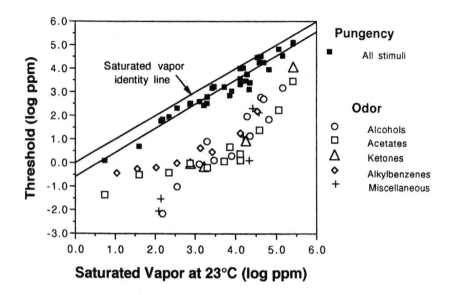

Saturated Vapor at 23°C (log ppm)

Figure 7   Pungency and odor thresholds for homologous alcohols, acetates, ketones, alkylbenzenes, and miscellaneous chemicals, depicted as a function of their saturated
vapor concentration at room temperature. In decreasing order of saturated vapor
concentrations, the alcohols include methanol, ethanol, 2-propanol; *tert*-butyl
alcohol, 1-propanol, 2-butanol, 1-butanol, 1-pentanol, 4-heptanol, 1-hexanol,
1-heptanol, and 1-octanol. The acetates include methyl, ethyl, propyl, *tert*-butyl,
butyl, *sec*-butyl, pentyl, hexyl, heptyl, octyl, decyl, and dodecyl acetate. The
ketones include acetone, 2-pentanone, 2-heptanone, and 2-nonanone. The alkylbenzenes include toluene through octyl benzene, and the miscellaneous substances include 1-octyne, 1-octene, pyridine, chlorobenzene, menthol, and
β-phenyl ethyl alcohol. The line representing pungency has a slope of 1.02 and
an r = 0.967. The saturated vapor identity line is shown for reference.

EYE IRRITATION THRESHOLDS

As mentioned in the introduction, CCS in the face includes the conjunctiva.
Eye irritation is another symptom often mentioned by occupants of spaces with
indoor air quality complaints. Also, from a more basic point of view, we are
interested in exploring the relative sensitivity of the CCS in the nasal and ocular
mucosae, both served by branches of the same trigeminal nerve but with different
mucus and epithelial layers.

A modification of the squeeze bottle's cap allowed us to produce ocular stimulation with vapors.[12] We, then, measured eye irritation thresholds for selected
members of all homologous series previously studied, using an analogous procedure
to that employed before (i.e., two-alternative forced choice with an ascending
concentration approach).[11] The results showed that eye irritation thresholds were
remarkably close to the nasal pungency thresholds measured before in anosmics
(Figure 8). This opens the possibility of learning about the nasal pungency-eliciting
potential of airborne VOCs (irrespective of odor) by simply testing them in the eyes.

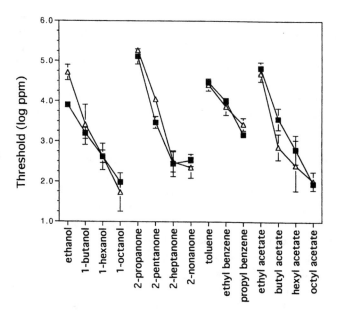

Figure 8   Comparison of eye irritation thresholds (triangles) and nasal pungency thresholds (squares) for selected members of homologous series of alcohols, ketones, alkylbenzenes, and acetates. Bars indicate standard deviations.

THRESHOLDS FOR MIXTURES OF SUBSTANCES

In the real world we are almost invariably faced with chemical stimulation from mixtures of compounds. Nevertheless, few investigations have addressed the issue of how odor thresholds for individual substances compare to odor thresholds for mixtures of them.[17-20] None addressed that issue for CCS thresholds (nasal pungency, eye irritation).

We designed an experiment where we prepared five mixtures of varying complexity: two three-component mixtures, two six-component mixtures, and one nine-component mixture. The constituents of such mixtures were selected compounds from the series studied before. The reason behind having two mixtures with three constituents and two with six was that one of each pair contained relatively high vapor pressure, highly water soluble substances while the other member of the pair contained relatively low vapor pressure and highly lipid soluble substances. This was done in an attempt to explore trends between the sensory thresholds in mixtures and general physicochemical properties.

Based on our previously measured average thresholds, we prepared the mixtures so they would be odor balanced, i.e., all components would be present at the same multiple or submultiple of their particular odor threshold. Odor (in normosmics), nasal pungency (in anosmics), and eye irritation thresholds were measured not only for the mixtures but also for the single compounds. At present we are exploring different ways to process the data. One of the problems we have encountered is that the new individual subject odor thresholds measured in this study depart from the odor threshold for the "average" subject on which the balanced mixtures were prepared. Thus, for each particular subject, the components in the mixtures were not necessarily odor balanced as intended (and described above).

A preliminary look at the results indicates that as mixtures get more complex (higher number of components) the concentration of each individual constituent necessary for a sensory threshold (odor, pungency, or eye irritation) to be elicited from the mixture as a whole becomes lower. Figure 9 depicts the average sensory threshold for each of nine substances. The nine sections of the graph correspond to the nine compounds. In each section, the first value (e.g., 1-propanol) represents the threshold for the substance alone. The following values represent the concentration at which the substance was present in mixtures of increasing complexity (e.g., in mixture A, C, and E), when the mixture *as a whole* achieved threshold. A and B are the three-component mixtures, C and D are the six-component mixtures, and E is the nine-component mixture. Note that, for every chemical, the concentration at which the chemical was present when the stimulus reached threshold is always lower in mixtures vs. when single, and also tends to be even lower if the mixture is more complex (i.e., if it has a higher number of components). This trend holds for all three sensations: odor, pungency, and eye irritation.

A comparison of the results obtained with mixtures of an equal number of substances but of contrasting general physicochemical properties (i.e., mixture A vs. mixture B, and C vs. D) suggest that for the attributes nasal pungency and eye irritation, but not for odor, mixtures of substances with higher lipid solubility drive down the threshold for the mixture to a larger extent than those with lower lipid solubility.

CONCLUSIONS AND NEEDS FOR FUTURE RESEARCH

Odor thresholds are always below (most times well below) nasal pungency and eye irritation thresholds. A common feature found in all series studied is the decline in all three sensory thresholds with increasing carbon chain length. The decline in odor thresholds tends to be steeper than that in pungency thresholds, at least for the first few members of each series, so, up to that point, the gap between odor and pungency grows larger as the series progresses. It should be pointed out that substances not usually regarded as irritants (e.g., 1-heptanol, 2-nonanone) not only can be detected by anosmics (i.e., they have pungency), but their pungency threshold is lower than that of more typical irritants (i.e., methanol, acetone).

In general, the decline in odor thresholds across each series reaches a plateau. The exception is the alcohol series where, up to 1-octanol, no tendency to plateau is observed in the odor thresholds. On the other hand, pungency thresholds across the series do not plateau, but a member is reached where the ability to evoke pungency is lost for one or more of the anosmics tested (1-octanol, octyl acetate, propylbenzene). Interestingly, the anosmics describe the quality of the pungency

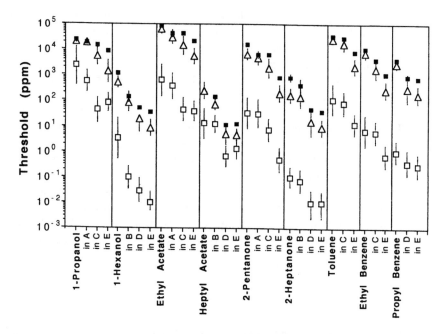

Figure 9 Thresholds (ppm ± SD) for nasal pungency (filled squares), eye irritation (triangles), and odor (empty squares). The nine sections of the graph correspond to nine substances. Each section lists, first, the threshold for the substance by itself (e.g., 1-propanol), then, consecutively, the level at which that substance was present when the threshold was achieved for mixtures of increasing complexity (e.g., 1-propanol *in* mixture *A* (3 components) when A achieved threshold, 1 propanol *in* mixture *C* (6 components) when C achieved threshold, 1-propanol *in* mixture *E* (9 components) when E achieved threshold).

evoked by the first members of each series as "sharp" or "biting," as opposed to that of the last members capable of being detected whose pungency is more "dull" or "pastel." This raises the question for future investigations of the discriminative capability of the human CCS, in the absence of olfaction, to distinguish among chemical vapors.

One issue that needs to be explored further due to its direct implications on indoor air is that of the sensory thresholds for mixtures of substances. Up to what extent can sensory thresholds for mixtures be driven down by adding more components? Are there certain compounds that add their sensory effects better than others? Is there a sensory channel that integrates the signals from different chemicals more completely than the others? What implications might this have regarding the mechanisms of the reception process(es) by which these airborne substances are perceived?

There are also at least two additional important parameters that need to be better understood. One of them is the effect of the presentation procedure on the sensory thresholds obtained. All the results described here were obtained with squeeze bottles. Our recent work indicates that the absolute value of the thresholds — although most likely *not* the relative value among compounds — changes if the stimuli are presented through face masks fed from small chambers or in environmental chambers. Results show that thresholds are lower from face masks and even lower in whole-body chambers. An important question is whether there is a constant factor that would allow comparison of thresholds measured with these different procedures (at least within related chemicals) or if the relationship among the procedures is very different from one chemical to another and cannot be generalized.

The second parameter, very relevant to indoor air quality, is the time-course characteristics of the various sensory impressions from airborne chemicals. All the results described here represent very short-term exposure responses (typically one or two sniffs). It is well known that odor sensations tend to diminish with time (olfactory adaptation) while common chemical sensations can build up for 30 or more minutes (temporal integration or summation) before adaptation starts to be produced.[21,22] Unveiling the temporal properties of all these sensations is central to understanding the adverse sensory reactions produced in indoor spaces.

ACKNOWLEDGMENT

Preparation of this article supported by NIH Grant DC00284.

REFERENCES

1. Cain, W.S. and J.E. Cometto-Muñiz: Irritation and odor: Symptoms of indoor air pollution. In *Indoor Air '93. Vol. 1: Health Effects (Proceedings of the 6th International Conference on Indoor Air Quality and Climate)*, edited by J.J.K. Jaakkola, R. Ilmarinen, and O. Seppänen. Helsinki: Indoor Air '93, 1993. pp. 21-31.
2. Møhave, L.: Volatile organic compounds, indoor air quality and health. *Indoor Air* 4: 357-376 (1991).
3. Møhave, L., B. Bach, and O.F. Pedersen: Human reactions to low concentrations of volatile organic compounds. *Environ. Int.* 12: 167-175 (1986).
4. Møhave, L., J. Grønkjær Jensen, and S. Larsen: Subjective reactions to volatile organic compounds as air pollutants. *Atmos. Environ.* 25A: 1283-1293 (1991).
5. Cain, W.S.: Olfaction. In *Stevens' Handbook of Experimental Psychology,* edited by R.C. Atkinson, R.J. Herrnstein, G. Lindzey, and R.D. Luce. New York: John Wiley & Sons, 1988. pp. 409-459.
6. Green, B.G., J.R. Mason, and M.R. Kare, ed.: *Chemical Senses. Vol. 2: Irritation.* New York: Marcel Dekker, 1990.
7. Chen, Z., and D. Lancet: Membrane proteins unique to vertebrate olfactory cilia: Candidates for sensory receptor molecules. *Proc. Natl. Acad. Sci. U.S.A.* 81: 1859-1863 (1984).
8. Rhein, L.D., and R.H. Cagan: Role of cilia in olfactory recognition. In *Biochemistry of Taste and Olfaction,* edited by R.H. Cagan, and M.R. Kare. New York: Academic Press, 1981. pp. 47-68.

9. Cometto-Muñiz, J.E. and W.S. Cain: Influence of airborne contaminants on olfaction and the common chemical sense. In *Smell and Taste in Health and Disease,* edited by T.V. Getchell, R.L. Doty, L.M. Bartoshuk, and J.B. Snow Jr. New York: Raven Press, 1991. pp. 765-785.

10. Cometto-Muñiz, J.E., and W.S. Cain: Sensory irritation. Relation to indoor air pollution. *Ann. N.Y. Acad. Sci.* 641: 137-151 (1992).

11. Cometto-Muñiz, J.E., and W.S. Cain: Thresholds for odor and nasal pungency. *Physiol. Behav.* 48: 719-725 (1990).

12. Cometto-Muñiz, J.E., and W.S. Cain: Nasal pungency, odor, and eye irritation thresholds for homologous acetates. *Pharmacol. Biochem. Behav.* 39: 983-989 (1991).

13. Cometto-Muñiz, J.E., and W.S. Cain: Efficacy of volatile organic compounds in evoking nasal pungency and odor. *Arch. Environ. Health* 48: 309-314 (1993).

14. Cometto-Muñiz, J.E., and W.S. Cain: Sensory reactions of nasal pungency and odor to volatile organic compounds: The alkylbenzenes. *Am. Ind. Hyg. Assoc. J.* 55:811-817 (1994).

15. Cain, W.S.: Testing olfaction in a clinical setting. *Ear, Nose & Throat* 68: 316-328 (1989).

16. Devos, M., F. Patte, J. Rouault, P. Laffort, and L.J. van Gemert, ed.: *Standardized Human Olfactory Thresholds.* Oxford: IRL Press, 1990.

17. Guadagni, D.G., R.G. Buttery, S. Okano, and H.K. Burr: Additive effect of subthreshold concentrations of some organic compounds associated with food aromas. *Nature* 200: 1288-1289 (1963).

18. Rosen, A.A., J.B. Peter, and F.M. Middleton: Odor thresholds of mixed organic chemicals. *J. Water Pollut. Control Fed.* 34: 7-14 (1962).

19. Laska, M., and R. Hudson: A comparison of the detection thresholds of odour mixtures and their components. *Chem. Senses* 16: 651-662 (1991).

20. Patterson, M.Q., J.C. Stevens, W.S. Cain, and J.E. Cometto-Muñiz: Detection thresholds for an olfactory mixture and its three constituent compounds. *Chem. Senses* 18: 723-734 (1993).

21. Cometto-Muñiz, J.E., and W.S. Cain: Olfactory adaptation. In *Handbook of Clinical Olfaction and Gustation,* edited by R.L. Doty. New York: Marcel Dekker, 1995. pp. 257-281.

22. Cain, W.S., L.C. See, and T. Tosun: Irritation and odor from formaldehyde: Chamber studies. In *IAQ'86. Managing Indoor Air for Health and Energy Conservation,* Atlanta, GA: American Society of Heating, Refrigerating and Air-Conditioning Engineers, Inc., 1986. pp. 126-137.

CHAPTER 3

The Potency of Gases and Vapors: QSARs — Anesthesia, Sensory Irritation, and Odor

Michael H. Abraham

INTRODUCTION

The first studies of biological potency of gases and vapors were carried out from 1920 onward by K.H. Meyer and his colleagues, who investigated anesthesia on mice and salamanders. Right from the start, they were interested in the relationship between the potency of gases and vapors and the solubility of the anesthetic gases and vapors in organic liquids such as olive oil. This was by no means an arbitrary choice, but was influenced by the earlier work of H. Meyer, and especially Overton, on aqueous anesthesia of tadpoles. Around 1900, Overton had put forward his "lipoid" theory of aqueous anesthesia and had shown that the potency of aqueous anesthetics was related to their partition between water and lipid-like phases such as olive oil. This relationship was later put on a quantitative basis by K.H. Meyer, and the lipoid theory of anesthesia (both aqueous and gaseous) has remained in force until relatively recently.

In this chapter, the work of H. Meyer, Overton, and K.H. Meyer, on aqueous anesthesia will be discussed briefly as an introduction to the work of K.H. Meyer on gaseous anesthesia. The Ferguson principle, put forward in 1939, was a useful suggestion that the biological potency of gases and vapors could be correlated with saturated vapor pressure, but the principle has recently been shown by Abraham, Nielsen, and Alarie to be purely empirical in nature, as will be discussed. No further advances in the understanding of the biological activity of gases and vapors were really made until the advent of new quantitative methods to relate activity to the chemical structure or to physicochemical properties of the anesthetics. Quantitative structure-activity relationships (QSARs) have

1-56670-144-9/96/$0.00+$.50

recently been applied to sensory irritation in mice and in humans and also to odor intensities and thresholds in humans. The general principles of these QSARs will be set out, and an account will be given of their application, stressing not only their use in the prediction of further values of gaseous potencies, but also their use in the understanding of the nature of biological processes.

EARLY WORK

The use of partition coefficients in the interpretation of narcotic activity was suggested by Overton and Meyer around the turn of the century.[1-3] Overton showed for numerous series of compounds that the narcotic concentration, C', of aqueous solute toward the tadpole was related to the water-olive oil partition coefficient with the latter defined as:

$$P' = (\text{concentration in organic phase})/(\text{concentration in water}) \qquad (1)$$

The narcotic concentration, C', in units of mol/l is that required just to narcotize tadpoles in aqueous solution.* Typical results of Overton and Meyer were set out by K.H. Meyer and Gottlieb-Billroth,[4] and in Table 1 are given averaged data from the list given by Meyer and Gottlieb-Billroth.[4]

There is an obvious connection between biological activity and partition between olive oil and water. As the value of C' decreases, i.e., as the narcotic substance becomes more potent, so does the water-olive oil partition coefficient increase. There are a few obvious exceptions in Table 1 to the general rule; for example, chloralhydrate and bromalhydrate are more potent than expected from their partition coefficients, no doubt because in water they are in equilibrium with the reactive free aldehydes.

This illustrates the general point that connections between biological activity such as aqueous (and gaseous) narcosis, and some physicochemical property such as a partition coefficient, will only hold, in general, for "nonreactive" compounds.

Although Overton and H. Meyer clearly showed that there was a connection between aqueous narcotic concentration and water-olive oil partition coefficients, the connection seems never to have been formulated as a mathematical equation. It was not until 1920 that K.H. Meyer and Gottlieb-Billroth[4] expressed the relationship in the quantitative form,

$$C' \cdot P' = \text{constant} \qquad (2)$$

They showed that the data of Overton and of H. Meyer, Table 1, led to a reasonably constant value of 0.05 for the product $C' \cdot P'$. This work of K.H. Meyer and Gottlieb-Billroth[4] represents the first quantitative structure-activity relationship ever reported.

---

* Aqueous concentrations and water-solvent partition coefficients are denoted by a superscript ', to distinguish them from gaseous concentrations and vapor pressures.

Table 1   Tadpole Narcosis and Partition

Solute	C′ mol/l	P′ (olive oil)
Ethanol	0.40	0.03
Methylurethane	0.27[a]	0.04
Propanone	0.26[a]	0.14
t-Butanol	0.13	0.18
Propan-1-ol	0.11	0.13
t-Pentanol	0.057	1.00
Pentanamide	0.05	0.07
Ethylurethane	0.035	0.10
Diethylether	0.024	2.4
Paraldehyde	0.023	3.0
Ethylacetoacetate	0.019	4.0
Chloralhydrate	0.015	0.22
Acetal	0.012	8.0
Acetanilide	0.0094	2.0
Methacitin	0.009	2.0
Sulfonal	0.0075	2.8
Benzamide	0.005	0.44
Phthalide	0.0043	3.3
Trional	0.0042	10.2
Chloroethane	0.004	24.0
Vanillin	0.0033	3.0
Phenacetin	0.003	4.0
2-Methoxyphenol	0.003	30.0
Bromoethane	0.0027	37.0
Bromalhydrate	0.002	0.70
Butylchloralhydrate	0.002	1.6
Salicylamide	0.002	14.0
Piperonal	0.002	100.0
Tetronal	0.0018	4.0
Coumarin	0.0017	10.0
Chloroform	0.0012	70.0
1,4-Dimethoxybenzene	0.0009	230.0
Chloreton	0.0008	22.8
Phenylurethane	0.0006	150.0
Carbon disulfide	0.0005	50.0
Menthol	0.0001	250.0
Thymol	$5.5 \cdot 10^{-5}$	600.0
Phenanthrene	$3.7 \cdot 10^{-6}$	40000.0

[a] From the original work of Overton.[1]
Data from Reference 4.

Some years later, Meyer and Hemmi[5] showed that water-oleyl alcohol partition coefficients could be used in Equation 2, again with narcotic concentrations of aqueous solutes toward the tadpole given in units of mol/l. Their own results are in Table 2, where the product C′ · P′ averages out to 0.031 with a standard deviation of only 0.011 units; a very impressive result considering the nature of the experiments involved.

Equation 2 can be expressed in logarithmic form as Equation 3, or more generally as Equation 4, where SP can be a biological property such as 1/C′ for tadpole narcosis, or can be a physicochemical property, such as another partition coefficient.

$$\log(1/C') = \log P' + \text{constant} \qquad (3)$$

$$\log SP = a \cdot \log P' + c \qquad (4)$$

Table 2   Tadpole Narcosis and Partition

Solute	C' mol/l	P'[a]	C' · P'
Ethanol	0.33	0.10	0.033
Propan-1-ol	0.11	0.35	0.038
Butan-1-ol	0.03	0.65	0.020
Pentanamide	0.07	0.30	0.021
Antipyrin	0.07	0.30	0.021
Pyramidon	0.03	1.30	0.039
Benzamide	0.013	2.50	0.033
Dial[b]	0.01	2.40	0.024
Salicylamide	0.0033	5.90	0.021
Luminal[c]	0.008	5.90	0.048
2-Nitroaniline	0.0025	14.00	0.035
Thymol	0.000047	950.00	0.045
Veronal	0.03	1.38	0.041

[a] Water-olelyl alcohol.
[b] Diallylbituric acid.
[c] Ethylphenylbarbituric acid.

Data from Reference 5.

Equation 3 is now referred to as the Overton-Meyer relationship, or sometimes just as the Overton relationship or rule, and has been applied to numerous series of biological results in aqueous solution, see for example, the review of Dearden.[6]

The first quantitative analysis of gaseous narcosis was also carried out by K.H. Meyer and Gottlieb-Billroth.[4] They showed that a similar relationship to Equation 2 holds also:

$$C \cdot L = \text{constant} \tag{5}$$

Here, C is the gaseous concentration of the narcotic, and L is the Ostwald solubility coefficient of the narcotic in some particular solvent, defined by Equation 6,

$$L = (\text{concentration in solvent})/(\text{concentration in gas phase}) \tag{6}$$

As an example, they related the gaseous narcotic concentration toward mice, to the narcotic L-value in olive oil at 310 K, as shown in Table 3, slightly altered from the original table given by Meyer and Gottlieb-Billroth.[4] The product $C \cdot L$ is reasonably constant. It should be pointed out that the Overton rule and the Meyer-Gottlieb-Billroth equation apply only to "nonreactive" compounds. Those solutes that interact with the biological system in some specific way are nearly always more potent than calculated — sometimes by orders of magnitude.

The results in Table 3 can also be treated through the more usual logarithmic relationship to yield a quite reasonable QSAR,

$$\log C \text{ (gas, vol\%)} = 2.11 - 0.950 \log L \text{ (olive oil)} \tag{7}$$
$$n = 14 \quad \rho = 0.9761 \quad sd = 0.127 \quad F = 242.4$$

Table 3   Gaseous Narcotic Potency toward Mice
           and Gaseous Solubility

Solute	C (gas, vol%)	L (olive oil)	C · L
Nitrous oxide	100	1.40	140
Dimethyl ether	12	11.6	139
Chloromethane	6.5	14.0	91
Ethylene oxide	5.8	31.0	180
Chloroethane	5.0	40.5	202
Bromomethane	3.5	32.0	112
Pentene	4.0	65.0	260
Diethyl ether	3.4	50.0	170
Methylal	2.8	75.0	210
Bromoethane	1.9	95.0	180
Dimethylacetal	1.9	100.0	190
Diethylformal	1.0	120.0	120
1,2-Dichloroethene	0.95	130.0	124
Chloroform	0.44	265.0	117

Mean: 160   SD: 49

Data from Reference 4.

Here, and elsewhere, n is the number of data points, ρ is the overall correlation coefficient, sd is the standard deviation in the dependent variable, and F is the Fisher F-statistic.

Meyer and Hemmi[5] continued work on gaseous narcosis and gave further examples of the Meyer-Gottlieb-Billroth equation, again with olive oil as the model solvent. They pointed out that the solutes studied (e.g., see Table 3) were not very polar, and that olive oil might not be such a good model solvent if solutes such as alcohols and amides were examined. They suggested that with these solutes, oleyl alcohol might be a better model than olive oil, but did not demonstrate this for gaseous narcosis.

What K.H. Meyer and Hopff[7] did do, however, was to show that the connection between gaseous narcotic concentration and L (solvent) was not valid if water was chosen as the model solvent. Their analysis is in Table 4, where the L (water) values are those given by Meyer and Hopff. Unlike the results in Table 3, the product C · L in Table 4 is nowhere near constant, and hence Meyer and Hopff[7] concluded that water was not at all a good model for gaseous narcosis. This first attempt to use QSARs to understand the mechanism of gaseous narcosis has received little attention. Only much more recently have QSARs been used to discuss matters such as the site of action of gaseous anesthetics and irritants through the use of model solvents.

A few years after the work of K.H. Meyer, Ferguson[8] advanced what has been the most widely used principle in gaseous narcosis, namely that the gaseous narcotic concentration, P^{nar}, is inversely proportional to the saturated vapor pressure, P^o, where P^{nar} is the partial pressure of a series of compounds giving rise to a particular effect on a given system by a physical mechanism. There are cases where the Ferguson rule holds quite well, especially for homologous or related series, see for example, the alcohol series of irritants given by Nielsen and Alarie,[9] and is still in use as an empirical observation. Unfortunately, there have been various attempts to give the Ferguson rule a thermodynamic status that it does

Table 4   Gaseous Narcotic Potency toward Mice
           and Gaseous Solubility

Solute	C (gas, vol%)	L (water)[a]	C · L
Nitrogen	1000	0.02	20
Methane	370	0.028	10.4
Nitrous oxide	100	0.50	50
Ethylene	80	0.098	7.8
Acetylene	65	0.84	54.6
Bromoethane	1.9	2.47	4.7
Carbon disulfide	1.1	0.80	0.88
Chloroform	0.44	4.17	1.8

[a] At 303 K.

Data from Reference 7.

not have, with the result that there are very considerable exceptions[10] to a rule that is supposed to have a thermodynamic validity. Brink and Posternak[11] were the first to advance a thermodynamic argument for the rule, but other workers[12,13] have accepted the Brink-Posternak argument also. Only recently has it been shown[14] that the thermodynamic analysis of Brink and Posternak is incorrect, and that the Ferguson rule is a useful empirical observation, but not a thermodynamic necessity.[14]

Later workers applied the treatment of K.H. Meyer and Gottlieb-Billroth[4] to gaseous potency. The $\log ED_{50}$/atm values for the righting reflex of mice, Table 5, can be quite well correlated with the solubility of the gaseous solutes in olive oil, benzene, cyclohexane, or carbon disulfide,[13-16] with olive oil giving slightly the best correlation.[15]

In Table 5 are also values of $\log P^{\circ}$/atm for the compounds at 298 K, from various sources,[17-19] and the corresponding logL values in olive oil at 310 K, mostly from Reference 20, with data for the perfluoroalkanes from Reference 15. Except for the perfluoro compounds, Ferguson's rule holds quite well: plots of $\log ED_{50}$ against logL (oil) and $\log P^{\circ}$ are shown in Figures 1 and 2, and the corresponding equations are given below:

$$-\log ED_{50} = -0.344 + 1.036 \log L \text{ (oil)} \qquad (8)$$
$$n = 18 \qquad r = 0.9924 \qquad sd = 0.18 \qquad F = 1038.6$$

$$\log ED_{50} = -1.179 + 0.941 \log P^{\circ} \qquad (9)$$
$$n = 18 \qquad r = 0.8690 \qquad sd = 0.67 \qquad F = 49.3$$

However, if the $\log ED_{50}$ values can be correlated with logL values in various solvents, and with $\log P^{\circ}$ values as well, we might ask whether or not these various correlations have any significance, at least for the set of compounds in Table 5.

It has been shown that values of logL for a set of nonpolar solutes in a given nonaqueous solvent are linearly related to logL values in any other nonaqueous solvent.[21] Thus for a set of solutes that are not very polar, such as that given in Table 5, correlations between the logL values would be expected, especially if the various solvents themselves are not too different. This is a familiar situation in regression analysis; if two independent variables A and B are themselves

Table 5   LogED$_{50}$ Pressures for Righting Reflex in Mice

Compound	logED$_{50}$/atm	logP°/atm[a]	logL (oil)
Helium	>2.1		−1.76
Neon	>2.1		−1.66
Hydrogen	2.11		−1.31
Nitrogen	1.52	2.94[b]	−1.13
Tetrafluoromethane	1.28	2.27	−1.14
Perfluorethane	1.26	1.61	−0.90
Perfluoropropane	1.26	0.97	−0.58
Argon	1.18	2.72[b]	−0.82
Sulfur hexafluoride	0.78	0.59[b]	−0.58
Krypton	0.65	2.54[b]	−0.35
Nitrous oxide	0.18	1.76[b]	0.15
Ethylene	0.15	1.78[c]	0.10
Xenon	−0.02	1.83[d]	0.24
Difluorodichloromethane	−0.40	0.81	0.76[e]
CF$_2$Cl · CH$_3$	−0.60	0.52	
Difluorochloromethane	−0.80	1.01	0.87[f]
Cyclopropane	−0.80	0.85	1.07
Diethylether	−1.52	−0.15	1.81
Chloroform	−2.08	−0.59	2.58
Halothane	−2.11	−0.41	2.29
Methoxyfluorane	−2.66	−1.40	2.93

[a] Values at 298 K from Reference 17, unless shown otherwise.
[b] From data in Reference 18.
[c] From Reference 19.
[d] International Critical Tables, Vol III.
[e] From Reference 13.
[f] From References 13 and 20.

Data from References 13–16.

Plot of -logED50 vs logL(oil)

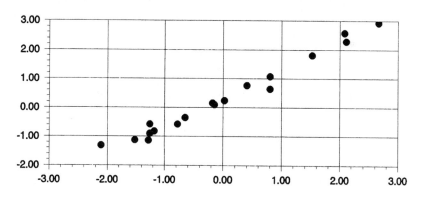

Figure 1 Righting reflex in mice. Plot of −logED$_{50}$ vs. logL (oil).

linearly related, so that a dependent variable X is linear with both A and B, it is impossible to tell which of the A/X and B/X correlations is fundamental. The only way that a choice between olive oil and benzene, for example, as suitable model solvents, could be made, would be to obtain further data using compounds that would break the olive oil/benzene correlation.

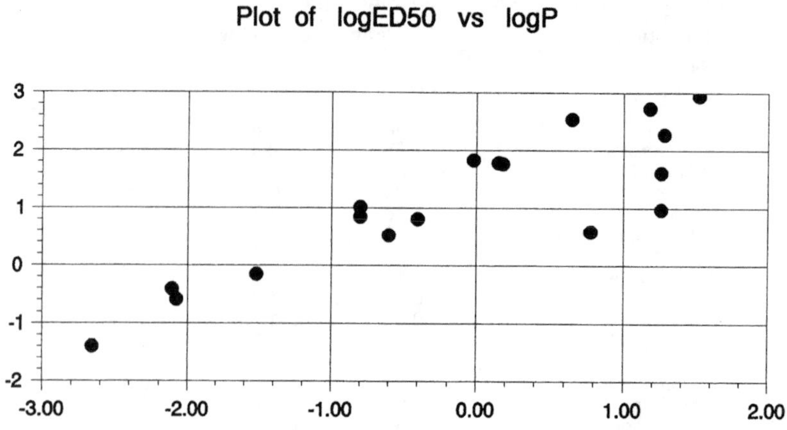

Plot of logED50 vs logP

Figure 2 Righting reflex in mice. Plot of $logED_{50}$ vs. logP.

In addition, for the less polar compounds, there is often a reasonable correlation between $logP^o$ and logL for solution in a not too polar solvent. Hence the quite good Ferguson rule plot in Figure 2. The perfluoro compounds are outliers here, because the compound/compound interactions in the bulk liquids that influence vapor pressure are now considerably different to the compound/olive oil or compound/benzene interactions that influence the logL values.

Miller and Smith[13] also showed that correlations of $logED_{50}$ values, Table 5, with hydrate dissociation pressure, or with a parameter related to the structuring of water, were rather poor, and concluded[13,15] that there was no evidence of a physicochemical nature to suggest that an aqueous phase of the central nervous system is the site of anesthetic action. This is really the same result that Meyer and Hopff[7] obtained, some 50 years earlier.

There have been other attempts to connect gaseous potency with various theoretical or empirical parameters. Mullins[12] used the old lipoid theory of Meyer and Overton to relate the so-called activity coefficient of a narcotic, γ^{nar}, in a membrane to the cohesive energy density of the narcotic and the membrane,

$$\ln(\gamma^{nar}) = V^{nar}\left(\delta^{nar} - \delta^{mem}\right)^2 \Big/ RT \qquad (10)$$

where V^{nar} is the molar volume of the narcotic as the bulk liquid. But since we have that

$$\gamma^{nar} = P^{nar} \Big/ P^o \cdot X^{nar} \qquad (11)$$

where X^{nar} is the mole fraction of the narcotic in the membrane, and since Mullins[12] suggests that

$$X^{nar} \cdot V^{nar} = k, \text{ a constant} \qquad (12)$$

it follows that

$$\ln\left(P^{nar} \cdot V^{nar}/P^{o}\right) - \ln k \; = \; V^{nar}\left(\delta^{nar} - \delta^{mem}\right)^{2}\Big/RT \qquad (13)$$

Now the right-hand side of Equation 13 reaches a minimum when $\delta^{nar} = \delta^{mem}$, so that no matter what value V^{nar} takes, the term P^{nar}/P^{o} would certainly not be expected to remain constant. Yet in many cases, this is found to be the case. We can only consider the analysis of Mullins[12] to be unhelpful.

On a more empirical basis, Krantz[22,23] observed that for seven anesthetic ethers, there was a linear relationship.

$$\log(C) = 1/S' \qquad (14)$$

where C is the concentration in mol/kg for respiratory arrest in dogs, and S' is the aqueous solubility of the liquid anesthetic in ml/100 ml water. This rather surprising result can be accounted for if we first follow the analysis of Nielsen, Hansen, and Alarie.[24] They noted that the water-octanol partition coefficient, P'oct, had been used as a descriptor for upper respiratory tract irritation in mice,[25-28] even though the water-octanol system cannot possibly be a good model for a gas-receptor (or receptor phase) process. However, if collinearities exist, for example, between the gas-water and gas-octanol logL values, then it is possible for the logP'oct values to parallel the gas-octanol logL values, for a particular restricted set of solutes. The relationship between the solubility of liquids in water, S' in mol/l, and logP'oct is well known,[29,30] and the Yalkowsky equation for the solubility of liquids and solids in water at 298 K is given by,[30]

$$\log S' = -\log P'oct - 0.01 \; MP + 1.05 \qquad (15)$$

where MP is the solid melting point in °C; for compounds that are liquid at 298 K, the MP term is zero. Hence if, by chance, there is a correlation between the biological activity of a series of gases or vapors and logP'oct, there could well be a relationship between the activity and logS'.

More important than such chance correlations is the observation of Nielsen, Hansen, and Alarie[24] that sensory irritation can be correlated with logL for the solubility of gases and vapors in wet octanol. Thus octanol is a possible model solvent for a receptor site or area, as will be discussed in detail below.

Other descriptors have been used to correlate sensory irritation; these include the compound normal boiling point,[25] an adjusted boiling point,[31] and the saturated vapor pressure[28] (in terms of the Ferguson rule). It is noteworthy, however, that all these descriptors, as well as the logL gas-solvent descriptors, are physicochemical parameters and do not involve the chemical structures of the compounds. The same is true for the descriptors used to correlate the $\log ED_{50}$ pressures for righting reflex in mice, Table 5, and the narcotic concentrations in Tables 3 and 4. There have been a few attempts to use structurally related descriptors in the correlation of biological activity of gases and vapors, however.

RECENT WORK ON QSARs

Laffort and co-workers[32,33] set out a gas-liquid chromatographic (GLC) procedure for the determination of solute descriptors. Using retention data on five stationary phases, they defined five descriptors:

α is an apolar factor, proportional to the molar volume at the boiling point.
ω is an orientation factor, proportional to the square of the dipole moment for simple molecules.
ε is an electron factor which for molecules with a regular distribution of electrons is proportional to the ratio between molar refraction and the molal volume at the boiling point.
π is a proton donor, or acidity, factor.
β is a basicity factor.

These solubility factors were determined for 240 solutes and were then used as descriptors in a number of linear free energy relationships (LFERs). Thus a set of gas-water logL values could be correlated[33] through the equation,

$$logL \text{ (water)} = -4.45 + 2.194\omega + 0.808\varepsilon + 2.947\pi + 3.406\beta - 2.556\pi\beta \quad (16)$$
$$n = 82 \qquad \rho = 0.96$$

Laffort and Patte[34] then examined the relative retention times as $logt_R$, of solutes across the frog olfactory mucosa, as set out in Table 6, together with the corresponding solubility factors.

Table 6 Values of $logt_R$ across Frog Olfactory Mucosa and the Laffort Solubility Factors

Solute	$logt_R$	α	β	ε	π	ω
Octane	0.644	3.33	0.00	0.00	0.00	0.00
Nonane	0.726	3.78	0.00	0.00	0.00	0.00
d-Limonene	0.877					
Heptanal	1.014	2.55	0.29	0.38	-0.22	1.11
Butyl acetate	1.055	2.61	0.35	-0.23	-0.20	1.26
Pentyl acetate	1.096	3.15	0.36	-0.29	-0.19	1.24
Heptan-4-one	1.178	2.38	0.27	0.42	-0.38	1.26
Geraniol	1.323					
Butan-1-ol	1.753	1.32	0.40	0.15	0.75	0.74
Benzaldehyde	1.874	1.85	0.42	1.42	-0.01	0.99
Methyl benzoate	2.006	1.78	0.61	1.71	0.71	0.73
Furfural	2.090	1.24	0.56	0.97	0.19	1.22
Carvone	2.647	4.30	0.48	0.78	-0.30	0.86
Diphenyl ether	2.666					
Isopentanoic acid	2.932	2.99	0.97	-1.94	1.77	1.40

Data from Reference 34.

As shown by Laffort and Patte,[34] the descriptors α and ε were not statistically significant, leaving,

$$logt_R = 0.71 + 4.80\beta - 0.63\pi - 0.99\omega \qquad (17)$$
$$n = 12 \qquad \rho = 0.951 \qquad sd = 0.27 \qquad F = 25.4$$

Interpretation of Equation 17 is by no means easy. It is not clear what the origin is of the negative coefficients of π the proton donor factor and ω the orientation factor, because any solute-mucosa interaction should lead to a positive coefficient. It is also not reasonable that the coefficient of the solute basicity factor in Equation 17 is larger than in Equation 16, since this implies that the solvent acidity in Equation 17 is larger than in Equation 16. Yet water is an extremely strong hydrogen-bond acid, and it is doubtful if any biological system is more acidic. We shall see later that other descriptors lead to a more rational explanation of the frog mucosa data.

More recently, Abraham and co-workers[35-40] have set out a number of new descriptors suitable for use in LFERs and QSARs. These are as follows:

R_2 is an excess molar refraction that can be determined simply from a knowledge of the compound refractive index.[35] Because R_2 is almost an additive property, it is quite straightforward to deduce values for compounds that are gaseous or solid at room temperature. Some 1500 R_2 values are available at present, and further values can be determined or estimated for any solute likely to be studied. The R_2 descriptor represents the tendency of a solute to interact with a phase through π- or n-electron pairs.

π_2^H is the solute dipolarity/polarizability,[36-38] it being impossible to devise descriptors for these properties separately. This descriptor can be obtained experimentally from GLC data and also from water-solvent partition coefficients.[39] At present, well over 1000 values of π_2^H are known, and it is reasonably easy to obtain further values by the GLC method or through water-solvent partitions.[39]

$\Sigma\alpha_2^H$ is the solute overall or effective hydrogen-bond acidity. For mono-acids, this descriptor was originally obtained directly from hydrogen-bond complexation constants, and in this way values were found for many types of solutes such as carboxylic acids, alcohols, and phenols.[41] Now that the acid scale is established, further values can be obtained by GLC or through the use of water-solvent partitions.[39]

$\Sigma\beta_2^H$ is the solute overall or effective hydrogen-bond basicity. Again, for mono-bases, this was first obtained from hydrogen-bond complexation constants.[42] Further values for mono-bases and all values for poly-bases can best be obtained from water-solvent partitions.[39]

$logL^{16}$ is a descriptor[20] based on the solute gas-liquid partition coefficient on hexadecane at 298 K. A database of over 1500 such values is available,[20,35-40] and additional values can easily be obtained by GLC on a variety of nonpolar stationary phases. The $logL^{16}$ descriptor is a measure of the lipophilicity of a solute.

These descriptors can be combined into a general linear equation,[40] either as an LFER or as a QSAR:

$$logSP = c + r \cdot R_2 + s \cdot \pi_2^H + a \cdot \Sigma\alpha_2^H + b \cdot \Sigma\beta_2^H + 1 \cdot logL^{16} \qquad (18)$$

Here, SP is a property for a series of solutes on a given phase. Thus, SP can be the gas-liquid partition coefficient for a number of solutes in a particular organic solvent, i.e., an LFER, or SP can be a biological property for a series of solutes, i.e., a QSAR. Then if SP is known for a series of solutes for which descriptors are available, Equation 18 can be solved by the method of multiple linear regression analysis (MLRA) to yield the constants c, r, s, a, b, and l. Not every term in Equation 18 may be significant, and each term is analyzed using Student t test. Usually, terms are retained only if the t test shows >95% significance.

The constants obtained by MLRA are important in that they can be used to characterize the phase (in LFERs) or receptor area (in QSARs) involved. In both cases, the r-constant gives the propensity of the phase to interact with solute π- and n-electron pairs, the s-constant is the phase-area dipolarity/polarizability (because a dipolar solute will interact with a dipolar phase and a polarizable solute will interact with a polarizable phase), the a-constant is the phase-area basicity (because a solute that is a hydrogen-bond acid will interact with a basic phase), similarly the b-constant is the phase-area acidity (because a solute that is a hydrogen-bond base will interact with an acidic phase), and finally, the l-constant is a measure of the phase-area lipophilicity. Note that by definition l = 1.00 for hexadecane at 298 K. Because the constants in Equation 18 can be used to characterize phases, they can be regarded as "characteristic constants," for solvents, for all kinds of phases, for receptor sites or areas, etc. Examples will follow of various phases that have been characterized through the constants in Equation 18.

Because the constants in Equation 18 represent quite specific properties of the phase or receptor area, they must follow correct chemical principles. Thus for a completely nonacidic phase/area, the b-constant must be zero, within some reasonable experimental error. Since for the highly acidic solvent water the b-constant is 4.84 at 298 K, it follows that for any biological phase/area, the b-constant must lie between these two limits, i.e., between zero and 4.84.

Numerous LFERs of gas-liquid partition coefficients, as logL values, on organic solvents have been carried out, but more relevant are those on biological phases. For example, logL values for organic compounds, mostly halogenated anesthetics, on human blood and fat at 310 K can be correlated[43] by the equations,

$$logL \text{ (blood)} = -1.27 + 0.61\ R_2 + 0.92\ \pi_2^H + 3.61\ \Sigma\alpha_2^H \qquad (19)$$
$$+ 3.38\ \Sigma\beta_2^H + 0.36\ logL^{16}$$
$$n = 82 \qquad \rho = 0.9884 \qquad sd = 0.20 \qquad F = 644.5$$

$$logL \text{ (fat)} = -0.29 - 0.17\ R_2 + 0.73\ \pi_2^H + 1.75\ \Sigma\alpha_2^H \qquad (20)$$
$$+ 0.22\ \Sigma\beta_2^H + 0.90\ logL^{16}$$
$$n = 36 \qquad \rho = 0.9940 \qquad sd = 0.12 \qquad F = 497.6$$

For a similar variety of solutes, gaseous solubility in water at 310 K gives rise[43] to a quite different set of constants in the regression Equation 21,

$$logL \text{ (water)} = -1.36 + 1.05 \ R_2 + 2.63 \ \pi_2^H + 3.74 \ \Sigma\alpha_2^H \quad (21)$$
$$+ 4.50 \ \Sigma\beta_2^H - 0.25 \ logL^{16}$$
$$n = 75 \qquad \rho = 0.9925 \qquad sd = 0.18 \qquad F = 912.8$$

The constants in Equation 18 for a number of gas-biological phase partitions, obtained by Abraham and Weathersby[43] are given in Table 7 below.

Table 7   Characteristic Constants in Equation 18
for some Biological Phases at 310 K

Phase	c	r	s	a	b	l
Water	−1.36	1.05	2.63	3.74	4.50	−0.25
Plasma	−1.48	0.49	2.05	3.51	3.91	0.16
Blood	−1.27	0.61	0.92	3.61	3.38	0.36
Brain	−1.07	0.43	0.29	2.78	2.79	0.61
Muscle	−1.14	0.54	0.22	3.47	2.92	0.58
Liver	−1.03	0.06	0.77	0.59	1.05	0.65
Fat	−0.29	−0.17	0.73	1.75	0.22	0.90
Olive oil	−0.24	−0.02	0.81	1.47	—	0.89

Data from Reference 43.

As noted above, these constants serve to characterize the phase in question. Thus water is highly dipolar (s = 2.63), is a strong hydrogen-bond base (a = 3.74), and a strong hydrogen-bond acid (b = 4.50), but is not lipophilic at all (1 = −0.25). At the other extreme, fat and olive oil are somewhat dipolar, are moderate bases, have only a weak or zero acidity, but are very lipophilic. The other biological phases can also be ranked with respect to these particular properties — the first time that any such properties have been assigned to biological phases.

Thus the LFERs generated through Equation 18 are not only useful as predictors of logL values, but can be used to assess the chemical nature of complicated systems. Of course, it must be recognized that the characteristic constants are average values for the systems taken as a whole, but even so, they are useful measures of the solubility-related properties of phases.

Retention times of solutes across the frog olfactory mucosa, see Table 6, can be analyzed using the general Equation 18. The resulting equation is:

$$logt_R = -0.97 + 1.35 \ \pi_2^H + 3.25 \ \Sigma\alpha_2^H + 0.37 \ logL^{16} \quad (22)$$
$$n = 14 \qquad \rho = 0.9680 \qquad sd = 0.22 \qquad F = 49.6$$

Although Equation 22 is statistically rather better than Equation 17, obtained by Laffort and Patte,[34] the real advantage of Equation 22 is that it can be interpreted much more easily. First of all, the s- and a-constants are positive, as required theoretically, and show that the mucosa is moderately dipolar and is quite a strong hydrogen-bond base. Compare s = 2.63 and a = 3.74 for water. But the mucosa does not behave quite like water, since b = 0, and 1 = 0.37, rather than −0.25 for water at 310 K. Of course the data set in Table 6 is by no means optimal for the application of Equation 18, but the advantage of descriptors that refer to specific chemical interactions is still evident.

The general Equation 18 can also be applied to the data on righting reflex in mice, Table 5, yielding Equation 23.

$$-logED_{50} = 0.614 + 0.991 \, \pi_2^H + 1.913 \, \Sigma\alpha_2^H + 0.894 \, logL^{16} \qquad (23)$$
$$n = 18 \qquad \rho = 0.9899 \qquad sd = 0.22 \qquad F = 226.6$$

Although Equation 23 is not as good, statistically, as the simple Equation 8, it does have the advantage that the structural factors that influence righting reflex can be elucidated. Thus solute dipolarity, hydrogen-bond acidity, and lipophilicity all increase the potency. (It should be noted that the limited structural range of the solutes studied leads to the a-constant, in particular, being subject to a very considerable error.) Furthermore, comparison of the characteristic constants in Equation 23 with those for olive oil, in Table 7, shows exactly why olive oil is such a good model solvent in this instance. A plot of observed values of $logED_{50}$ vs. those calculated through Equation 23 is in Figure 3, for comparison with Figures 1 and 2.

Plot of -logED50 obs vs calc

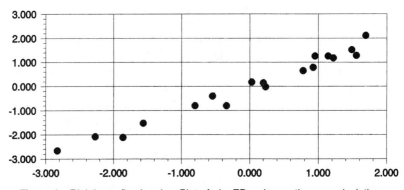

Figure 3 Righting reflex in mice. Plot of $-logED_{50}$ observation vs. calculation.

UPPER RESPIRATORY TRACT IRRITATION

A much more varied set of solutes was listed by Muller and Greff[25] for upper respiratory tract irritation of male Swiss OF1 mice by airborne chemicals, and an earlier version of Equation 18 was used to analyze the effect of nonreactive solutes.[44] A recalculation[14] using Equation 18 yields a similar equation, Equation 24. Potencies in Equation 24 are expressed as RD_{50} values with gaseous concentration in ppm. These refer to the gaseous concentration required to halve the respiratory rate. Note that in Reference 14 values were given in mol/l.

$$-\log[RD_{50}] = -6.45 + 0.83\ \pi_2^H + 2.58\ \Sigma\alpha_2^H + 0.77\ \log L^{16} \qquad (24)$$
$$n = 39 \qquad \rho = 0.9810 \qquad sd = 0.12 \qquad F = 440$$

A large set of nonreactive solutes is available from the fine review of Schaper[45] and leads to Equation 25, again for OF1 mice with RD_{50} values in ppm,

$$-\log[RD_{50}] = -6.71 + 1.30\ \pi_2^H + 2.88\ \Sigma\alpha_2^H + 0.76\ \log L^{16} \qquad (25)$$
$$n = 45 \qquad \rho = 0.9810 \qquad sd = 0.14 \qquad F = 350$$

The characteristic constants s and a are positive, as required by theory, and the QSAR is chemically reasonable. The gaseous solute factors that lead to increased potency are dipolarity (s = 1.30), hydrogen-bond acidity (a = 2.88), and lipophilicity (l = 0.76), and the QSAR Equation 24 or Equation 25 is good enough to use to estimate further values of $\log[RD_{50}]$. The constants indicate that the receptor site or area is moderately dipolar, quite basic, not acidic (because b = 0), and rather lipophilic. A plot of observed vs. calculated values on Equation 25 is in Figure 4.

Plot of -logRD50 obs vs calc

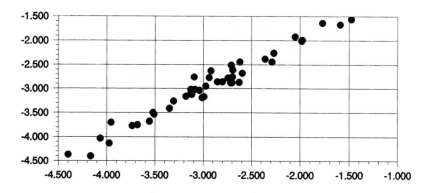

Figure 4 Sensory irritation of OF1 mice. Plot of $-\log RD_{50}$ observation vs. calculation.

Abraham, Nielsen, and Alarie[14] used data on the potencies of the 39 unreactive solutes toward OF1 mice, Equation 24, to compare with logL values for the solutes in various model solvents that had characteristic constants close to those in Equation 24. As can be seen from the similarity between Equation 24 and Equation 25 use of the data on the 45 solutes instead of the 39 solutes in Equation 24 would lead to essentially the same results. Values of $-\log[RD_{50}]$ for the 39 nonreactive irritants are in Table 8.

Before comparing the constants in Equation 24 with those for model solutes, it is interesting to see if the Ferguson rule holds for the 39 solutes. There is quite a reasonable quantitative connection between the $\log[RD_{50}]$ values and the logP° values, as shown by Equation 26. The corresponding plot is in Figure 5.

Table 8   Values of −log[RD$_{50}$] in ppm for
          the 39 Nonreactive Irritants in
          Equations 24 and 26–30

Irritant	−logppm
Propanone	−4.37
Butanone	−4.03
Pentan-2-one	−3.77
Cyclohexanone	−2.88
Hexan-2-one	−3.41
4-Methylpentan-2-one	−3.50
3,3-Dimethylbutan-2-one	−3.75
Heptan-2-one	−2.95
Heptan-4-one	−3.04
5-Methylhexan-2-one	−3.09
Octan-2-one	−2.68
5-Methylheptan-3-one	−2.88
Nonan-5-one	−2.44
2,6-Dimethylheptan-4-one	−2.51
Undecan-2-one	−1.56
Methanol	−4.40
Ethanol	−4.13
Propan-1-ol	−3.68
Propan-2-ol	−3.70
Butan-1-ol	−3.10
2-Methylpropan-1-ol	−3.26
Pentan-1-ol	−2.78
3-Methylbutan-1-ol	−2.86
Hexan-1-ol	−2.38
4-Methylpentan-2-ol	−2.63
Heptan-1-ol	−1.99
Octan-1-ol	−1.67
2-Ethylhexan-1-ol	−1.64
Toluene	−3.53
Chlorobenzene	−3.02
Bromobenzene	−2.61
1,2-Dichlorobenzene	−2.26
2-Chlorotoluene	−2.76
o-Xylene	−3.17
p-Xylene	−3.12
Styrene	−2.77
Ethylbenzene	−3.16
a-Methylstyrene,	−2.44
1,4-Divinylbenzene	−1.89

$$\log\left[RD_{50}\right] = -5.63 + 1.15 \; \log P°/atm \qquad (26)$$

$$n = 38 \qquad \rho = 0.9695 \qquad sd = 0.21 \qquad F = 565$$

One data point is missing in Equation 26, no value of logP° being available at 298 K for divinylbenzene.

Examination of the constants in either Equation 24 or Equation 25 can be used to select a solvent that closely resembles the receptor phase or area in upper respiratory tract irritation, without recourse to trial and error methods. In Table 9 are listed the characteristic constants for a number of solvents, obtained through the application of Equation 18 to logL values for the solubility of gases and vapors in the various given solvents.[14,43,48,49] In principle almost any solvent could be examined through Equation 18, and it should be possible in the future to construct a database of characteristic constants for a wide range and variety of solvents.

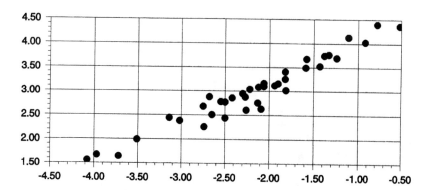

Figure 5 Sensory irritation of OF1 mice. Plot of $\log RD_{50}$ vs. logP/atm.

Table 9 Characteristic Constants in Equation 18 for Solvents

Phase	c	r	s	a	b	l
Water (310 K)[a]	−1.36	1.05	2.63	3.74	4.50	−0.25
NFM (298 K)[b]	−0.53	—	2.57	4.32	—	0.73
NFM (313 K)	−0.56	—	2.39	3.92	—	0.68
EHP (298 K)[c]	−0.07	−0.26	0.91	3.74	—	0.95
EHP (310 K)	−0.09	−0.19	0.83	3.41	—	0.89
OCT (298 K)[d]	−0.18	—	0.62	3.73	1.36	0.86
CH_2I_2 (298 K)[e]	−0.74	0.32	1.34	0.38	1.19	0.87
$CHCl_3$ (298 K)[e]	0.10	−0.35	1.26	0.60	1.18	0.99
CCl_4 (298 K)[e]	0.23	−0.20	0.35	0.07	0.27	1.04
DCE (298 K)[f]	−0.01	−0.28	1.72	0.73	0.59	0.93
3-Ethylphenol[g]	−1.08	−0.20	0.87	1.80	3.42	0.90
Oil (310 K)[h]	−0.24	−0.02	0.81	1.47	—	0.89
Hexadecane	0.00	0.00	0.00	0.00	0.00	1.00
OF$_1$ mice	−6.71	—	1.30	2.88	—	0.76
SW + CF$_1$ mice	−6.70	—	1.44	1.47	—	0.68
All mice	−6.97	—	1.50	2.51	—	0.77

[a] Equation 21.
[b] N-Formylmorpholine.[14]
[c] Tri-(2-ethylhexyl)phosphate.[14]
[d] Wet octanol.[14]
[e] Reference 48.
[f] 1,2-Dichloroethane.[48]
[g] Reference 49.
[h] Olive oil.[43]

Several of the listed solvents in Table 9 could be considered as good models, for example, the solvents N-formylmorpholine (NFM), tri-(2-ethylhexyl)phosphate (EHP), wet octanol (oct), and olive oil. The equation connecting potency (in ppm) with solubility in NFM is given by Equation 27, see Figure 6.

$$-\log[RD_{50}] = -6.25 + 0.88 \log L(NFM, 313\ K) \tag{27}$$
$$n = 38 \qquad \rho = 0.9645 \qquad sd = 0.17 \qquad F = 477$$

Plot of -logRD50 vs logL(NFM,313)

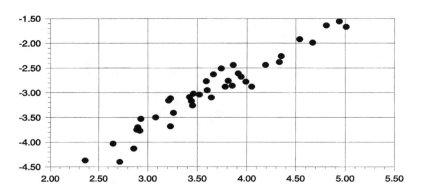

Figure 6 Sensory irritation in OF1 mice. Plot of $-\log RD_{50}$ vs. logL (NFM, 313).

Plot of -logRD50 vs logL(EHP,310)

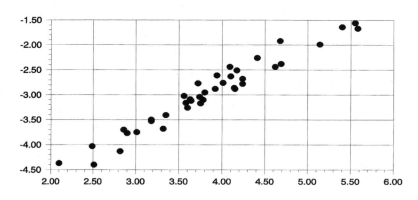

Figure 7 Sensory irirtation in OF1 mice. Plot of $-\log RD_{50}$ vs. logL (EHP, 310).

and that for EHP is shown in Figure 7 and is given by,

$$-\log[RD_{50}] = -6.25 + 0.86 \, \log L(EHP, 310 \, K) \tag{28}$$
$$n = 39 \qquad \rho = 0.9734 \qquad sd = 0.17 \qquad F = 673$$

Either NFM or EHP would be excellent model solvents, as obtained from the data on the 39 solutes. Wet octanol is not quite as good a model, see Figure 8 and Equation 29.

$$-\log[RD_{50}] = -6.38 + 0.81 \, \log L(oct, 298 \, K) \tag{29}$$
$$n = 36 \qquad \rho = 0.9193 \qquad sd = 0.27 \qquad F = 185$$

Plot of -logRD50 vs logL(oct,298)

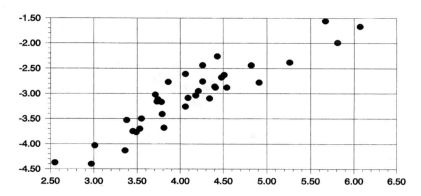

Figure 8 Sensory irritation in OF1 mice. Plot of $-\log RD_{50}$ vs. logL (oct, 298).

Plot of -logRD50 vs logL(water,298)

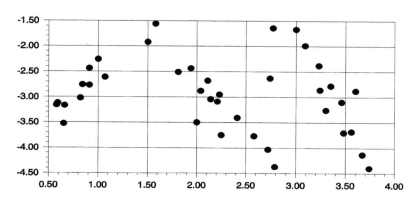

Figure 9 Sensory irritation in OF1 mice. Plot of $-\log RD_{50}$ vs. logL (water 298).

Water is far removed as a model solvent, see Figure 9 and Equation 30, and compare with the conclusions of Meyer and Hopff[7] on the narcotic activity of gases and vapors.

$$-\log[RD_{50}] = -2.60 - 0.17 \, \log L(\text{water, 298 K}) \tag{30}$$
$$n = 39 \qquad \rho = 0.2421 \qquad sd = 0.70 \qquad F = 2.3$$

Because olive oil has always been advocated as a model solvent, and because the characteristic constants for olive oil match reasonably well with those in Equation 24 or 25, this solvent was also tested. The resulting equation is given in Equation 31, and the corresponding plot is shown in Figure 10.

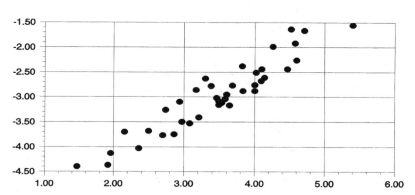

Figure 10 Sensory irritation in OF1 mice. Plot of $-\log RD_{50}$ vs. logL (oil, 310).

$$-\log[RD_{50}] = -5.70 - 0.78 \log L(oil, 310 \text{ K}) \qquad (31)$$
$$n = 39 \qquad \rho = 0.9437 \qquad sd = 0.24 \qquad F = 302$$

In this case, olive oil is not quite as good a model solvent as NFM or EHP. The practical application of being able to choose a good model solvent on a rational basis is that Equation 28 could be used to estimate further values of $\log[RD_{50}]$ for the effect of nonreactive solutes simply through an experimental determination of the vapor solubility in tri-(2-ethylhexyl)phosphate.

The review of Schaper[45] lists RD_{50} values for upper respiratory tract irritation in other strains of mice. Results on Swiss-Webster (SW) mice, mostly of Alarie and Nielsen[46,47] can be combined with those on CF_1 mice to yield the QSAR,

$$-\log[RD_{50}] = -6.70 + 1.44 \, \pi_2^H + 1.47 \, \Sigma\alpha_2^H + 0.68 \, \log L^{16} \qquad (32)$$
$$n = 28 \qquad \rho = 0.9669 \qquad sd = 0.21 \qquad F = 115$$

If the three strains are taken together, then for $-\log[RD_{50}]$ values in ppm for OF_1, SW, and CF_1 strains:

$$-\log[RD_{50}] = -6.97 + 1.50 \, \pi_2^H + 2.51 \, \Sigma\alpha_2^H + 0.77 \, \log L^{16} \qquad (33)$$
$$n = 73 \qquad \rho = 0.9468 \qquad sd = 0.25 \qquad F = 199$$

The constants in Equations 25, 32, and 33 are not too dissimilar, and probably EHP and NFM will be suitable model solvents here also.

The nasal pungency thresholds (NPT) in ppm of Cain and Cometto-Muñiz[50,51] on human subjects can be used to obtain a quite reasonable QSAR, again based on Equation 18, as follows,

$$-\log[NPT] = -8.80 + 2.58 \, \pi_2^H + 3.73 \, \Sigma\alpha_2^H + 1.42 \, \Sigma\beta_2^H + 0.88 \, \log L^{16} \qquad (34)$$
$$n = 32 \qquad \rho = 0.9744 \qquad sd = 0.28 \qquad F = 127$$

Interestingly, the equation for NPTs now contains a term in $\Sigma\beta_2^H$, so that wet octanol might be a very good model solvent judging by the constants in Table 9. A plot of observed vs. calculated –log[NPT] values is in Figure 11.

Plot of -log[NPT] obs vs calc

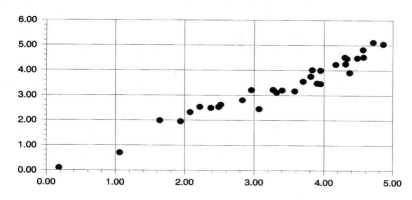

Figure 11 Nasal pungency thresholds. Plot of –log[NPT] observation vs. calculation.

ODOR THRESHOLDS

There have been so many investigations on the relationship between odor thresholds (OTs) and structural or physicochemical features, that any attempt to review the literature is beyond the scope of this chapter. In fact, most reported work has only dealt with a limited number of compounds, usually of similar structure.[52] For example, Greenberg[53] analyzed data on odor intensities for 59 varied compounds by dividing them up into various subsets. In one of the few attempts to analyze data on a varied set, Edwards and Jurs[54] managed to set up a QSAR for 58 of the 59 compounds studied by Greenberg. They used eight theoretically calculated descriptors, yielding an equation with n = 58, r = 0.9050, sd = 0.74 log units, and F = 27.8, certainly a very reasonable QSAR in view of the practical difficulties in obtaining the data. The QSAR of Edwards and Jurs[54] is not easy to interpret in terms of the chemical features that influence odor intensity, although Edwards and Jurs[54] stress that compound molecular weight is a significant descriptor. Jurs et al.[55] also examined odor thresholds, but now dealt with two separate series — one of 53 aliphatic alcohols and one of 74 pyrazine derivatives. They obtained good regression equations, again using theoretically calculated descriptors. For 49 of the alcohols they found n = 49, ρ = 0.929, sd = 0.37 log units, and F = 69 using four descriptors: CPSA1 the sum of all of the partial positively charged surface areas on the molecule, MOLC8 the path 4 cluster molecular connectivity index, MOMI1 the first principle moment of inertia, and ENVR59 the path 1 molecular connectivity index for tertiary carbon substructures. For the pyrazines they found n = 73, ρ = 0.941, sd = 0.59 log units, and F = 104 with five descriptors: WTPT3 the sum of the atomic IDs for all heteroatoms, FRAG12 the number of single bonds, MOMI1 as above, WTPT1 the sum of the atomic IDs for all the atoms in the molecule, and MOLC3 the path 1 valence and ring connected molecular connectivity index. As seems to be usually the case, two sets of different descriptors are needed to model the two different classes of odorants, with a third set altogether being used for the 58 varied compounds. As Jurs et al.[55] point out, many of the descriptors are related to compound size, compared to the molecular weight descriptor used before.

Cometto-Muñiz and Cain[56,57] have determined OTs in human subjects, in ppm using a standardized test procedure, for a variety of compounds, as shown in Table 10. They noted that there was only a rather poor correlation with the saturated vapor pressure,[58] thus the Ferguson rule is not well obeyed. Application of Equation 18 to the OTs in ppm, as –log[OT] values, yielded:

Table 10   Observed and Calculated (Equation 35) Values of OTs as Log[OT] in ppm

Solute	Observation	Calculation
Methanol	3.180	1.932
Ethanol	1.850	1.844
Propan-1-ol	1.150	1.404
Propan-2-ol	2.700	2.016
Butan-1-ol	0.300	0.945
Butan-2-ol	1.980	1.553
2-Methylpropan-2-ol	2.780	2.169
Pentan-1-ol	0.110	0.538
Hexan-1-ol	−0.050	0.132
Heptan-1-ol	−1.000	−0.275
Heptan-4-ol	0.910	0.335
Octan-1-ol	−2.150	−0.681
Pyridine	0.110	0.538
Methyl acetate	3.460	2.199
Ethyl acetate	2.240	1.951
n-Propyl acetate	1.390	1.621
n-Butyl acetate	0.380	1.191
n-Pentyl acetate	0.070	0.795
n-Hexyl acetate	−0.200	0.387
n-Heptyl acetate	−0.010	−0.027
n-Octyl acetate	−0.410	−0.429
n-Decyl acetate	−0.500	−1.242
n-Dodecyl acetate	−1.360	−2.054
Pentan-2-one	0.930	1.366
Heptan-2-one	−0.150	0.557
Nonan-2-one	−0.030	−0.229
Toluene	2.190	1.520
Ethylbenzene	1.260	1.193
n-Propylbenzene	0.470	0.867
n-Butylbenzene	0.630	0.426
n-Pentylbenzene	0.000	0.023
n-Hexylbenzene	−0.190	−0.333
n-Heptylbenzene	−0.250	−0.658
n-Octylbenzene	−0.430	−1.057
Oct-1-ene	2.310	3.009
Oct-1-yne	2.130	1.973
Chlorobenzene	1.110	0.755
2-Phenylethanol	−2.050	−2.273
s-Butyl acetate	0.670	1.547
Propanone	4.070	1.976
t-Butyl acetate	0.110	1.865

Data from References 50 and 51.

$$-\log[OT] = -6.190 + 3.83\ \pi_2^H + 4.17\ \Sigma\alpha_2^H + 0.81\ \log L^{16} \qquad (35)$$
$$n = 39 \qquad \rho = 0.8912 \qquad sd = 0.64 \qquad F = 45$$

The goodness-of-fit of Equation 35 is about the same as that obtained by Edwards and Jurs[54] for the varied set of compounds of Greenberg.[53] Possibly, this is the best that can be done in terms of a general QSAR, but further QSAR studies are needed.

Of the original data set, two compounds, propanone and t-butyl acetate, were outliers and are not included in Equation 35. As well as propanone, other small compounds, such as methyl acetate and methanol are not well correlated, but have been left in the correlation equation. By the usual standards, Equation 35 is rather poor and can correlate the log[OT] values to only 0.64 log units. However, the data cover a range of over five log units, and the experimental error in the obtained values should not be overlooked. Furthermore, the data set covers a range of compound types. Also given in Table 10 are the log[OT] values calculated on Equation 35. A graph of observed vs. calculated –log[OT] values is shown in Figure 12. What can be deduced from Equation 35 is that solute dipolarity, hydrogen-bond acidity, and lipophilicity all increase the perceived odor. It is tempting to compare the characteristic constants in Equation 35 with those for solvents in Table 9, where N-formylmorpholine now seems a reasonable model. However, as Cometto-Muñiz and Cain[58] point out, a realistic model for OTs needs to be more finely tuned to the features of the solute molecules.

Plot of -log[OT] obs vs calc

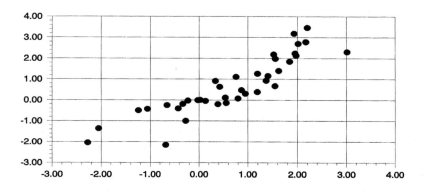

Figure 12 Odor thresholds. Plot of –log[OT] observation vs. calculation.

Finally, in an interesting development, Okahata et al.[59] have shown that the odor intensity of a number of odorants and perfumes could be related to their partition between air and a synthetic lipid, used as a coating on a quartz crystal microbalance. Unfortunately, no numerical data were reported, and as is often the case, the compounds investigated were divided into small subsets. What is clear, however, is that a very wide range of phases is available as candidate model solvents for odor intensity or OTs. All the more relevant is some method of systematically selecting solvents or phases as set out above, rather than relying on trial and error.

ACKNOWLEDGMENTS

It is a pleasure to acknowledge over the years the fruitful discussions with, and help given by, Dr. Gunnar D. Nielsen, Dr. Yves Alarie, Dr. William S. Cain, and Dr. J. Enrique Cometto-Muñiz. Much of this work at UCL was carried out with Dr. Gary S. Whiting who initiated the computing programs, and I thank also Dr. Harpreet S. Chadha and Dr. Marti Roses who have subsequently helped with the development of these programs.

REFERENCES

1. E. Overton, *Studien uber die Narkose,* Fischer, Jena, Germany,1901.
2. H. Meyer, *Arch. Exp. Pathol. Pharmakol. (Naunyn-Schmiedebergs)* 1899, 42, 109-118.
3. H. Meyer, *Arch. Exp. Pathol. Pharmakol. (Naunyn-Schmiedebergs)* 1901, 46, 338-346.
4. K.H. Meyer and H. Gottlieb-Billroth, *Z. Physiol. Chem.,* 1920, 112, 55-79.
5. K.H. Meyer and H. Hemmi, *Biochem. Z.,* 1935, 277, 39-71.
6. J.C. Dearden, *Environ. Health Persp.,* 1985, 61, 203-228.
7. K.H. Meyer and H. Hopff, *Z. Physiol. Chem.,* 1923, 126, 281-298.
8. J. Ferguson, *Proc. R. Soc. (London),* 1939, B127, 387-404.
9. G.D. Nielsen and Y. Alarie, *Toxicol. Appl. Pharmacol.,* 1982, 65, 459-477.
10. J.C. Kranz and F.G. Rudo, *Handb. Exp. Pharmacol.,* 1966, 20, 501-564.
11. F. Brink and J.M. Posternak, *J. Cell. Comp. Physiol.,* 1948, 32, 211-233.
12. L.J. Mullins, *Chem. Rev.,* 1954, 54, 289-323.
13. K.W. Miller and E. Brian Smith, in "A Guide to Molecular Pharmacology-Toxicology, Part II," ed R.M. Featherstone, Marcel Dekker, New York, 1973.
14. M.H. Abraham, G.D. Nielsen and Y. Alarie, *J. Pharm. Sci.,* 1994, 83, 680-688.
15. K.W. Miller, W.D.M. Paton, E.B. Smith and R.A. Smith, *Anesthesiology,* 1972, 36, 339-351.
16. E.I. Eger, C. Lundgren, S.L. Miller and W.C. Stevens, *Anesthesiology,* 1969, 30, 129-135.
17. R.M. Stephenson and S. Malanowski, "Handbook of the Thermodynamics of Organic Compounds," Elsevier, New York, 1987.
18. *Handbook of Chemistry and Physics,* CRC Press, Boca Raton, FL, 1975.
19. M.H. Abraham, *J. Chem. Soc., Faraday Trans. 1,* 1984, 80, 153-181.
20. M.H. Abraham, P.L. Grellier and R.A. McGill, *J. Chem. Soc., Perkin Trans. 2,* 1987, 797-803.
21. M.H. Abraham, *J. Am. Chem. Soc.,* 1979, 102, 5477-5484; 1982, 104, 2085-2094.
22. N.M. Cone, S.E. Forman and J.C. Kranz, Jr., *Proc. Soc. Exp. Biol. Med.,* 1941, 48, 461-463.
23. G. Lu, F.K. Bell, C.J. Carr and J.C. Kranz, Jr., *Arch. Exp. Pathol. Pharmakol.,* 1953, 219, 115-118.

24. G.D. Nielsen, L.F. Hansen and Y. Alarie, in "Chemical, Microbiological, Health and Comfort Aspects of Indoor Air Quality — State of the Art in SBS," ed H. Knoppel and P. Wolkoff, Kluwer Academic Publishers, Dordrecht, The Netherlands, 1992.

25. J. Muller and G. Greff, *Food Chem. Toxicol.,* 1984, 22, 661-664.

26. G.D. Nielsen and A.M. Vinggaard, *Pharmacol. Toxicol.,* 1988, 63, 293-304.

27. G.D. Nielsen and M. Yamagawi, *Chem. Biol. Interact.,* 1989, 71, 223-244.

28. G.D. Nielsen, E.S. Thomsen and Y. Alarie, *Acta Pharmacol. Nord.,* 1990, 2, 31-44.

29. C. Hansch, J.E. Quinlan and G.L. Lawrence, *J. Org. Chem.,* 1968, 33, 347-350.

30. S.H. Yalkowsky and S.C. Valvani, *J. Pharm. Sci.,* 1980, 69, 912-922.

31. D.W. Roberts, *Chem. Biol. Interact.,* 1986, 57, 325-345.

32. P. Laffort and F. Patte, *J. Chromatogr.,* 1976, 126, 625-639.

33. F. Patte, M. Etcheto and P. Laffort, *Anal. Chem.,* 1982, 54, 2239-2247.

34. P. Laffort and F. Patte, *J. Chromatogr.,* 1987, 406, 51-74.

35. M.H. Abraham, G.S. Whiting, R.M. Doherty and W.J. Shuely, *J. Chem. Soc. Perkin Trans. 2,* 1990, 1451-1460.

36. M.H. Abraham, G.S. Whiting, R.M. Doherty and W.J. Shuely, *J. Chromatogr.,* 1991, 587, 213-228.

37. M.H. Abraham and G.S. Whiting, *J. Chromatogr.,* 1992, 594, 229-241.

38. M.H. Abraham, *J. Chromatogr.,* 1993, 644, 95-139.

39. M.H. Abraham, *J. Phys. Org. Chem.,* 1993, 6, 660-684.

40. M.H. Abraham, *Chem. Soc. Rev.,* 1993, 22, 73-83; *Pure Appl. Chem.,* 1993, 65, 2503-2512.

41. M.H. Abraham, P.L. Grellier, D.V. Prior, P.P. Duce, J.J. Morris and P.J. Taylor, *J. Chem. Soc. Perkin Trans. 2,* 1989, 699-711.

42. M.H. Abraham, P.L. Grellier, D.V. Prior, J.J. Morris and P.J. Taylor, *J. Chem. Soc. Perkin Trans. 2,* 1990, 521-529.

43. M.H. Abraham and P.K. Weathersby, *J. Pharm. Sci.,* 1994, 83, 1450-1456.

44. M.H. Abraham, G.S. Whiting, Y. Alarie, J.J. Morris, P.J. Taylor, R.M. Doherty, R.W. Taft and G.D. Nielsen, *Quant. Struct. Act. Relat.,* 1990, 9, 6-10.

45. M. Schaper, *Am. Ind. Hyg. Assoc. J.,* 1993, 54, 488-544.

46. G.D. Nielsen and Y. Alarie, unpublished results.

47. G.D. Nielsen, U. Kristiansen, L. Hansen and Y. Alarie, *Arch. Toxicol.,* 1988, 62, 209-215.

48. M.H. Abraham, J. Andonian-Haftvan, J.P. Osei-Owusu, P. Sakellariou, J.S. Urieta, M.C. Lopez and R. Fuchs, *J. Chem. Soc. Perkin Trans. 2,* 1993, 299-304.

49. M.H. Abraham, I. Hamerton, J.B. Rose and J.W. Grate, *J. Chem. Soc. Perkin Trans. 2,* 1991, 1417-1423.

50. J.E. Cometto-Muñiz and W.S. Cain, *Physiol. Behav.,* 1990, 48, 719-725.

51. J.E. Cometto-Muñiz and W.S. Cain, *Pharmacol. Biochem. Behav.,* 1991, 39, 983-989.

52. D. Ottoson, *Acta Physiol. Scand.,* 1958, 43, 167-181.

53. M.J. Greenberg, *J. Agric. Food Chem.,* 1979, 27, 347-352.

54. P.A. Edwards and P.C. Jurs, *Chem. Senses,* 1989, 14, 281-291.

55. P.A. Edwards, L.S. Anker and P.C. Jurs, *Chem. Senses,* 1991, 16, 447-465.

56. J.E. Cometto-Muñiz and W.S. Cain, *Physiol. Behav.,* 1990, 48, 719-725.

57. J.E. Cometto-Muñiz and W.S. Cain, *Pharmacol. Biochem. Behav.,* 1991, 39, 983-989.

58. J.E. Cometto-Muñiz and W.S. Cain, *Arch. Environ. Health,* 1993, 48, 309-314.

59. Y. Okahata, O. Shimizu and H. Ebato, *Bull. Chem. Soc. Jpn,* 1990, 63, 3082-3088.

Electrophysiological Indices of Human Chemosensory Functioning

Gerd Kobal

SUMMARY

Based on the experiments of Hosoya and Yoshida (1937),[2] and Ottoson (1956),[3] various attempts were made, in order to record electrophysiological activities elicited by chemical stimuli in humans, such as electro-olfactograms (EOG) from the human olfactory mucosa and cerebral event-related potentials (ERPs) from the electroencephalogram (EEG). Electro-olfactograms are generally considered to be summated generator potentials of olfactory receptor cells. They are recorded by aid of macro-electrodes from the surface of the olfactory mucosa. Therefore, it is of paramount importance that the process of applying the odorants does not produce artifacts at the mucosa. Consequently, an olfactometer must be utilized, which does not alter the mechanical or thermal conditions at the mucosa during stimulation.

Similar requirements, albeit for different reasons, apply to the recording of ERPs. Because these potentials are highly integrated responses of the human brain, an additional activation of thermo- and mechanoreceptors would lead to synchronous cortical somatosensory responses, which would, virtually, not be distinguishable from the olfactory responses. Another type of olfactorily induced brain wave activity is an endogenous component called the late "cognitive component" (or P300). It is recorded as part of an ERP during a discrimination task where subjects are requested to discriminate low probability stimuli from high probability stimuli. This late ERP component only appeared after application of the low probability stimulus. Technically, the recording of the P300 components adds another difficulty to the olfactory stimulation procedures. Because the high probability stimuli must be repeated frequently, interstimulus intervals of less than 8 s are required. Therefore, the olfactometer needs the option to rapidly deliver two or more different odorants through the same outlet with very little residues of the preceding substance in the tubing system.

Recently, magnetoencephalography (event-related magnetic field, ERMF) became available as a better means for localization of cortical activity. CO_2 stimuli were applied to the nasal mucosa activating free nerve endings of the trigeminal nerve. The measurements were made in a magnetically shielded chamber using a first order SQUID gradiometer. Sources of the responses were located within a 2.5×2.5 cm^2 area in a region where the central sulcus and the Sylvian fissure are contiguous with each other. This corresponds to an activation of the secondary somatosensory projection area (SII). Similar experiments using a 37 channel magnetoencephalograph (Krenikon) investigating odorants, such as vanillin, are in progress. There are first indications that the generators of the olfactory ERMF are located in the temporal lobe.

INTRODUCTION

A variety of techniques have been used in an effort to reliably measure the influence of odorants on electrophysiological responses. For example, the first recording of an ERP evoked by chemical stimulation was obtained in 1890 by Fleischl von Marxow,[1] who presented ammonia to the nasal mucosa of a rabbit, eliciting a negative electrical deflection on the surface of the brain. Following the pioneering animal experiments of Hosoya and Yoshida[2] and Ottoson,[3] Osterhammel et al.[4] provided the first report of the human EOG (arising from the olfactory epithelium in response to odorant stimulation). Cerebral potentials in response to odorants were obtained from the human EEG by Finkenzeller[5] and Allison and Goff.[6] In 1979, Hughes and Andy,[7] using implanted electrodes in patients, recorded the spontaneous EEG from the olfactory bulb, as well as from the amygdala.

STIMULATION TECHNIQUES

To record chemosensory event-related potentials (CSERP), considerable sophistication in stimulus presentation is needed. Small fluctuations in stimulus concentration can lead to variations in latencies which distort or even cancel components of the CSERP. For the same reason the quality of the temporal presentation of the stimuli has to be very precise, and stimulus onset has to be exceedingly steep (≤ 50 ms) to guarantee the simultaneous excitation of a sufficiently large number of cortical neurons. Importantly, the synchronous excitation of other sensors (e.g., mechano- or thermosensors) has to be avoided. Only recently have olfactometers been developed which meet these requirements.[8,9] Monomodal chemical stimulation is achieved by embedding pulses of the odorant into a constantly flowing air stream (140 ml/s) of controlled temperature (36.5°C) and humidity (80%).

Figure 1 is a diagram of the olfactometer's odor switching device delivering pulses of odorants to the subjects. Similar to the functioning of a railway switch two gaseous streams are switched. Switching is induced by means of a vacuum so that during stimulation odorant pulses of pre-established concentrations *(odor + dilution)* reach the olfactory region, and during interstimulus intervals only nonodorous air *(control)* reaches the nose. This insures that the subjects have no tactile, thermal, or acoustic cues to provide extraneous information about the timing of stimulus presentations (for more details, see References 8–12).

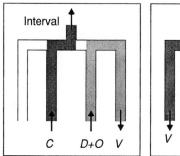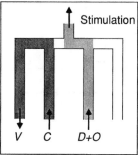

Figure 1   Schematic diagram of the switching device (*C*ontrol air stream, *D*ilution + *O*dorant, *V*acuum). When the odorant is switched on or off, the subject is unable to discern turbulences or changes in flow rate or pressure. The carrier gas (air) is thermostabilized (36.5°C) and humidified to 80% relative humidity.

ELECTROPHYSIOLOGICAL RESPONSES TO CHEMOSENSORY STIMULI

The Human Electro-Olfactogram

The EOG is considered to consist of summated generator potentials from olfactory receptor cells.[3] The EOG is recorded from the surface of the olfactory mucosa using macroelectrodes. For this reason, it is of paramount importance that the process of applying the odorants does not produce artifacts at the mucosal surface. Clearly, the recording of EOGs from the human nasal mucosa is difficult. Placing of electrodes is no easy task, since the intrusion of any foreign matter into the nose may lead to sneezing and to mucous discharge. Local anesthesia also has to be avoided or used carefully, because it may render the subject temporarily anosmic.

Kobal[8] performed an EOG study which utilized the stimulation method described above. The odorants amyl acetate, H_2S, and eugenol were presented to four subjects. The responses, all of which were characteristic negative electrical potentials at the surface of the olfactory mucosa, were dependent upon stimulus concentration. When stimuli (H_2S) of longer durations were applied, temporal integration over a 10-s period of time was observed. The same phenomenon also occurred when pairs of 1-s stimuli were successively presented. Although the response to the second stimulus was of smaller amplitude, it was superimposed upon the remnant of the preceding one (Figure 2). Little adaptation was noted in the EOG, supporting the view that adaptation following repeated or continuous stimulation is due the habitutation of central nervous system (CNS) mechanisms.[13]

Relation of EOG to Repetitive Stimulation: Example of a Single Subject
Hydrogen Sulphide - Stimulus Duration 0.5 s

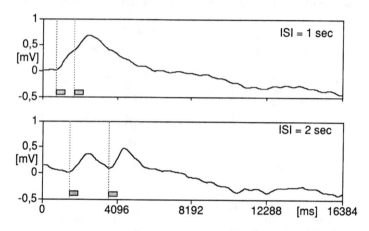

Figure 2 Electro-olfactograms (recordings from the olfactory epithelium; example of a single subject) after repetitive olfactory stimulation with H_2S at an interstimulus interval (ISI) of 1 and 2 s, respectively. Stimulus duration was 500 ms as indicated by the filled rectangular boxes. At an ISI of 1 s responses are superimposed on each other. For the second response no major decrease of amplitudes could be observed when compared to the response to the first stimulus.

When recording EOGs from the olfactory mucosa, Kobal[8] also recorded from a second electrode placed on the respiratory epithelium. This electrode registered additional negativities after application of odorants which also produced trigeminal activity. These negative mucosal potentials (NMPs) were not correlated with the olfactory sensations of the subjects, but were rather correlated with the degree to which they experienced pain. In animals, NMPs can be eliminated by the systemic injection of capsaicin[14] implicating their relationship to excitation of nociceptive C- or Adelta-fibers. Additionally, in human subjects the NMP was shown to be independent of autonomic reflexes.[15] Thus, these potentials appear to represent peripheral events closely related to the activation of nociceptors of the trigeminal nerve.[9]

More research is needed before EOG or NMP recordings will be of practical use. Nevertheless, because EOG and NMP represent the input signals into the olfactory and trigeminal channels, respectively, their measurement may be useful in the future for the interpretation of more centrally generated responses (e.g., CSERP and subjective ratings), including the determination of the site of modulatory processes of such diverse phenomena as hyperosmia, parosmia, or peripheral sensitization.

The CSERP — General Considerations

ERPs are due to changes in electrical fields generated by large populations of neurons occurring before, during, or after a sensory or internal psychological event (for review see Reference 16). When recording such potentials from the surface of the skull, their amplitudes are very small (~50 μV), and it is necessary to isolate them from the background activity by averaging and/or filtering. For averaging, a number of stimulus-synchronous EEG records of 1- to 2-s duration are digitized. Averaging of the stimulus-locked array of numbers results in visualized waveforms which represent responses of synchronously reacting cortical neurons.

Because CSERPs are highly integrated responses of the human brain, additional activation of thermo- or mechanoreceptors by odorants or the stimulus airstream can lead to synchronous cortical somatosensory responses which are virtually indistinguishable from olfactory responses. Therefore, odorants must be presented in a manner which does not contaminate the CSERP with other neural responses. The major technical problem with stimulus presentation is that steep stimulus onsets are required to produce synchronous activation of a sufficient number of cortical neurons to result in a measurable CSERP. However, even when most causes responsible for artifacts are eliminated, a major problem still remains; namely, the problem of distinguishing between responses elicited by activation of the chemoreceptors of the olfactory nerve and those of the trigeminal nerve.[17] This problem is discussed later in more detail.

If a CSERP study is to yield reliable data, several additional prerequisites have to be met:

1. The background activity of the EEG has to be stationary, and, at the same time, stochastic in relation to the concealed response.
2. The stimulus-specific response has to be stable, especially if there are a number of consecutive positive and negative components. Even the slightest phase-shifting can cause the positive and negative components of the stimulus-related responses to cancel each other out, in a manner analogous to the canceling that occurs from background activity. As mentioned earlier, positive and negative waves are only separated by a fraction of a second, making it absolutely essential that the temporal presentation of the stimulus is accurate and reproducible.
3. A minimum flow rate is needed to obtain reliable CSERP recordings; no responses have been obtained with flow rates below 85 ml/s.[8]

Finally, it should be emphasized that suprathreshold odorant concentrations are needed to distinguish CSERPs from background activity; therefore, measurement of responses within the perithreshold region cannot be achieved using olfactory event-related potentials (OERPs). Nevertheless, the amplitude of the OERP increases and its latency decreases as stimulus intensity is increased,[8] demonstrating its sensitivity to stimulus changes within the suprathreshold domain.

ERPs are often grouped into early, intermediate, late, and ultra-late potential components on the basis of their time of appearance after the onset of an event. Another way to classify ERPs is to determine the distances of their generators from the recording electrode (e.g., generators of nearfield potentials are located in the cortex, relatively close to the recording electrode; generators of farfield recordings are located subcortically).[18]

To date, only late nearfield CSERPs have been recorded. The first responses of this kind were obtained by Finkenzeller[5] and Allison and Goff.[6] However, Smith et al.[17] contested the olfactory nature of these responses on the grounds that they did not see them in patients who had lost trigeminal sensitivity. Nevertheless, Kobal and Hummel[11] found such responses could be obtained using the odorant vanillin, an agent that is not perceived by anosmic patients.[19] Importantly, such responses failed to occur in anosmic patients whose filae olfactoriae were presumably damaged by trauma or other causes (Figure 3). In contrast, CO_2, a nonodorous trigeminal stimulant, is perceived by such subjects and elicits ERPs via the trigeminal nerve.

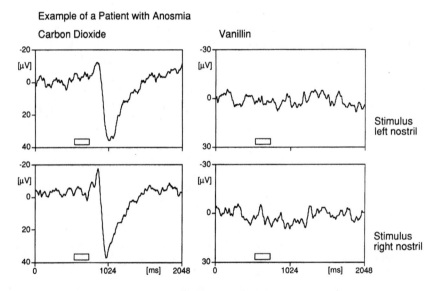

Figure 3   CSERP in a 22-year-old woman with anosmia (recording position Pz). No responses could be observed after olfactory stimulation with vanillin. In contrast, the patient clearly responds to trigeminal stimulation with CO_2.

Coding of Stimulus Quality

Attempts to experimentally determine differences in the shape of CSERPs have not been successful. It became apparent in earlier studies[11] that the shape of ERPs elicited by activation of the olfactory and trigeminal systems (i.e., by vanillin or CO_2, respectively) were the same. Thus, the hypothesis put forward by Herberhold[20] and Tonoike et al.[21] that different size amplitudes of odorant evoked potentials have their origin in different sensory systems is probably incorrect; the differential responses observed in these studies were likely due to the extent of afferent fiber activation.

Cortical Sources of the CSERP

The exact location of the cortical generators of CSERPs is still unknown. One approach to localizing their foci is to topographically analyze the registered activity. Unfortunately, few studies have performed such analyses and surface potential distributions have to be interpreted cautiously. Realistic models must take into account the fact that ERP wave forms are usually not produced by a single intracranial generator, and that such waves can be compatible with a variety of intracerebral charge configurations.[22] Research using multichannel EEG recordings, in combination with current source density techniques,[23-25] is needed to better define details of the underlying processes.

Recently, magnetoencephalography (ERMF) has become available. This technology allows for better localization of cortical activity than the EEG.[26-28] Although both the ERP and the ERMF are epiphenomena of the transmembrane currents, it has been suggested that the ERMF is almost entirely due to membrane currents rather than to the extracellular volume currents that are responsible for the ERP.[27] Importantly, the ERMFs are not modified by the physical properties of the conductive medium.[29]

In the field of chemosensation, only one study was performed on recording of ERMFs.[30] CO_2 stimuli at concentrations which produced pain were applied to the nasal mucosa.[9] Measurements were made using a first order SQUID gradiometer. Simultaneous recording was made from three channels, and responses were measured from 15 to 87 locations on the scalp, covering an area of 60 to 200 cm^2. ERMF waves had maximal deflections at 350 to 400 ms after stimulus onset coinciding with the negative potential recorded from the vertex. Because approximately 100 to 150 ms utilization times at the receptor site have to be subtracted,[9] the corrected latency of this component was in the range of 200 to 300 ms. Isofield maps were found to be dipolar at the peak latencies of the electrical potentials in all subjects. The equivalent dipoles, which caused a negative potential deflection at the vertex, were found within a 2.5×2.5 cm^2 area in a region where the central sulcus and the Sylvian fissure are contiguous. This corresponds to an activation of the secondary somatosensory projection area (SII). According to Chudler et al.,[31] this area is closely connected to the processing of nociceptive information. This observation is in accordance with the point made earlier that CO_2 specifically activates nocisensors of the trigeminal nerve.[14,32] Similar experiments using odorants with little or no trigeminal activity, such as vanillin, are in progress.[33] They suggest that the generators of the olfactory ERMF are located in the temporal lobe.

Topographical Distribution of the CSERP

Topographical ERP mappings have allowed the differentiation between olfactory stimulation (OERP) and chemosomatosensory (trigeminal) stimulation (CSSERP). Hummel and Kobal[34] found a different topographical distribution of

CSERPs after stimulation with vanillin or acetaldehyde than after stimulation with sulfur dioxide (SO_2) or ammonia (NH_3). When acetaldehyde and vanillin were used as stimulants, N1P2 amplitudes were largest at the parietal site Pz and the central site Cz, and smallest at lateral frontal positions. When SO_2 and NH_3 were used, amplitudes obtained at Cz were significantly larger than those obtained at any other recording position (Figure 4). These data have been confirmed by a number of other studies.[35-37] They suggest that objective information as to the intranasal chemosensory system that has been activated by a nasally inhaled stimulus can be obtained by using the technique of "brain mapping."

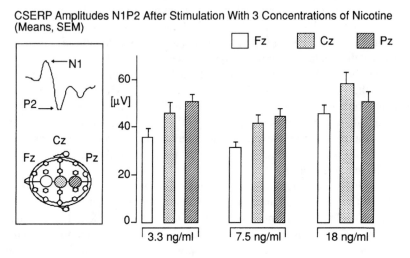

CSERP Amplitudes N1P2 After Stimulation With 3 Concentrations of Nicotine (Means, SEM)

Figure 4   Measurement of CSERP amplitudes N1P2 (means, standard errors, n = 9) at midline recording positions Fz, Cz, and Pz. Responses are shown to three concentrations of nicotine which produces odorous (3.3 ng/ml), burning (7.5 ng/ml), and stinging (18 ng/ml) sensations in a concentration-related manner. At low concentrations the distribution of ERP amplitudes clearly revealed an "olfactory" pattern, i.e., largest amplitudes were obtained at the parietal recording site. Following stimulation with the strongest stimulus the distribution exhibited a "trigeminal" pattern with largest amplitudes at central sites, i.e., this particular pattern emerged only when the concentration of nicotine was above recognition threshold for stinging sensation.

Clinical applications

An important application of OERPs is in the clinical diagnosis of some forms of olfactory dysfunction. The test procedure standardized by Kobal and Hummel[12] includes the recording of responses to CO_2 (52% v/v), H_2S (0.78 ppm), and vanillin (2.06 ppm). All three stimulants are applied 15 times, both to the left and the right nasal chambers. The interstimulus interval is approximately 40 s. The session, including preparation, lasts 80 min. The EEG (filtering 0.2 to 30 Hz, electrode impedance 2 to 4 kΩ) is recorded from 6 positions of the international 10/20 system (Fz, Cz, Pz, C3, C4, and Fp2; referenced to A1 + A2). EEG records of 2048-ms duration are digitized (sampling frequency 250 Hz) and averaged in 6 groups according to the 3 stimulants and the binasal stimulation. EEG records contaminated by eye blinks or motor artifacts are discarded from the average. In all anosmic patients tested thus far no OERPs could be detected after stimulation with vanillin and H_2S.[38] However, CSSERPs were obtained in these cases after stimulation with the trigeminal stimulant CO_2 (Figure 3).

OERP recording may prove to be helpful in the diagnosis of individuals suffering from neurological diseases associated with changes in olfactory function (e.g., Alzheimer's disease,[39] Parkinson's disease,[40] multiple sclerosis,[41] progressive supranuclear palsy,[42] and amyotrophic lateral sclerosis[43]). In a recent study, CSERPs to presentations of CO_2, vanillin, and H_2S (16 presentations of each stimulus to each nostril) were examined in 8 patients with Parkinson's disease (PD) and 9 controls.[44] Longer latencies were found in the PD patients. No difference was observed for amplitudes. There were also no differences in the responses to the trigeminal stimulant CO_2.

Only recently a study was published on patients with temporal lobe epilepsy.[45] Twenty-two patients were investigated, 12 of whom suffered from epilepsy with a focus located in the left temporal lobe; 10 patients had a right-sided temporal lobe focus. In both groups prolonged CSERP latencies were found after left-sided stimulation with CO_2 compared to stimulation of the right nostril. In contrast, a different pattern emerged for olfactory stimuli. After right-sided olfactory stimulation latencies were prolonged in patients with right-sided epileptical foci. Similarly, when the left nostril was stimulated in patients with a left-sided focus, CSERP latencies were prolonged. Thus, it was suggested that the neocortical processing of olfactory, but not trigeminally mediated information is affected by functional lesions of the temporal lobe. Moreover, analyses revealed nonoverlapping 95% confidence intervals for latency N1 when vanillin was applied to the right nostril. These results indicated that the right temporal lobe may play a different role in the processing of olfactory information compared to the left temporal lobe.

Investigation of Olfactory-Induced Emotions

There are indications that correlates of odor-induced emotions are observable in the CSERP. In one study,[36] in which CO_2, menthol, H_2S, and vanillin were presented to each side of the nose, differences between ERPs in relation to the stimulated side were observed. Responses to CO_2, menthol, and H_2S showed significantly shorter latencies and smaller amplitudes after stimulation of the left nasal chamber, whereas in responses to vanillin latencies of the late components were shorter and the N1P2 amplitude was significantly smaller after stimulation of the right nasal chamber. This cannot be explained by differences in subjectively perceived intensities, because none of the substances elicited statistically significant differences in intensity estimates after stimulation of the right or the left side. With regard to the grouping of ERPs into CO_2, menthol, and H_2S vs. vanillin, it became evident that the hedonic reactions to both groups of odorants were different. Vanillin caused a pleasant sensation in all subjects. The substances of the other groups were associated with more or less negative experiences. CO_2

was always perceived as painful. Menthol induced positive as well as negative sensations. On one occasion it was termed as pleasant and cooling, whereas on another it caused slightly painful sensations. H_2S reminded all subjects of rotten eggs. Because vanillin was the only substance to be associated with pleasant emotions, it was hypothesized that the topographical ERP differences caused by stimulation of the different sides was related to the manner in which the two hemispheres process emotional information on different qualities, in accordance with previous studies on hemispheric differences in emotional information processing.[46,47] Interestingly, the finding of hedonic-related latency shifts was also observed when the odorants phenyl ethyl alcohol and acetaldehyde served as stimuli.[48,49]

The Late Positive Component of the OERP

Another way of classifying ERPs, in addition to "nearfield vs. farfield" and "early vs. late" potentials (see above), relates to the information encoded within them. By definition, "exogenous" ERPs are determined by the characteristics of the afferent input. In contrast, "endogenous" ERPs vary with the physiological state of the subject, the meaning of the stimulus to the subject, and/or demand of the task related to the respective stimulus. These classes are not always mutually exclusive. ERP components are usually less subject to endogenous processes the earlier they appear.[50] For example, in the case of audition, the late "cognitive component P300" (i.e., the positive wave appearing approximately 300 ms after the stimulus onset) is recorded in ERPs during a discrimination task where subjects are requested to discriminate low probability stimuli from high probability stimuli.[51] Visual P300 recordings have also been performed in an odor-labeling task.[52] Here, the effect of correct and incorrect visual labeling of a preceding odor presentation was investigated. As expected, P300 was significantly larger during the rare nonmatching trials.

Recently, a late positive component (LPC) analogous to the P300 has been successfully recorded following the application of an odorant.[53] To observe this response, stimulus presentation times had to be carefully worked out. Because the recording of this response requires an interstimulus interval of only a few seconds, rapid habitutation processes can alter the response if the interstimulus interval is too short. On the other hand, if it is too long, the high probability stimulus would not be "frequent" anymore and would also elicit an LPC. The lower limit of the interstimulus interval was found to be 6 to 8 s when vanillin, H_2S, or a mixture of vanillin with CO_2, of 200 ms duration, was used.

In a similar study, the LPC was assessed in 16 subjects using the odorants vanillin and H_2S.[53] Unlike the earlier CSERP components, LPCs were not related hedonically to the side of stimulation. This implies that the latency effect related to the olfactory-induced emotion is generated subcortically, since it is already seen in the first ERP wave, P1.

To summarize, considerable promise is being realized in the development of electrophysiological responses to odorants. When the location of cortical generators becomes better known, measures such as the LPC may be useful in helping to determine the focus of various brain lesions associated with olfactory dysfunction, including foci in temporal lobe epilepsy patients. Such measures may also be of value in the exploration of the physiological underpinings of odor-related hedonic and cognitive processes. In addition, CSERPs are likely to be of value in the understanding of diseases such as multiple chemical sensitivities where exposition to odors may play the crucial role.

ACKNOWLEDGMENTS

This research was supported by DFG grant Ko812/5-1.

REFERENCES

1. E. Fleischl von Marxow, Mittheilung, betreffend die Physiologie der Hirnrinde, *Zbl. Physiol.* 4 (1890) s37-s4.
2. Y. Hosoya and H. Yoshida, Über die bioelektrischen Erscheinungen an der Riech-schleimhaut, *J. Med Sci. III Biophys.* 5 (1937) 22.
3. D. Ottoson, Analysis of the electrical activity of the olfactory epithelium, *Acta Physiol. Scand.* 35 Suppl. 122 (1956) 1-83.
4. P. Osterhammel, K. Terkildsen and K. Zilsdorff, Electro-olfactograms in man, *J. Laryngol.* 83 (1969) 731-733.
5. P. Finkenzeller, Gemittelte EEG-Potentiale bei olfactorischer Reizung, *Pflügers Arch.* 292 (1965) 76-85.
6. T. Allison and W.R. Goff, Human cerebral evoked responses to odorous stimuli, *Electroencephalogr. Clin. Neurophysiol.* 23 (1967) 558-560.
7. J.R. Hughes and O.J. Andy, The human amygdala. I. Electrophysiological responses to odorants, *Electroencephalogr. Clin. Neurophysiol.* 46 (1979) 428-443.
8. G. Kobal, Elektrophysiologische Untersuchungen des menschlichen Geruch-ssinnes. Thieme, Stuttgart, 1981.
9. G. Kobal, Pain-related electrical potentials of the human nasal mucosa elicited bychemical stimulation, *Pain* 22 (1985) 151-163.
10. G. Kobal, Process for measuring sensory qualities and apparatus terefor, United States Patent Number 4,681,121 (1987).
11. G. Kobal and C. Hummel, Cerebral chemosensory evoked potentials elicited by chemical stimulation of the human olfactory and respiratory nasal mucosa, *Electroencephalogr. Clin. Neurophysiol.* 71 (1988) 241-250.
12. G. Kobal and T. Hummel, Olfactory evoked potentials in humans, in Smell and Taste in Health and Disease (ed. T.V. Getchell, R.L. Doty and J.B. Snow, Jr.), Raven Press, New York 1992, p. 255-275.
13. H. Zwaardemaker, The sense of smell, *Acta Oto-Laryngol.* 11 (1927) 3-15.
14. N. Thürauf, I. Friedel, C. Hummel and G. Kobal, The mucosal potentials elicited by noxious chemical stimuli: is it a peripheral nociceptive event? *Neurosci. Lett.* 128 (1991) 297-300.
15. N. Thürauf, T. Hummel, B. Kettenmann and G. Kobal, Nociceptive and reflexive responses recorded from the human nasal mucosa, *Brain Res.* 629 (1993) 293-299.
16. A.S. Gevins and A. Remond, Handbook of Electroencephalography and Clinical Neurophysiology, Volume 1: Methods of Analysis of Brain Electrical and Magnetic Signals. Elsevier, Amsterdam 1987.

17. D.B. Smith, T. Allison, W.R. Goff and J.J. Princitato, Human odorant evoked responses: effects of trigeminal or olfactory deficit, *Electroencephalogr. Clin. Neurophysiol.* 30 (1971) 313-317.

18. J.E. Desmedt, Somatosensory Evoked Potential, in Human Event-Related Potentials (ed. T.W. Picton) Elsevier, Amsterdam 1988, p. 245-261.

19. R.L. Doty, W.E. Brugger, P.C. Jurs, M.A. Ornsdorff, P.J. Snyder and L.D. Lowry, Intranasal trigeminal stimulation from odorous volatiles: Psychometric responses from anosmics and normal humans, *Physiol. Behav.* 20 (1978) 175-185.

20. C. Herberghold, Funktionsprüfungen und Störungen des Geruchssinnes, *Arch. Oto-Rhino-Laryngol.* 210 (1975) 67-164.

21. M. Tonoike, N. Seta, T. Maetani, I. Koizuka and M. Takebayashi, Measurements of olfactory evoked potentials and event related potentials using odorant stimuli, *Ann. Int. Conf. IEEE Eng. Med. Biol. Soc.* 12 (1990) 912-913.

22. H.G. Vaughan and J.C. Arezzo, The neural basis of event-related potentials, in Human Event-Related Potentials (ed. T.W. Picton) Elsevier, Amsterdam 1988, pp. 45-96.

23. U. Mitzdorf, Current source-density method and application in cat cerebral cortex: Investigation of evoked potentials and EEG phenomena, *Physiol. Rev.* 65 (1985) 37-100.

24. M. Scherg and D. von Cramon, Evoked dipole source potentials of the human auditory cortex, *Electroencephalogr. Clin. Neurophysiol.* 65 (1986) 344-360.

25. Z. Zhang and D.L. Jewett, Insidious errors in dipole localization parameters at a single time-point due to model misspecification of the number of shells, *Electroencephalogr. Clin. Neurophysiol.* 88 (1993) 1-11.

26. R. Hari, E. Kaukoranta, K. Reinikainen, T. Huopainiemi and J. Mauno, Neuromagnetic localization of cortical activity evoked by painful dental stimulation in man, *Neurosci. Lett.* 42 (1982) 77-82.

27. Y. Okada, Discrimination of localized and distributed current dipole sources and localized single and multiple sources, in Biomagnetism: Applications and Theory (ed. H. Weinberg, G. Stroink and T. Katila) Pergamon Press, New York 1985, p. 266-272.

28. S.J. Williamson and L. Kaufman, Analysis of neuromagnetic signals, in Handbook of Electroencephalography and Clinical Neurophysiology, Volume 1: Methods of Brain Electrical and Magnetical Signals (ed. A.S. Gevins and A.A. Remond) Elsevier, Amsterdam 1987, p. 405-448.

29. R. Hari and O.V. Lounasmaa, Recording and interpretation of cerebral magnetic fields, *Science* 244 (1989) 432-436.

30. H. Huttunen, G. Kobal, E. Kaukoronta and R. Hari, Cortical responses to painful CO_2 stimulation of nasal mucosa: a magnetencephalographic study in man, *Electroencephalogr. Clin. Neurophysiol.* 64 (1986) 347-349.

31. E.H. Chudler, W.K. Dong and Y. Kawakami, Tooth pulp evoked potentials in the monkey: cortical surface and intracortical distribution, *Pain* 22 (1985) 221-223.

32. K.H. Steen, P.W. Reeh, F. Anton and H.O. Handwerker, Protons selectively induce lasting excitation and sensitization of nociceptors in rat skin, *J. Neurosci.* 12 (1992) 86-95.

33. G. Kobal, T. Hummel and B. Kettenmann, Chemosensory event-related potentials in temporal lobe epilepsy and first recordings of olfactory event-related magnetic fields, *Chem. Senses* 18 (1993) 582.

34. T. Hummel and G. Kobal, Differences in human evoked potentials related to olfactory or trigeminal chemosensory activation, *Electroencephalogr. Clin. Neurophysiol.* 84 (1992) 84-89.

35. T. Hummel, A. Livermore, C. Hummel and G. Kobal, Chemosensory event-related potentials: relation to olfactory and painful sensations elicited by nicotine, *Electroencephalogr. Clin. Neurophysiol.* 84 (1992) 192-195.

36. G. Kobal, T. Hummel and S. Van Toller, Differences in chemosensory evoked potentials to olfactory and somatosensory chemical stimuli presented to left and right nostrils, *Chem. Senses* 17 (1992) 233-244.

37. A. Livermore, T. Hummel and G. Kobal, Chemosensory evoked potentials in the investigation of interactions between the olfactory and the somatosensory (trigeminal) systems, *Electroencephalogr. Clin. Neurophysiol.* 83 (1992) 201-210.

38. G. Kobal, Olfactory event-related potentials in the diagnosis of smell disorders, in Olfaction and Taste XI (ed. K. Kurihara, N. Suzuki and H. Ogawa) Springer, Berlin 1994, p. 626-630.

39. R.L. Doty, P. Reyes and T. Gregor, Presence of both odor identification and detection deficits in Alzheimer's disease, *Brain Res. Bull.* 18 (1987) 597-600.

40. R.L. Doty, D. Deems and S. Steller, Olfactory dysfunction in Parkinsons' disease: A general deficit unrelated to neurologic signs, disease stage, or disease duration, *Neurology* 38 (1988) 1237-1244.

41. C.H. Hawkes, B.C. Shephard and G. Kobal, Assessment of olfaction in multiple sclerosis: evidence of dysfunction by olfactory evoked response and identification tests, *Brain* (in press).

42. R.L. Doty, R. Smith, D. Mckeown and J. Raj, A comparison of tests of olfactory function: Factor analysis and reliability studies, *Percept. Psychophys.* (in press).

43. A. Sajjadian, R.L. Doty, D.N. Gutnick, R.J. Shirugi, M. Sivak and D. Perl, Olfactory dysfunction in amyotrophic lateral sclerosis (submitted).

44. T. Hummel, G. Kobal and T. Mokrusch, Chemosensory evoked potentials in patients with Parkinson's disease, in New Developments in Event-Related Potentials (ed. H.-J. Heinze, T. Münte and G.R. Mangun) Birkhäuser Verlag, Boston 1993, p. 275-281.

45. T. Hummel, E. Pauli, H. Stefan, B. Kettenmann, P. Schüler and G. Kobal, Chemosensory event-related potentials in patients with temporal lobe epilepsy, *Epilepsia* 36 (1995) 79-85.

46. J.D. Davidson, Affect, cognition, and hemispheric specialization, in Emotions, Cognition, and Behavior (ed. C.E. Izard, J. Kagan and R.B. Zajonc), Cambridge University Press, Cambridge 1984, p. 320-365.

47. S.J. Dimond, L. Farrington and P. Johnson, Differing emotional response from right and left hemispheres, *Nature* 261 (1976) 690-692.

48. G. Kobal, T. Hummel and E. Pauli, Correlates of hedonic estimates in the olfactory evoked potential, *Chem. Senses* 14 (1989) 718.

49. T. Hummel and G. Kobal, Chemosensory event-related potentials: effects of dichotomous stimulation with eugenol and dipyridyl, in Olfaction and Taste XI (ed. K. Kurihara, N. Suzuki and H. Ogawa), Springer, Berlin 1994, p. 659-663.

50. A.C.N. Chen, C.R. Chapman and S.W. Harkins, Brain evoked potentials are functional correlates of induced pain in man, *Pain* 6 (1979) 365-374.

51. S. Sutton, M. Braren and J. Zubin, Evoked-potential correlates of stimulus uncertainty, *Science* 150 (1965) 1187-1188.

52. T.S. Lorig, Cognitive and non-cognitive effects of odor exposure: electrophysiological and behavioral evidence, in Fragrance — the Psychology and Biology of Perfume (ed. S. Van Toller and G.H. Dodd), Elsevier Applied Science, London 1992, p. 161-173.

53. M. Durand-Lagarde and G. Kobal, P300: a new technique of recording a cognitive component in the olfactory evoked potentials *Chem. Senses* 16 (1991) 379.

CHAPTER 5

Human Responses to Ambient Olfactory Stimuli

Susan C. Knasko

INTRODUCTION

It has been estimated that Americans spend over 90% of their time indoors and that elderly people and children spend even more time in buildings.[1,2] This means that average Americans live over 90% of their lives in settings with artificial lighting, furnishings, and temperature control. Because so much time is spent in built environments, it makes sense to learn as much as possible about how physical aspects of these environments affect people, so that homes, worksites, and recreational settings can be designed to best suit human needs.

The effects of certain environmental factors, such as temperature, have already been studied to a great extent. For example, it has been determined what temperature and humidity ranges are comfortable for sedentary young men wearing light clothing.[3] It would be beneficial to know this kind of information for other types of environmental factors such as air quality, in particular, odors. Research is needed to address such questions as whether certain ambient odors can affect a person's performance, health, or sense of well-being in positive or negative directions.

Environmental scenting is currently an issue at both the national and international level. Some buildings in Japan have air-conditioning systems designed to deliver a variety of odors to sections of the building throughout the day.[4] This is done with the hope that the odors will improve work performance or instill a sense of well-being in the occupants. In the U.S. and Europe there are numerous odor products on the market that claim to be beneficial to one's emotional state or health, or to be able to influence behavior.

At the same time there are people who do not want any odorants added to their environment. Magazine scent strips are deplored by many individuals who have pressed for legislation in several states to control this form of advertising.[5] There are also groups pressing for legislation to establish fragrance-free zones.[6,7] These would be areas in buildings where people would not be allowed to wear perfume or cologne. Much of the concern is due to beliefs that exposure to odorants can have negative health effects, or play a role in the development of multiple-chemical sensitivity.

Although there is much interest in how ambient odors affect humans, there is relatively little scientific research on the topic. This chapter will give an overview of what is known about ambient odors and their effect on human well-being and behavior. The effects of odors on human mood and health will be addressed first, followed by a look at the effects of ambient odors on task performance.

MOOD AND HEALTH

Odor Hedonics

Most studies concerning the effects of exposure to ambient odors have focused on the hedonic value of the odors (i.e., whether they were pleasant or unpleasant).[8,9] This focus is not unexpected since hedonics is one of the qualities most frequently mentioned when individuals describe odors. Historically, pleasant odors were believed to be beneficial while malodors were viewed as harmful. The ancient Romans and Greeks had scented public baths that were thought to improve one's health,[10] while the early Egyptians placed scented cones on the heads of banquet guests to enhance their sense of well-being.[11] On the other hand, during the plagues of the Middle Ages whole towns were fumigated, not to get rid of disease-infested rodents, but to purge the cities of malodors that people thought were causing the plagues.[11] In a similar vein, homes of sick Maya Indians were scented with pleasant smelling plants to expel the malodors of evil believed to have caused the illness.[12]

This raises the question of how people today think exposure to pleasant or unpleasant odors affects them. Are their beliefs similar to those of ancient peoples? To address this issue a placebo study was conducted.[13] Water vapor was sprayed into a room but subjects were told that it was an odorant and that the majority of people who could smell the odor thought it was pleasant, neutral, or unpleasant. The subjects worked on several tasks then completed mood, health, and environmental quality questionnaires. The results of the study showed that subjects who had been told the odor was pleasant reported being in a better mood compared to those in the other two conditions. On the other hand, subjects who were told that the odor was unpleasant, reported having more health symptoms than subjects in the other conditions. Because no odor actually was present, these results most likely indicate the underlying beliefs people have about the role of hedonics in the effect of odors on mood and health.

The above experiment implies that the ancient beliefs concerning the effects of pleasant and unpleasant odors are still prevalent. The enduring nature of these beliefs suggests that they may be grounded in actual experience, i.e., there may be some relationship between odor hedonics and mood and health in daily life. To examine this premise a field study was conducted.[14] Forty-five men and women between the ages of 18 and 35 were given a questionnaire to complete six times a day for five consecutive weekdays. The subjects carried an electronic pager so that they could be prompted at random times to complete the questionnaire. The questionnaire asked what activities they were engaged in, who they were with, how they felt physically and emotionally, and questions about the physical environment. The results indicate that when people reported being the happiest, if they also reported an odor, that odor was most likely to be rated as pleasant, strong, controllable, and natural. When people reported having the most physical symptoms, if they also reported smelling an odor, that odor was most likely to be rated somewhat unpleasant, weak, artificial, and uncontrollable. Thus it seems that odor hedonics are to some extent related to mood and health in daily life. It also appears that controllability and naturalness are other odor characteristics that play a role in that relationship.

Odor and Mood Studies

Olfaction is often referred to as the emotional sense, implying that it elicits more emotional responses from people than the other senses.[15] The fact that the olfactory system is the sense most directly connected to the limbic system[16] is often cited as a reason that this might be the case. There is, however, little research to back this notion. In fact, one of the few studies conducted to test this claim found visual stimuli to be rated more emotional than other types of sensory stimuli.[9]

Although olfaction may not be the main sensory system to elicit emotions, and a substantial number of studies have found no mood effects associated with odor exposure,[17-20] in general, the literature indicates that pleasant odors improve mood while unpleasant odors have a negative effect.[9,21-23] It should be noted, however, that few studies concerning self-reported mood have employed both pleasant and unpleasant odors in the same experiment.[19,20,24] This makes it difficult to determine if, in a given study, odor hedonics played a role in the mood effects or if any odor would have lead to the same results. Although the emphasis of a particular experiment may warrant the use of odorants of limited hedonic value, the question of the role of odor hedonics in mood induction will only be answered with any certainty when more experiments include both pleasant and unpleasant odors.

Some studies that looked at a variety of pleasant odorants found reports of good moods during exposure to all of the odors but for certain ones there were stronger responses on specific mood scales.[25] Studies concerning the physiological effects of odor exposure also indicate that various pleasant odors can elicit quite different physiological responses.[26,27] This suggests that they have dissimilar effects on arousal level. In such cases factors other than odor hedonics are probably involved. Similar tests with various malodors may lead to different results since it has been suggested that good and bad mood might not fall along a continuum.[28] It has also been suggested that pleasant and unpleasant odors may be judged differently[29] or may differentially motivate behavior.[30] In a review of odor and mood[9] it was noted that malodors may be more important to mood induction than pleasant ones.

Odor and Health Studies

Unpleasant odors

The health consequences of exposure to high concentrations of air pollutants have been the focus of numerous studies. This research often involves human epidemiological studies or toxicological tests with animals. Typically, these investigations focus on serious illnesses such as cancer, respiratory disease, reproductive problems, or death.[31-35] The pollutants are often malodorous, but it is difficult to tease out the effects of odor from the toxicological effects of the chemicals. If odors do contribute to the health effects of high levels of air pollutants, their role is most likely a small one compared to the physiological damage of the chemicals on body tissue.

Other studies have focused on chemicals that are in the air at levels believed to be below toxic concentrations. Because many of the chemicals are odorous at these levels, negative exposure effects may be due, in part, to the olfactory nature of the chemicals. The health symptoms reported in these investigations usually involve relatively minor symptoms such as headache, eye, skin, and upper respiratory tract irritation, lethargy, dizziness, irritability, and concentration difficulties.[36] These are the same types of symptoms that are reported in sick-building syndrome (SBS).[37-41]

The term SBS is applied when a large percentage of people in a building, or in a specific part of the structure, are experiencing the above symptoms at the same time.[38,42,43] Because these symptoms are often similar to those known to be caused by air contaminants, building air-quality is usually considered suspect.[43,44] Pollutants in an office-type environment may arise from multiple sources.[45] In the majority of cases, however, SBS complaints have not been linked to any particular pollutant or group of pollutants,[46] and the chemicals are found at levels far below existing occupational health standards.[38,40,45,47,48] It is possible that because the majority of threshold limit values pertain to manufacturing environments, they are not appropriate for office workers, who may be concerned primarily with comfort[40] or for individuals, especially unhealthy ones, who may spend more than eight hours in a setting.[49]

It is noteworthy that a large percentage of SBS complaints have been attributed to ventilation systems. Ventilation standards were originally set to help control odors,[50] and most indoor air pollutants in office space are odorous.[43,51,52] In a large percentage of sick-building incidents, the victims complain of odors as well as health symptoms, and the two are believed to be related.[40,53-56] Symptomatic employees report unusual odors more frequently than well employees.[57] Odors, therefore, may help explain SBS in cases where no pollutants are found above the occupational health standards.

The relationship of odors to SBS, or to the health effects of low concentrations of chemicals, may have both physiological and psychological components.[56,58] On the physiological side, discomfort and functional changes, primarily of the vegetative system, may be induced by malodors.[59] Odorants are known to cause many of the symptoms associated with SBS (e.g., eye irritation, stuffy or runny noses, coughs),[60] mainly due to chemical irritation of the olfactory or trigeminal nerves.[61] It is possible that people who complain of SBS symptoms are more sensitive to odors or perceive them differently.

On the psychological side, the relationship between odors and SBS-type symptoms may center around beliefs in the effects of odors. It has been suggested that in the past few decades people have become more suspicious of the health effects of their environment due to media coverage of such events as chemical spills or release of toxic industrial fumes.[48,62] Because air pollution is often defined by sensory qualities, such as odor, people may assume that if odors are present, so are pollutants that can negatively affect their health.[53,63] These beliefs can lead to odors becoming the scapegoat for vague symptoms experienced by workers that have no known etiology.[38,56,64,65]

Exposure to odors can be very annoying[43,66] leading to stress responses that are included among SBS symptoms (e.g., headaches, fatigue, irritability).[38,60] Stress is also a factor that can connect beliefs about odors to health symptoms. When people are exposed to an odorous environment, belief in the negative health effects of odors can cause a stress response, thereby actually triggering symptoms or exacerbating those that people already have.[38,57,67] This may be especially true when people smell odors in environments that have been known to have dangerous levels of pollution in the past.[68] Stress is known to reduce resistance to biological pathogens,[69] therefore, people under any type of stress may be more susceptible to the negative effects of air pollution. Over a period of time, annoying, stressful odors may thereby reduce resistance to air pollutants, even at levels that do not normally affect healthy people.

Research shows that animals and possibly humans can adapt to air pollution, building up tolerance with sustained exposure;[31,36,70] psychological adaptation to pollutants has even been observed.[71] People are also known to be able to adapt to odors. When individuals are continuously exposed to an odorant, their sensitivity to the odor decreases sometimes to the extent of not being able to smell it. The extent of adaptation depends on the duration of exposure and concentration of the odor.[72,73] In one study, individuals who rarely or never adapted to odors during eight sessions held at weekly intervals were found to have the highest incidence of stress recorded in their daily diaries on test days. It was suggested that olfactory adaptation occurs less in persons under stress.[74] Complaints of SBS are often associated with a stressful work and/or home environment.[55,75] If stress and the ability to adapt to stimuli are related, people who report SBS may be adapting less to the olfactory stimuli in their environment.

Finally, it should be recognized that little is known about the health effects of trace concentrations of pollutants[40] or about the synergistic effect of low levels of multiple pollutants.[33,38,76] It has been found that the detection threshold of single odorants can be lower in a mixture than by themselves.[77] There is concern that there also may be synergism among odorants in mixtures that can produce adverse health effects even though for the concentrations of the individual odorants, such effects would not be expected.[78]

Pleasant Odors

A few scientific studies found that individuals reported fewer SBS-type health symptoms when they were exposed to certain pleasant odors.[24,79] In general, however, there is a scarcity of research investigating the effects of pleasant odors on human health. This is surprising since historically people have used odorants to improve health, and aromatherapy, an art that has regained popularity in recent years, makes many claims in this area. Aromatherapy is based on the premise that oils from natural products (essential oils) have healing powers when inhaled or rubbed into the skin. It usually involves the use of essential oils in a therapeutic massage. Different essential oils are thought to have various physiological, pharmacological, or emotional effects. Books on aromatherapy often list dozens of diseases or conditions that a specific oil is thought to help heal.[80] It should be pointed out that little scientific research has been done to test the claims of aromatherapy. At this point the assertions are based almost entirely on anecdotal evidence, tradition, or folk medicine. Because much of the medicine used in modern Western society was developed from natural substances, it is possible that some of the claims of aromatherapy may be supported by scientific research in the future.

Psychological Factors

Mood and health effects occurring during odor exposure can be due to a number of factors. Particular odorants may have specific pharmacological properties or physiological effects. For example, odorants inhaled or rubbed into the skin, as in aromatherapy, may be absorbed into the bloodstream through the lungs or subcutaneously. Once in the blood stream these chemicals can have an effect on a number of physiological systems. As noted in a previous section, psychological factors may also play a role in the mood and health effects of odor exposure. Several possible psychological determinants are considered below.

Associations

Some of the effects of odor on human mood and health may be due to associations people have with the particular odors. In one study scents of lemon and lavender were both rated as pleasant. Only when the room was scented with lemon scent, however, did people report significantly fewer health symptoms compared when they were exposed to no odor.[24] Similarly, in another study, chocolate and baby powder scents were both rated as very pleasant, but only in the baby powder condition did people report fewer health symptoms compared to a no odor condition.[79] It is possible these health effects were due to associations people had with the odors. Lemon is often marketed as a clean fresh scent so people may associate it with a clean environment that is beneficial to their health. The subjects in the second study were 18 to 35 years old and it is likely that they were raised with baby powder. Therefore, they may associate being physically cared for with the odor.

The scent of spiced apple was found to induce EEG responses that are normally associated with relaxation.[27] There was speculation that perhaps this odorant had some special properties that could directly affect the central nervous system. Later studies, however, revealed that other food odors gave similar EEG responses,[81] as did the imagining of food in the absence of actual food stimuli.[82] These later results suggest that the relaxation effect may be due more to cognitive influences, such as people associating the odor with food, rather than a direct effect on the central nervous system.[82]

Odor Conditioning

Another psychological factor that can determine odor's influence on human mood and health is odor conditioning. Sometimes, after a person has been exposed to a high dose of odorous chemicals, lower levels of the chemicals will trigger symptoms similar to those normally associated with the high dosage. This has been labeled "behavioral sensitization to an odorant."[83] It is thought by some to be a type of involuntary odor conditioning with the only remedies in some cases being: removal of individuals from the setting, reduction of the chemical concentrations to levels below odor detection,[68] or deconditioning.[84]

There is evidence that odor conditioning can occur even when people are not consciously aware of the odors. In one study, while subjects worked on a stressful task, half were exposed to no experimental odor while the other half were exposed to an odorant.[85] Several days later the subjects were seated in another room that was scented with the odor used in the first session. Women who had been exposed to the odor while working on the stressful task reported more anxiety during their second exposure compared to women who had not been exposed to the odor in the first session. This was true even though the subjects were not consciously aware of an odor on either day. The possibility of unconscious odor conditioning is given additional support by studies which suggest that people can respond physiologically to odors that they are not aware of consciously[81,86] or that odorants that are perceived as smelling very similar can cause large differences in EEG responses.[27]

Placebo Effects

As was mentioned previously, placebo effects may be influencing the consequences of odor exposure. People's beliefs that odors can influence their mood or health may lead them to perceive such consequences when they are exposed to an odorant; such beliefs may even help trigger actual effects. The potential for placebo effects is high in an area such as aromatherapy where various essential oils are promoted as having specific beneficial mood and health effects and the people using the odorants desire such outcomes.

Another area in which placebo effects have the potential to play a major role is in indoor air quality complaints. If a few employees say they smell an odor, even if no odor actually exists, other people may start to think that they smell an odor too. These employees, along with those who do not smell an odor but believe that others do, may react as they would to an actual air contaminant. This spreading of beliefs concerning the presence of an odorous pollutant, along with health reports associated with such beliefs, may be a factor in incidents of mass psychogenic illness.[38,42,54]

Recall Bias

It has been suggested that some of the reported effects of odors may be due to recall or reporting bias.[67] Certain circumstances may make people recall events in a slanted way. People who think they are exposed to dangerous substances may have a lowered threshold for noticing mild symptoms, or for reporting them, and be more likely to have "value-laden recollections."[68] It has been suggested that this is what is happening in many air pollution studies.[68,87]

Findings in a laboratory study support this theory.[20] In that study, individuals were exposed to either no odor, intermittent bursts of two pleasant odors, or intermittent bursts of two malodors, while they worked on four performance tasks and then completed a mood, health, and environmental quality questionnaire. Subjects were then taken into an unscented room and asked how they thought the odor of the testing room had affected them. The results indicate that odor did not have an effect on mood, health, or performance while people were actually exposed to the odors. Subjects in the malodor condition, however, retrospectively reported that the malodor had had a negative effect on all of the variables. This type of recall bias may occur in other odor studies and in air quality investigations where individuals are asked about the effects of an odor exposure after the fact.

Expectations

Some odor effects may be due to people expecting to smell a particular odor. In a study that measured contingent negative variation (CNV), subjects were asked to sniff six odorants.[88] They were led to believe that these were three different odors each presented at a high and a low concentration. Actually, there were three different odors and one mixture of the three. The mixture was used as the low concentration of all three odors. CNV measures for the odor mix differed depending on the specific odor subjects thought it was (i.e., the response for the "low-level odor" usually matched the response for the corresponding high-level odor). These results imply that the expectation of a particular odor can lead to brain wave responses similar to those obtained during exposure to the actual odor.

Findings that might be attributable to odor expectations also were found in another study.[24] Subjects were exposed to dimethyl sulfide (DMS) either on the first test day or the second test day, one week later. During the first session subjects who were exposed to no odor were in a good mood while those who were exposed to DMS were in a less positive mood. During the second session those who were exposed to DMS were in a less positive mood then they had been the first day, but those who were exposed to no odor in the second session were in the least positive mood of all. It may be that being in the same environment where they had previously experienced a foul odor, and not knowing if or when an odor would appear, may have caused a decrease in mood. This brings in the issue of control which could be important in buildings or communities where malodors are occasionally present but people do not know when to expect the smells and feel they have no control over the quality of their air.

EFFECT OF ODOR ON HUMAN BEHAVIOR

Besides mood and health, behavior also may be influenced by ambient odors. Investigations of odor's effect on many types of human behavior is lacking, while its effect on other behaviors (e.g., memory, social interactions and evaluations, identification of individuals) have been examined to some extent. This section will focus on the influence of ambient odor on task performance (i.e., the execution of tasks often encountered in offices, factories or schools, such as verbal and math problems, vigilance, computer keyboard typing, or creativity). While the idea of adding odors to the environment in order to improve performance has captivated the interest of people in many fields, research in the area is limited (see Table 1).

Table 1   Studies on Odor and Task Performance

Tasks	Odor hedonics	Performance effect	Ref.
No effect			
Block puzzle	Neutral	None	85
Psychology exam	Pleasant	None	89
Clerical coding	Pleasant	None	21
Creativity	Pleasant/unpleasant	None	24
Addition	Pleasant/unpleasant	None	20
Multiplication		None	
Odd-Word		None	
Proofreading		None	
Negative effect			
Addition	Unpleasant	None	23
Proofreading		↓ Errors detected	
Memory	Pleasant	None	18
Vocabulary		None	
Analogy		None	
Arithmetic		↓ Correct answers	
Letter detection	Pleasant	↑ Time	90
Positive effect			
Data entry	Pleasant	↓ Errors and speed	91
Vigilance	Pleasant	↑ Signals detected	17
Anagrams	Pleasant	↑ Words made	92
Word construction	Pleasant	↑ Words completed	93
Message decoding		↑ Letters decoded	

Task Performance Studies

No Odor Effect

Five studies that tested the influence of ambient odor on task performance showed odor to have no effect on performance.[20,21,24,85,89] This was the case even though the tasks varied in type and in their degree of difficulty. In two of these studies the odors used were unpleasant, indicating that people can be exposed to malodors (which air pollutants often are) and not necessarily have their performance negatively influenced.

Negative Odor Effect

A negative effect of odor on performance was detected in three studies.[18,23,90] In one of the studies[23] only the more complex of the two tasks employed (proof-reading) was negatively effected during odor exposure. This was the same task that was used in a previously noted study where no odor effect was detected.[20] Both studies employed malodors but in one the odor was presented continuously[23] while in the other it was presented intermittently.[20] Therefore, a factor that may play an important role in the effect of odor on task performance is the odor's delivery schedule.

Another study[18] found only one of four tasks (arithmetic problems) to be affected by odor exposure. Other studies have found no effect of odor on math problems[20,23] and the negative effect in this study arose only during a second exposure to one of the odorants. This indicates that the frequency with which people are exposed to an odor can play a role in its effect on task performance. Since most of the interest in environmental scenting involves scenting work environments over lengthy periods of time, the effect of odor on task performance over numerous exposures becomes an important question that warrants further investigation.

Also of note is the fact that two of the studies which found negative effects on performance used pleasant odors. This indicates that it is unwise to assume that scenting a room with pleasant odors will lead to either no effect or a positive one.

Positive Odor Effect

A positive effect of odor on task performance has been found in several papers.[17,91-93] In the first of these studies the findings were mixed. During a data entry task, the number of errors made decreased when subjects were exposed to pleasant odors, but so did the speed with which subjects completed the task. This suggests that the results may have been due to a speed and accuracy trade-off.

Of note is the fact that pleasant odors were used in all of the studies that showed a positive effect of odor on task performance. Also of note is the length of time people worked on the tasks that were positively effected by odor exposure. The time frames varied from a 5 min anagram task, to a 40 min vigilance task and a data entry task performed over several weeks. This implies that pleasant odors can positively influence the performance of tasks that vary greatly in the amount of time they involve.

Other Factors Involved in Performance

The above review of ambient odor's effect on task performance attests to the fact that not only has a limited number of studies investigated the role of odor on performance but that the results of these studies are mixed and sometimes contradictory. To a large extent these differences may be due to variables that have not been examined systematically. For example, there are many odor characteristics that might influence how odors affect task performance such as the odor's delivery schedule, intensity, familiarity or controllability. Similarly, various task characteristics (e.g., duration, amount of attention required, stressfulness, physical requirements) or subject characteristics (e.g., age, motivation, personality, odor sensitivity) may play a role in the effect of odor on task performance. Odors may have strong effects on task performance under the right mix of odor, subject, and task characteristics.

It should be noted that even when odors do not affect actual performance, they might influence perception of performance[20] which, in itself, can be important in many types of work. This was found to be the case in several music studies. Individuals exposed to music while they worked thought it improved their performance on various tasks but, in reality, it did not.[94-96] The less complex the task, the greater the perceived effect of music.[97] Perceived effects were also found in a pilot study involving gum chewing (Knasko, unpublished data). Subjects chewed mint flavored gum while they worked on five physical performance tasks. Performance was not affected by the gum chewing, but in the gum-chewing condition, the perception of task difficulty and stressfulness decreased for several of the tasks.

From the research that is currently available it seems premature to add odors to environments to improve task performance. If in the future more research determines that odors can be beneficial for certain types of performance tasks, another issue will need to be addressed — people's willingness to have their work environments scented. Already individuals frequently air their concern that they would consider it manipulative if employers scented work environments in order to improve performance. It does not seem to matter that lighting, wall color, acoustics, and spacial layout are all designed with the end goal of maximizing performance. People seem to fear that odors can make them do things against their will. Perhaps this is due to the ancient beliefs that odors are powerful. If some individuals feel that environmental scenting is manipulative it could become an environmental stressor for them. Stress research has found that if people feel they have control over a stressor, even if they do not exercise that control, they often show weaker autonomic responses or do not experience any negative effects.[96-100] Therefore, developing scenting systems that are under each individual's control (e.g., a small desk unit or a switch to prevent scented air from entering a particular office) may be important if environmental scenting is incorporated into a worksite. This would not only be effective for people who feel manipulated by environmental odors, but also for people who like the idea of environmental scenting, but may be working on a task that is negatively influenced by odor exposure.

REFERENCES

1. Committee on Indoor Pollutants. *Indoor Pollutants*. Report by the National Research Council. (Washington, DC: National Academy Press, 1981).
2. Samet, J.M., M. C. Marbury, and J. D. Spengler. Health effects and sources of indoor air pollution. Part II. *Am. Rev. Respir. Dis.,* 137:221-242 (1988).
3. Frisancho, A. R. *Human Adaptation. A Functional Interpretation.* (Ann Arbor: The University of Michigan Press, 1981).
4. Shimizu Corporation. *Shimizu Insight,* 2:6 (1988).
5. Givhan, R. D. A sniff penalty. Detroit Free Press, (March 18, 1990).
6. National Center for Environmental Health Strategies. Center urges comprehensive indoor pollution legislation to break current logjam on chemical sensitivity question. (Press release, April 10, 1991).
7. Bishop, K. Stamping Out Perfume: It's Blowing in the Wind. (New York Times, September, 12, 1991).
8. Van Toller, S. Odours, emotion and psychophysiology. *Int. J. Cosmet. Sci.,* 10:171-197 (1988).
9. Ehrlichman, H. and L. Bastone. Olfaction and emotion, in *Science of Olfaction.* Serby, M. J. and K. L. Chobor, Eds. (New York: Springer-Verlag, 1992) pp. 410-438.
10. Gibbons, B. The intimate sense of smell. National Geographic, 170:324-360 (1986).
11. Stoddart, D.M. *The Scented Ape. The Biology and Culture of Human Odour.* (Cambridge: Cambridge University Press, 1990).
12. Nash, J. The logic of behavior: Curing in a Maya Indian town. *Hum. Organ.,* 26:132-140 (1967).
13. Knasko, S. C., A. N. Gilbert, and J. Sabini. Emotional state, physical well-being, and performance in the presence of feigned ambient odor. *J. Appl. Soc. Psychol.,* 20:1345-1357 (1990).
14. Knasko, S. C. The relationship of odors to mood, health and behavior in daily life. Manuscript in preparation.
15. Lieff, B. and J. Alper. Aroma driven: On the trail of our most emotional sense. *Health,* 20:62-67 (1988).
16. Castellucci, V. F. The chemical senses: smell and taste, in *Principles of Neural Science,* 2nd ed. Kandel, E. R. and J. H. Schwartz, Eds. (New York: Elsevier Science Publishing Co., Inc., 1985) pp. 34-57.
17. Warm, J. S., W. N. Dember, and R. Parasuraman. Effects of olfactory stimulation on performance and stress in a visual sustained attention task. *J. Soc. Cosmet. Chem.,* 42:199-210 (1991).
18. Ludvigson, H. W. and T. R. Rottman. Effects of ambient odors of lavender and cloves on cognition, memory, affect and mood. *Chem. Senses,* 14:525-536 (1989).
19. Cann, A. and D. A. Ross. Olfactory stimuli as context cues in human memory. *Am. J. Psychol.,* 102:91-102 (1989).

20. Knasko, S. C. Performance, mood and health during exposure to intermittent odors. *Arch. Environ. Health,* 48:305-308 (1993).
21. Baron, R. A. Environmentally induced positive affect: its impact on self-efficacy, task performance, negotiation, and conflict. *J. Appl. Soc. Psychol.,* 20:368-384 (1990).
22. Rotton, J., T. Barry, J. Frey and E. Soler. Air pollution and interpersonal attraction. *J. Appl. Soc. Psychol.,* 8:57-71 (1978).
23. Rotton, J. Affective and cognitive consequences of malodorous pollution. *Basic Appl. Soc. Psychol.,* 4:171-191 (1983).
24. Knasko, S. C. Ambient odor's effect on creativity, mood and perceived health. *Chem. Senses,* 17:27-35 (1992).
25. Warren, C. and S. Warrenburg. Mood benefits of fragrance. *Perfum. Flavorist,* 18:9-10, 12-14, 16 (1993).
26. Torii, S., H. Fukuda, H. Kanemoto, R. Miyanchi, Y. Hamauzu, and M. Kawasaki. Contingent negative variation (CNV) and the psychological effects of odour, in *Perfumery: The Psychology and Biology of Fragrance,* Van Toller, S. and G. H. Dodd, Eds. (New York: Chapman and Hall, 1988) pp. 107-120.
27. Lorig, T. S. and G. E. Schwartz. Brain and odor: I. Alteration of human EEG by odor administration. *Psychobiology,* 16:281-284 (1988).
28. Bradburn, N. M. *The Structure of Psychological Well-being.* (Chicago: Aldine, 1969).
29. Yoshida, M. Studies in psychometric classification of odors. *Jpn. Psycholog. Res.,* 6:145-154 (1964).
30. Engen, T. *The Perception of Odors.* (New York: Academic Press, 1982).
31. Coffin, D. and H. Stokinger. Biological effects of air pollutants. In *Air Pollution* (3rd ed, Vol III). Stern, A. C., Ed. (New York: Academic Press, 1977).
32. Shy, C. M. Epidemiological studies of neurotoxic, reproductive, and carcinogenic effects of complex mixtures. *Environ. Health Perspect. Suppl.,* 101:183-186 (1993).
33. Samet, J. M. and F. E. Speizer. Assessment of health effects in epidemiologic studies of air pollution. *Environ. Health Perspect. Suppl.,* 101:149-154 (1993).
34. Sexton, K., H, Gong, Jr., J. C. Bailar, III, J. G. Ford, D. R. Gold, W. E. Lambert, and M. J. Utell. Air pollution health risks: Do class and race matter? *Toxicol. Ind. Health,* 9:843-878 (1993).
35. Lave, L., and E. Seskin. Air pollution and human health. *Science,* 169:723-733 (1970).
36. Evans, G. W. and S. V. Jacobs. Air pollution and human behavior. *J. Soc. Issues,* 57:95-125 (1981).
37. Sterling, E., T. Sterling, and D. McIntyre. New health hazards in sealed buildings. *AIA,* April:64-67 (1983).
38. Letz, G. A. Sick building syndrome: Acute illness among office workers — The role of building ventilation, airborne contaminants and work stress. *Allergy Procedures,* 11:109-116 (1990).
39. Finnegan, M. J., C. A. C. Pickering, and P. S. Burge. The sick building syndrome: Prevalence studies. *Br. Med. J.,* 289:1573-1575 (1984)
40. National Institute for Occupational Safety and Health. *Guidance for Indoor Air Quality Investigations.* (Cincinnati, OH: Hazard Evaluations and Technical Assistance Branch. Division of Surveillance, Hazard Evaluations and Field Studies; Department of Health and Human Services, 1987).
41. Skov, P. The sick building syndrome. *Ann. N.Y. Acad. Sci.,* 641:17-20 (1992).

42. Colligan, M. J. and M. J. Smith. A methodological approach for evaluating outbreaks of mass psychogenic illness in industry. *J. Occup. Med.,* 20:401-402 (1978).

43. Lindvall, T. Assessing the relative risk of indoor exposures and hazards and future needs, in *Proceedings of the 4th International Conference on Indoor Air Quality and Climate, Vol. 4.* Seifert, B., H. Esdorn, M. Fischer, H. Rüden and J. Wegner, Eds. (West Berlin: Institute for Water, Soil and Air Hygiene, August 17-21, 1987) pp. 117-133.

44. Kjærgaard, S., L. Mølhave, and O. F. Pedersen. Human reactions to indoor air pollution: *N*-decane, in *Proceedings of the 4th International Conference on Indoor Air Quality and Climate, Vol. 1.* Seifert, B., H. Esdorn, M. Fischer, H. Rüden and J. Wegner, Eds. (West Berlin: Institute for Water, Soil and Air Hygiene, August 17-21, 1987) pp. 97-101.

45. Hughes, R. T. and D. M. O'Brien. Evaluation of building ventilation systems. *Am. Ind. Hyg. Assoc. J.,* 47:207-213 (1986).

46. Sterling, E. and T. Sterling. The impact of different ventilation levels and fluorescent lighting types on building illness: An experimental study. *Can. J. Public Health,* 74:385-392 (1983).

47. Fanger, P. O. A solution to the sick building mystery, in *Proceedings of the 4th International Conference on Indoor Air Quality and Climate, Vol. 4.* Seifert, B., H. Esdorn, M. Fischer, H. Rüden and J. Wegner, Eds. (West Berlin: Institute for Water, Soil and Air Hygiene, August 17-21, 1987) pp. 49-55.

48. Bardana, Jr., E. J. Office epidemics: Why are Americans suddenly allergic to the workplace? *Sciences,* Nov/Dec, 1986, pp. 39-44.

49. van der Wal, J. F. Guidelines for a healthy indoor environment and impact on the acceptance of the actual situation. *Toxicol. Environ. Chem.,* 40:111-119 (1993).

50. Hodgson, M. J. Environmental tobacco smoke and the sick building syndrome. *Occup. Med. State of the Art Rev.,* 4:735-740 (1989).

51. Berglund, B., U. Berglund, T. Lindvall, and H. Nicander-Bredberg. Olfactory and chemical characterization of indoor air. Towards a psychophysical model for air quality. *Environ. Int.,* 8:327-332 (1982).

52. Berglund, B. and T. Lindvall. Olfactory evaluation of indoor air quality, in *Indoor Climate: Effects of Human Comfort, Performance and Health in Residential, Commercial and Light-Industry Buildings.* Fanger, P. O. and O. Valbjørn, Eds. (Copenhagen: Danish Building Research Institute, 1979) pp. 141-156.

53. Beller, M. and J. P. Middaugh. Results of an indoor air pollution investigation. *Alaska Med.,* 31:148-155 (1989).

54. Faust, H. S. and L. B. Brilliant. Is the diagnosis of "mass hysteria" an excuse for incomplete investigation of low-level environmental contamination? *J. Occup. Med.,* 23: 22-26 (1981).

55. Cohen, B. G. F., M. J. Colligan, W. Wester, II, and M. J. Smith. An investigation of job satisfaction factors in an incident of mass psychogenic illness at the workplace. *Occup. Health Nursing,* 26:10-16 (1978).

56. Colligan, M. J. and L. R. Murphy. Mass psychogenic illness in organizations: An overview. *J. Occup. Psychol.,* 52:77-90 (1979).

57. Donnell, H. D. Jr., J. R. Bagby, R. G. Harmon, J. R. Crellin, H. C. Chaski, M. F. Bright, M. Van Tuinen, and R. W. Metzger. Report of an illness outbreak at the Harry S Truman state office building. *Am. J. Epidemiol.,* 129:550-558 (1989).

58. Norbäck, D., I. Michel, and J. Widström. Indoor air quality and personal factors related to the sick building syndrome. *Scand. J. Work Environ. Health,* 16:121-128 (1990).

59. Rothweiler, H. and C. Schlatter. Human exposure to volatile organic compounds in indoor air — a health risk? *Toxicol. Environ. Chem.,* 40:93-102 (1993).

60. Mølhave, L. Human reactions to controlled exposures to VOC's and the "total-VOC" concept, in *Chemical, Microbiological, Health and Comfort Aspects of Indoor Air Quality — State of the Art in SBS,* Knoppel H. and P. Wolkoff, Eds. (Brussels: ECSC, 1992) pp. 247-261.

61. Berglund, B. and T. Lindvall. Sensory reactions to "sick buildings." *Environ. Int.,* 12:147-159 (1986).

62. Shusterman, D. J., A. L. Casey, and R. R. Neutra. The health significance of odorous air emissions from stationary sources: Investigation of odor and symptom complaints near a hazardous waste site. Presented at the 82nd Annual Meeting and Exhibition of the Air and Waste Management Association (Anaheim, CA, June 25-30, 1989).

63. Evans, G. W., S. V. Jacobs, and N. B. Frager. Behavioral responses to air pollution, in *Advances in Environmental Psychology,* Vol. 4. Baum, A. and J. Singer, Eds. (Hillsdale, NJ: Erlbaum, 1982) pp. 237-269.

64. Gochman, I. R. Arousal, attribution and environmental stress, in *Stress and Anxiety.* Sarason, I. G. and C. D. Speilberger, Eds. (Washington, DC: Hemisphere, 1979) pp. 67-92.

65. Keating, J. P. Environmental stressors: Misplaced emphasis, in *Stress and Anxiety.* Sarason, I. G. and C. D. Speilberger, Eds. (Washington, DC: Hemisphere, 1979) pp. 55-66.

66. Berglund, B., U. Berglund, and T. Engen. Do "sick buildings" affect human performance? How should one assess them? (Department of Psychology, University of Stockholm, Report No. 609, 1983).

67. Shusterman, D., J. Lipscomb, R. Neutra, and K. Satin. Symptoms prevalence and odor-worry interaction near hazardous waste sites. *Environ. Health Perspect.,* 94:25-30 (1991).

68. Neutra, R., J. Lipscomb, K. Satin, and D. Shusterman. Hypotheses to explain the higher symptom rates observed around hazardous waste sites. *Environ. Health Perspect.,* 94:31-38 (1991).

69. Dubos, R. *Man Adapting.* (New Haven, Conn: Yale University Press, 1965).

70. Evans, G. W., S. Jacobs, and N. Frager. *Human Adaptation to Photochemical Smog.* (New York: American Psychological Association, 1979).

71. Wohlwill, J. F. Human response to levels of environmental stimulation. *Hum. Ecol.,* 2:127-147 (1974).

72. Pryor, G. T., G. Steinmetz, and H. Stone. Changes in absolute detection threshold and in subjective intensity of suprathreshold stimuli during olfactory adaptation and recovery. *Percept. Psychophys.,* 8:331-335 (1970).

73. Stone, H., G. T. Pryor, and G. A. Steinmetz. A comparison of olfactory adaptation among seven odorants and their relationship with several physicochemical properties. *Percept. Psychophys.,* 12:501-504 (1972).

74. Schneider, R. A. and J. P. Costiloe. Limitation of olfactory adaptation in subjects under stress. *Fed. Proc.,* 28: 829 (1969).

75. Smith, M. J., M. J. Colligan, and J.J. Hurrell. Three incidents of industrial mass psychogenic illness. A preliminary study. *J. Occup. Med.,* 20:399-400 (1978).

76. Lundholm, M., G. Lavrell, and L. Mathiasson. Self-leveling mortar as a possible cause of symptoms associated with "sick building syndrome." *Arch. Environ. Health,* 45:135-140 (1990).

77. Patterson, M. Q., J. C. Stevens, W. S. Cain, and J. E. Cometto-Muniz. Detection thresholds for an olfactory mixture and its three constituent compounds. *Chem. Senses,* 18:723-734 (1993).

78. Samet, J. M. and F. E. Speizer. Introduction and recommendations: Working group on indoor air and other complex mixtures. *Environ. Health Perspect. Suppl.,* 101:143-147 (1993).

79. Knasko, S. C. Pleasant odors and congruency: Effects on approach behavior. *Chem. Senses,* 20:479-487 (1995).

80. Tisserand, R. *The Art of Aromatherapy,* Eleventh Ed. (Frome, Somerset, Great Britain: Hillman Printers Ltd., 1990).

81. Lorig, T. S. Human EEG and odor response. *Prog. Neurobiol.,* 33:387-398 (1989).

82. Lorig, T. S. and G. E. Schwartz. EEG activity during relaxation and food imagery. *Imag., Cognit. Personal.,* 8:201-208 (1988-89).

83. Shusterman, D., J. Balmes, and J. Cone. Behavioral sensitization to irritants/odorants after acute overexposures. *J. Occup. Med.,* 30:565-567 (1988).

84. Bolla-Wilson, K., R. J. Wilson, and M. L. Bleecker. Conditioning of physical symptoms after neurotoxic exposure. *J. Occup. Med.,* 30:684-689 (1988).

85. Kirk-Smith, M. D., S. Van Toller, and G. H. Dodd. (1983) Unconscious odour conditioning in human subjects. *Biol. Psych.,* 17:221-231 (1983).

86. Lorig, T. S., G. E. Schwartz, K. B. Herman, and R. D. Lane. Brain and odor: II. EEG activity during nose and mouth breathing. *Psychobiology,* 16:285-287 (1988).

87. Shusterman, D. Critical review: The health significance of environmental odor pollution. *Arch. Environ. Health,* 47:76-87 (1992).

88. Lorig, T. S. and M. Roberts. Odor and cognitive alteration of the contingent negative variation. *Chem. Senses,* 15:537-545 (1990).

89. Rottman, T. R. The effects of ambient odor on the cognitive performance, mood, and activation of low and high impulsive individuals in a naturally arousing situation. *Diss. Abstr. Int.,* 50:364B (1989).

90. Lorig, T. S., E. Huffman, A. DeMartino, and J. DeMaro. The effects of low concentration odors on EEG activity and behavior. *J. Psychophysiol.,* 5:69-77 (1991).

91. Hashimoto, S., N. Yamayuchi, and M. Kawasaki. Experimental research on the aromatherapeutic effects of fragrances in living environments. *Proc. Jpn. Arch. Soc.,* Oct 1988) pp. 83-84.

92. Baron, R. A. and J. Thomley. A whiff of reality: Positive affect as a potential mediator of the effects of pleasant fragrances on task performance and helping. *Environ. Behav.,* 26:766-784 (1994).

93. Baron, R. A. and M. I. Bronfen. A whiff of reality: Empirical evidence concerning the effects of pleasant fragrances on work-related behavior. *J. Appl. Soc. Psychol.,* 24:1179-1203 (1994).

94. Newman, R. I., D. L. Hunt, and F. Rhodes. Effect of music on employee attitude and productivity in a skateboard factory. *J. Appl. Psychol.,* 50:493-496 (1966).

95. Lucaccini, L. F. and L. H. Kreit. Music, in *Ergogenic Aids and Muscular Performance,* Morgan, W. P., Ed. (New York: Academic Press, 1972) pp. 235-262.

96. Deverux, G. A. Commercial background music: Its effect on workers' attitudes and output. *Personnel Practice Bull.,* 25:24-30 1969.

97. Jacoby, J. Work music and morale: A neglected but important relationship. *Personnel J.,* 47:882-886 (1968).

98. Glass, D. C. and J. E. Singer. *Urban Stress.* (New York: Academic Press, 1972).

99. Geer, J. H., G. C. Davison, and R. I. Gatchel. Reduction of stress in humans through nonverbal perceived control of aversive stimulation. *J. Personal. Soc. Psychol.,* 16:731-738 (1970).

100. Corah, N. L. and J. Boffa. Perceived control, self observation, and response to aversive stimulation. *J. Personal. Soc. Psychol.,* 16:1-4 (1970).

101. Puente, A. E. and I. Berman. The effects of behavior therapy, self-relaxation and TM on cardiovascular stress response. *J. Clin. Psychol.,* 36:291-295 (1980).

102. Baum, A., J. E. Singer, and C. S. Baum. Stress and the environment. *J. Soc. Issues,* 37:4-35 (1981).

Part II
Allergy and Respiratory

Overview: Indoor Air and Human Health Revisited

Cecile Rose

INTRODUCTION

Outbreaks of building-associated illness occur in many settings including offices, factories, healthcare facilities, and residences. Complaints are often non-specific and results of investigations inconclusive. There are, however, circumstances in which specific symptoms point to diagnoses which have profound implications for the affected individual and for those sharing the affected environment. Probably the most important category of symptoms are those that reflect respiratory tract responses to exposures. The respiratory tract is both an important route of entry and a major end organ for the effects of indoor inhalant exposures. Identification of building-related respiratory symptoms should prompt careful investigation for relevant causal exposures and their control.

This portion of the book is designed to organize respiratory effects from indoor exposures into broad categories and to examine some current state-of-the-art issues for each. The broad categories of exposure are irritants, infections, and sensitizers.

Understanding the effects of irritant exposures in buildings is complicated by the usual finding of complex mixtures and by host variability in irritant response. Bascom addresses these challenges by selecting subjects who self-identified as symptomatic to a particular irritant, environmental tobacco smoke (ETS), which is often found in indoor environments. Using posterior nasal rhinomanometry, physiologic measurements of nasal resistance demonstrated biological variability following controlled exposure to the vapor phase of ETS, with the ETS-sensitive subjects developing vascular congestion. Further mechanistic studies will explore the role of substance P and mucociliary clearance in mediating these effects.

Devlin and his colleagues in Chapter 9 examine the effect of the common pollutant ozone on allergic asthmatics, a population at increased risk from low level irritant exposures. Following exposure to low level ozone for two hours, they used the creative "split nose" design to challenge allergic asthmatics with allergen on one side of the nose while the other side was challenged with saline. Subsequent nasal lavage demonstrated that ozone alone induces an inflammatory effect and augments allergen-induced cellular influx into the nasal mucosa. These experimental studies are essential to our understanding of upper airway allergic and irritant effects from low level indoor air exposures.

Both the more recently recognized infectious agents such as the bacterium causing Legionnaires' disease and the infections dating back to antiquity such as tuberculosis are relevant to the discussion of building-related respiratory disease. In Chapter 7 Kubica summarizes the current state-of-the-art for tuberculosis risk from indoor exposures to infected individuals, particularly nosocomial transmission, as well as the three methods currently available to control airborne mycobacterial droplet nuclei. The compelling need for further research efforts into the control of this increasing and potentially deadly disease are well documented. Streifel, in Chapter 6, describes the prevention and control of opportunistic infection in hospitals from exposure to the environmental microbial agents *Aspergillus* and *Legionella*. He explores the roles of building design, construction methods, engineering controls including ventilation rates, and procedures to minimize release of these agents during maintenance and housekeeping.

Indoor exposure to microbial sensitizers is the focus of Rose's discussion of building-related hypersensitivity diseases. These common illnesses can cause devastating effects on the upper and lower respiratory tracts if unrecognized, and probably contribute significantly to the lost productivity and absenteeism which can affect symptomatic occupants of contaminated buildings. In particular, failure to identify cases of mild or early building-related hypersensitivity pneumonitis, which often has a high attack rate among those exposed, leads to a poor prognostic outcome and significant respiratory impairment. One of the most fascinating and least well-understood categories of bioaerosols which play a role in building-related respiratory diseases (having both airway and interstitial effects) are the bacterial endotoxins, the focus of Milton in Chapter 11. As with most of the data on indoor air and respiratory outcomes, the small number of studies and methodological issues in endotoxin measurement point to the need for further research in this area.

CHAPTER 6

Controlling Aspergillosis and Legionella in Hospitals

Andrew J. Streifel

ABSTRACT

Spores from common airborne fungi are essential for the natural decomposition of organic material. Certain immunocompromised patients are extremely susceptible to opportunistic infection from environmental microbes such as airborne *Aspergillus* or waterborne *Legionella* species.

ASPERGILLOSIS

Patients requiring immune system manipulation for treatment of malignancy or solid organ transplantation are particularly susceptible to saprophytic airborne fungi which are capable of growth at body temperature. The majority of environmental fungi are mesophilic making temperature a selection factor for human pathogenic fungi. *Aspergillus fumigatus* and *A. flavus* are commonly identified as pathogens in the patients requiring long term hospitalization during bone marrow transplants (BMT) or long term chemotherapy for solid organ transplants. Other thermotolerant fungi may also be infectious given the opportunity to colonize susceptible hosts.

INTRODUCTION

Advances in medical technology have provided successful treatment protocols for the management of malignant diseases while using severe immunocompromising treatments such as toxic radiation treatment and/or chemotherapy. Patients

1-56670-144-9/96/$0.00+$.50

with solid organ failure may require long term immune suppressive therapy after transplantation for survival. Cawley[1] described the unique disease aspergillosis in 1947 and since then this fungal genus has become a nemesis for many large transplantation centers and immunocompromised patients worldwide.[2-10] Patients with susceptibility to the opportunistic *Aspergillus* species must be identified and methods provided for the protection of these specific hospitalized patients. As an example of institution problems, Petersen[6] retrospectively showed that 20% of BMT patients in minimally ventilated patient rooms developed aspergillosis (see Table 1). In this retrospective evaluation, 17 of 19 episodes of these fungal infections were fatal.

Table 1   ***Aspergillus fumigatus* Air Concentration and Infection Rate**

Air concentration (cfu/m³)	(%) Infection rate	Ref.
≤0.2	0.3	17
1.1–2.2	1.2	17
0.9	5.4	11
2.0	20	6
0.009	0	28

Special ventilation measures using high efficiency particle air (HEPA) filters (99.97% efficient at 0.3-μm particles) were instituted at the University of Minnesota Hospital and Clinic (UMHC) to minimize exposure to this cosmopolitan microbe. Initially, at the University of Minnesota a sterile laminar flow clean room environment was thought to be necessary for control of common environmental microorganisms. The infection problems associated with endogenous microbes required application of broad-spectrum regimens of antibiotics[6] which provided survival to increasing numbers of patients undergoing radical allogenic BMT. Transition into new space and little experience with aspergillosis in this patient group brought about an epidemic outbreak in 1979 to 1981. The evaluation of ambient air[11] in that BMT unit did not seem to have excessive levels of *A. fumigatus* at 2.0 cfu/m³. The implementation of in-room HEPA filters provided a reduction to about 0.9 cfu/m³ and a subsequent reduction of *Aspergillus* infections. The ongoing construction around this BMT unit during a renewal project was most likely responsible for the endemic level of 5.4% after the epidemic period between 1979 and 1981 when the level was close to 20%.[6] The higher infection rate from a seeming low average *A. fumigatus* air content may instead be the result of the extreme airborne spore levels caused by environmental disturbances and a lack of ventilation control.

Construction of a specially designed hospital, UMHC, provided for design considerations for the control of airborne fungi for the entire building.[12] Table 2 lists the features specific to the whole building. One can determine that even with special ventilation in place, the levels of infection vary to exposure potential. The patients do leave their rooms and in the hospital corridors levels for *A. fumigatus* are 0.4 cfu/m^3 while the special ventilated BMT in room levels for *A. fumigatus* are 0.08 cfu/m^3.[13] This experience has allowed for a more complete understanding of the nature of the buoyant spore.

Table 2   Features Specific to the Whole Building

- Incorporation of a fungal inhibitor in the structural steel insulation.
- A smoke evacuation system which permits a total seal of windows in building.
- Special increased sealing of the windows hospital wide.
- Minimal (<1%) carpet floor cover in the building.
- Filtration for complete hospital space at 90-95% efficiency dust spot ASHRAE 52-76 except air to stairwells during fire alarm conditions.
- Transfer fan recirculation with filtration at entrances to the building and connections to adjacent buildings.
- Greater supply air volume to exhaust ratio for the building to allow over pressurization.
- Specially designed BMT unit with HEPA (99.97% efficiency at 0.3 μm particles) filtration in rooms >15 ac/h, >supply vs. exhaust room air volumes, directed airflow, controlled ventilation, and sealed rooms.

Thermophilic *A. fumigatus* is important for the decomposition of organic material. This trait allows for growth at relatively high temperatures to about 115°F incubation for production of aerodynamic spores.[14] *A. fumigatus* spores have echinulations on a 2.0 to 3.5 μm diameter spore. These protuberances may enhance buoyancy or adhesion to vectors assisting in the dissemination of this soil microorganism. The cosmopolitan nature of the genus *Aspergillus,* especially species *A. fumigatus*, *A. flavus,* or *A. terreus,* are of particular concern to a critical hospital environment.[15] These microbes have been found in hospital environments often apart from airflow accumulations of airborne spores or from a loci of growth caused by prolonged uncontrolled water. These opportunistic saprophytic spores require active liberation in order to find a nutrition source for additional spore propagation. In the hospital environment these thermotolerant spores seem to be especially adapted to being present in a sensitive environment.

SOURCES

The decomposition of organic material by saprophytic fungi is essential for recycling organic material into reusable material. Such beneficial but problematic fungal spores then become common in most soils rich in organic material. The indoor air levels have been recorded to levels in excess of 10^6 spores/m^3.[14] Such elevated levels of airborne fungi can often be associated with transient levels due to some disturbance of soil, demolition, or simple movement of wood.[16] Airflow accumulations from entrained spores can attach to lint or other airborne particles and eventually gather significant numbers of spores which when disturbed disseminate into the local environment. The recognition and control of such sources becomes a factor essential if susceptible patients are nearby in the hospital environment.[17,18]

Air becomes the mode of transportation for the aerodynamic spore. Because
A. fumigatus has surface echinulations aiding spore buoyancy a settling rate of
0.03 cm/s[19] increases the inhalation potential. Other larger spore thermotolerant
fungi settle rapidly which may be a factor for diminishing infection potential.
The accumulated spores may be redistributed by vacuum cleaners,[11] moving
ceiling tile, stopping and starting heating ventilating air-conditioning systems,[22]
or brushing clothing worn outside.[20] These ubiquitous spores are a challenge to
control through measures which are not cost or patient care prohibitive. Clean-
room sterile environments are expensive to maintain and operate. It is difficult
to provide consistent protection for increasing numbers of patients reqiring such
an enviornment.[21] A whole hospital approach for fungal spore control has been
attempted at the University of Minnesota. Air samplings at that institution provide
data that *A. fumigatus* levels are low compared to outdoor and minimally filtered
areas.[13] It should be recognized that the transient nature of disturbed spores means
a high concentration initially dissipating into the local environment[22] presenting
a greater risk for acquisition during the susceptible patient condition. The low
levels of inpatient rooms and low endemic levels of *A. fumigatus* may only
indicate the need to control behavior which may generate spore bursts in critical
patient care areas by behavior modification through procedural controls.

SPECIAL VENTILATION ENVIRONMENTS

Laminar flow clean rooms with high efficiency filtration at 99.97% removal
of 0.3-μm particles and procedural practice to maintain sterile conditions were
part of the pioneering efforts to provide a microbial safe BMT environment.[23]
The uncertainty of acquisition of infection brought on a concept for clean-room
aseptic patient care which proved to be labor intensive. The early success of
marrow transplantation procedures along with increasing numbers of diseases[6]
amenable for treatment with BMT meant that not enough BMT clean rooms were
available to treat all patients with leukemia. The stringent aseptic requirements
for patient care were a labor intensive effort which required extensive resources
for sterilizing patient care items.[24] It was soon discovered that microorganisms
initially problematic to the patient were endogenous and controllable with anti-
microbial therapy. As BMT became a more common treatment and clean-room
type beds were not available, regular patient beds were used with environmental
microbes beginning to cause nosocomial infection. Long term extreme immune
suppression up to six weeks is needed for BMT. The time and nature of the
immune suppression is different for solid organ transplants; regardless, both
patient groups should be protected while in the hospital.

The modern hospital environment is very dynamic. Ongoing utility upgrade
means the hospital environment is undergoing continuous remodel and new
construction. In the 90 articles published on aspergillosis, construction and natural
ventilation are often cited as the source of opportunistic fungi.[10] Noteworthy, the

number of patients susceptible to opportunistic infection is also increasing due to the availability of immune suppressing technologies. The need to protect such acute care patient is necessary to assure survival of patients requiring certain modern lifesaving medical resources.

The University of Minnesota designed and built a hospital from 1980 to 1986 which included design features which should protect all of the building's patient care[12,25] environments from opportunistic environmental microbes (see Table 2). Air sampling averages indicated significant differences in the averages of 25°C incubated total airborne fungal levels when comparing the outside (788 cfu/m^3) and minimally (40%) filtered areas (96 cfu/m^3) with the 95% (17.7 cfu/m^3) to 99.97% (14.1 cfu/m^3) efficient filtration in the hospital.[13] The difference in total fungal averages for 37°C incubation isolates from the hospital ward corridors (3.0 cfu/m^3) seems insignificant when compared to the HEPA-filtered corridors of the BMT unit (2.0 cfu/m^3). The rooms have even lower levels of airborne fungi averaging 0.8 cfu/m^3 at 37°C incubation and a level of 0.08 cfu for *A. fumigatus*/m^3.[13] The endemic level of infection caused by *Aspergillus* organisms is still about 3.2% in 883 patients. While the filtration and ventilation system provides a disinfected air supply with 99+% removal of airborne fungi,[26,27] the in-room behavior and patient restriction to the room becomes a factor. The 3 seperate times when >2 cfu/plate of *A. fumigatus* were recovered from a patient room it was the same patient with unrestricted visitors. A visitation policy was established from the observation. The patients should be masked with a highly filtered properly fitted mask when out of their rooms. Unless the patient can be continually confined in a clean-room environment[28] allowing for complete protection of the susceptible host acquisition still occurs. The control provided in the true laminar flow clean room includes a generally >100 air changes per hour which will translate into a perceivable draft. Patient comfort should be a priority during convalescence; patient response to such an extreme environment may have detrimental effects due to the confinement. The control of ambient fungi need not be that complex and burdensome to the patient care staff and physical plant requirements.

Table 3 lists the elements of a protected fungal spore control environment. The guideline for prevention of nosocomial pneumonia in 1994[29] includes these and additional factors needed for such control, measures especially focusing on disruptive activity control of construction projects and educating hospital personnel.

Table 3   Elements of Fungal Spore Control Ventilation

- A sealed room with self-closing door and airtight windows will assure ventilation control.
- Use directed airflow while providing increased room air changes to greater than 15/h.
- More supply air vs. exhaust/return air volume to create positive pressure.
- Point of use HEPA (99.97% efficiency @ 0.3-μm particles) to prevent maintenance-derived release of opportunistic spores.
- Procedural practice measures to avoid spore release during cleaning, maintenance, or visitation.

PROCEDURAL PRACTICE

Construction, renovation, and maintenance procedures create local disturbances from disturbed opportunistic spores which potentially migrate throughout
structures.[14] The need to control construction activities during fungal spore loci
disruption is essential for contamination control.[22] The planning and implementation of outdoor vs. indoor projects differs and specifications should be made
for each type of construction. Regardless, responsibility should be delegated to
assure that the projects are closely supervised. Table 4 includes factors which
should be considered for control in hospital construction projects. By far the most
important factor is the contractual language which obliges the vendor with an
economic incentive attached. External projects require emphasis sealing the skin
of the building to prevent unwanted infiltration while internal projects require
containment using clean to dirty airflow. Some institutions have imposed fines
as a means for assuring compliance.

Table 4   Construction Control Measures

Outdoor Construction:

- Seal windows and obvious openings to prevent infiltration from nearby construction.
- Provide filter maintenance and assure appropriate installation before construction.
- Assure ventilation and infiltration control when the buildings or additions are joined.
- Anticipate effects of construction on the protected structure such as leaking roof.

Indoor Renovation:

- Provide contractual language for bid submittal to include the following:
- Maintain clean to dirty airflow using asbestos-like containment methods in specific areas.
- Provide emergency procedures to handle air supply outage or water spillage.
- Specify the barrier types with circumstances for use and monitoring daily.
- Control construction traffic to avoid construction worker/debris contact with patients.
- Assign responsibility to communicate and maintain the conditions of the contract.
- Verify through commissioning the design specification and air balance for ventilation systems.

Recognition of the critical care areas and the various utility interconnections
is essential. For example, a general exhaust ventilation system may be shut off
for maintenance or renovation and the loss of airflow may compromise the special
ventilation area several floors distant. The institution providing transplantation
services requires trained maintenance and planning/construction personnel with
foresight into the effects of maintenance and renovation activity. Such awareness
on a proactive basis will help to prevent exposure during construction-related
ventilation interruptions. Simple procedures such as moving a dirty ceiling tile
may disrupt significant numbers of opportunistic fungi and appropriate procedural
practice must be considered a part of the training for maintenance and contracted
personnel. During maintenance common sense measures, such as minimizing the
presence of personnel and patients and a closed door policy for the sensitive

positive pressure rooms, should help prevent the spores from reaching patients. Ceiling tile should be carefully cleaned before moving; however, they should be periodically cleaned as part of routine maintenance.

Recognition of water damaged "mildew odor" sources by personnel and subsequent careful cleanup requires basic decontamination and planned control measures. Water in the form of leaks or uncontrolled humidity with condensation will create conditions of fungal sporulation on most organic substrates when relative humidity exceeds 70%. The control of this available water becomes an essential factor for minimizing spore forming fungal growth. Handwashing facilities essential for infection control can become growth areas if maintenance of sink areas is overlooked or delayed. Unfortunately, installation of sinks with seamed covered laminated particle board is problematic, especially in the areas requiring ongoing handwashing in critical care environments.[30] Such handwashing sinks can become reservoirs of growth from such leakage and subsequently pollute the patient care environment if the growth is disturbed. In one situation it was observed that $>6.0 \times 10^5$ spores were released in an hour through passive liberation from a moldy handwashing sink.[31] Also, fungal sources can be anticipated when splashing occurs in airhandling systems designed with internal insulation creating conditions for sporulation. Uncontrolled disturbances of these sources cause the subsequent "seeding" of the environment with the spores.[17,32,33] In addition, neglected cleaning and improper drainage of water collection pans create conditions for growth. Humidification methods are also a problem if filters are after humidifiers. Some evaporative cooling devices can cause microbial growth producing spores. Odors from volatile organic chemicals released during microbial growth also cause occupant complaints.[34]

It is difficult to minimize spore release during HVAC system maintenance because as fans are stopped and restarted, the cake dust in the filters becomes disrupted with the release of captured spores.[22] The variety of situations which create conditions with sufficient water activity are numerous in a modern hospital building. Uncontrolled water leakage or humidity is more common than most medical personnel realize. The leaks caused by holes in a roof, washers, ice machines, water treatment devices, or beverage service can all contribute to sources of indoor air microbial pollution.

Prevention of *Aspergillus* infections in immunocompromised patients requires insightful planning of engineering control with patient and environment surveillance, along with careful maintenance and personnel training for medical facilities housing severely immunocompromised patients. Preventing epidemic outbreaks with detail to planning includes facility design, procedural practice in patient care areas, and the definition of the nosocomial disease. Definition for disease aquisition location (nosocomial vs. community) will assist the epidemiologist and medical staff in determining if the mycosis was acquired in the community or in the hospital. Environmental reservoirs of opportunistic microbes in immunocompromised patient environments should be avoided in hospital buildings. In addition to *Aspergillus,* other microbes must also be controlled, such as *Legionella* which is also seen in the immunocompromised patients.[35]

LEGIONELLA

Legionella is an opportunistic environmental bacteria found commonly in natural water environments. Sometimes conditions in hospital water systems are conducive to the presence of this opportunistic bacteria and when discovered in hospital patient specimens, it can be an indication of a potential nosocomial problem. The appropriate reaction required after discovery may require disinfection of the hospital water supply distribution system using chemical or physical methods.[36] Such efforts are not completely efficacious because most institutions were not built with disinfectant application methods which provide ease of water supply disinfection. In addition, hospital space requirements continually change which in turn require water utility changes. This often causes discontinuing water service by capping off the water service line which then causes a dead end connection. Such stagnation often can produce the ideal conditions needed for *Legionella* proliferation in large multi-aged buildings, often part of a medical complex. Methods for disinfecting include the usage of chlorine, ozone, heat, or metal ions.[37,38]

Legionella might be associated with the design of drinking water distribution[39] system or the hospital water supply having residual vs. free available chlorination. If the hospital water distribution has dead end connections due to discontinued water utility, especially on the hot water line, and the municipal water supply has gaseous chlorination, the heating will dissipate the gaseous chlorination. Without residual chlorination and a sufficient water temperature, opportunistic water bacteria such as *Legionella* can proliferate. The surveillance of water or high risk patients for the isolation of these microbes should be provided[29] in order to provide a clinical protocol for recognition of potential epidemic outbreak of this nosocomial infection. Table 5 lists methods for recognizing and controlling *Legionella* in hospitals. Cooling towers may also serve as reservoirs for these opportunistic bacteria. Several outbreaks of Legionellosis have recently been associated with hospital cooling towers especially in debilitated patients visiting outpatient clinics and in community residents.[40] Maintenance and disinfection requirements of hospital water systems are essential for minimizing the presence of *Legionella* species from their natural environment.

CONCLUSION

The future control of opportunistic environmental microbes must at this stage rely on prevention of acquisition through appropriate surveillance methods. This includes the elaborate construction measures incorporated into hospital building contracts. Recognizing this sensitivity of the patient requires vigilance. Factors such as construction or maintenance in the workplace or in any building may produce conditions after disrupting the loci of spores which may contaminate the airway of sensitive persons. The proper design of the ventilation and water distribution system along with maintenance will assist in the control of opportunistic environmental microbes.

Table 5   Prevention and Control of *Legionella* in Hospitals

Option 1

- Culture hospital water for *Legionella*
  - If found, culture high risk patients
  - Provide retrospective epidemiology
  - Provide water system decontamination

Option 2

- Follow high risk in hospital patients
  - If *Legionella* isolated as nosocomial pneumonia
  - Initiate search for water source
  - Maintain cooling towers and hot water systems
  - Use sterile water for nebulization

Maintenance of potable water should include:

- Recirculation systems with >50°C water temperature
- Heated water should maintain 1-2 mg/l free residual chlorine
- Eliminate dead end connections especially in the hot water system
- Maintain hot water heaters and storage tanks by flushing residual minerals
- Convert to on demand hot water heaters with no storage

REFERENCES

1. Cawley, E., Aspergillosis and the Aspergilli. *Arch. Intern. Med.,* 1947. 80(4): p. 423-434.
2. Bodey, G.P., The Emergence of Fungi as Major Hospital Pathogens. *J. Hosp. Infect.,* 1988. 11 (Suppl. A): p. 411-426.
3. Bodey, G. and S. Vartivarian, Aspergillosis. *Eur. J. Clin. Microbiol. Infect. Dis.,* 1989. 8(5): p. 413-437.
4. Kyriakides, G., Zinneman, H., Hall, W., Arora, V., Lifton, J., DeWolf, W., and Miller, J., Immunologic Monitoring and Aspergillosis in Renal Transplant Patients. *Am. J. Surg.,* 1976. 131: p. 246-252.
5. Levenson, C., Wohlford, P., Djou, J., Evans, S., and Zawacki, B., Preventing Postoperative Burn Wound Aspergillosis. *J. Burn Care Rehab.,* 1991. 12(2): p. 132-135.
6. Petersen, P.K., McGlave, P., Ramsay, N.K., and Rhame, F.S., et al., A Prospective Study of Infectious Diseases Following Bone Marrow Transplantation: Emergence of Aspergillus and Cytomegalovirus as the Major Causes of Mortality. *Infect. Control,* 1983. 4(2): p. 81-89.
7. Rinaldi, M.G., Invasive Aspergillosis. *Rev. Infect. Dis.,* 1983. 5(6): p. 1061-1077.
8. Rotstein, C., Cummings, K.M., Tidings, J., Killan, K., Powell, E., Gustafson, T.L., and Higby, D., An Outbreak of Invasive Aspergillosis Among Allogeneic Bone Marrow Transplants: A Case-Control Study. *Infect. Control,* 1985. 6(9): p. 347-355.
9. Rubin, R., The Compromised Host as Sentinel Chicken. *New Engl. J. Med.,* 1987. 317(18): p. 1151-1153.
10. Walsh, T. and D. Dixon, Nosocomial Aspergillosis: Environmental Microbiology, Hospital Epidemiology, Diagnosis and Treatment. *Eur. J. Epidemiol.,* 1989. 5(2): p. 131-142.

11. Rhame, F., Streifel, A., Kersey, J., and McGlave, P., Extrinsic Risk Factors for Pneumonia in the Patient at High Risk of Infection. *Am. J. Med.,* 1984. p. 42-51.

12. Streifel, A.J., Vesley, D., Rhame, F.S., and Murray, B., Control of Airborne Fungal Spores in a University Hospital. *Environ. Int.,* 1989. 15: p. 221-227.

13. Streifel, A.J. and F.S. Rhame. *Hospital Air Filamentous Fungal Sore and Particle Counts in a Specially Designed Hospital,* in 6th International Conference on Indoor Air Quality and Climate. 1993. Helsinki, Finland.

14. Marsh, P., P. Millner, and J. Kla, A Guide to the Recent Literature on Aspergillosis as Caused by Aspergillus fumigatus, a Fungus Frequently Found in Self-Heating Organic Matter. *Mycopatholgia* 69, 1979. 1-2: p. 67-81.

15. Sarubbi, F.A., Kopf, H.B., Wilson, M.B., McGinnis, M.R., and Rutala, W.A., Increased Recovery of Aspergillus Flavus from Respiratory Specimens during Hospital Construction. *Am. Rev. Respir. Dis.,* 1982. 125: p. 33-38.

16. Reoponen, T., Lehtonen, M., Raunemaa, T., and Nevalainen, A., Effect of Indoor Sources on Fungal Spore Concentrations and Size Distribution. *J. Aerosol. Sci.,* 1992. 23: p. 663-666.

17. Arnow, P.M., Sadigh, M., Costas, C., Weil, D., and Chudy, R., Endemic and Epidemic Aspergillosis Associated with In-Hospital Replication of Aspergillus Organisms. *J. Infect. Dis.,* 1991. 164(November): p. 998-1002.

18. Streifel, A., Aspergillosis and Construction. Architectural Design and Indoor Microbial Pollution, ed. R. Kundsin. 1988, New York: Oxford University Press.

19. Gregory, P.H., The Microbiology of the Atmosphere. 2nd ed. 1973, New York: John Wiley & Sons.

20. Staib, F., Abel, T., Mishra, S., Grosse, G., Frocking, M., Occurrence of Aspergillus fumigatus in West Berlin — Contribution to the Epidemiology of Aspergillosis. *Zbl. Bakt. Hyg.,* 1978. I. Abt. Orig. A 241: p. 337-357.

21. Rhame, F., Nosocomial Aspergillosis: How Much Protection for Which Patients. *Infect. Control Hosp. Epidemiol.,* 1989. 10(7): p. 296-298.

22. Streifel, A.J., Vesley, D., and Rhame, F.S. *Occurrence of Transient High Levels of Airborne Fungal Spores,* in 5th International Conference on Indoor Air Quality and Climate. 1990. Toronto, Canada.

23. Solberg, C., Matsen, J., Vesley, D., Wheeler, D., Good, R., and Meuwissen, H., Laminar Airflow Protection in Bone Marrow Transplantation. *Am. Soc. Microbiol.,* 1971. 21(2): p. 209-216.

24. Michaelsen, G.S., D. Vesley, and M.M. Halbert, Laminar Flow Studied as Aid in Care of Low-Resistance Patients. *Hospitals,* 1967. 41: p. 91-106.

25. Murray, W., Streifel, A., O'Dea, T., and Rhame, F., Ventilation for Protection of Immune Compromised Patients. *ASHRAE Trans.,* 1988. 94(1): p. 1185-1191.

26. Rhame, F.S., Mazzarella, M., Streifel, A.J., and Vesley, D. *Evaluation of Commercial Air Filters for Fungal Spore Removal Efficiency,* in 3rd International Conference on Nosocomial Infections. 1990. Atlanta, GA.

27. Rhame, F., Prevention of Nosocomial Aspergillosis. *J. Hosp. Infect.,* 1991. 18(Suppl. A): p. 466-472.

28. Sherertz, R.J., Belani, A., Kramer, B.S., Elfenbein, G.J., Weiner, R.S., Sullivan, M.L., Thomas, R.G., and Samsu, G.P., Impact of Air Filtration on Nosocomial Aspergillus Infections. *Am. J. Med.,* 1987. 83: p. 709-718.

29. Guideline for Prevention of Nosocomial Pneumonia. *Infect. Control Hosp. Epidemiol.,* 1994. 15: p. 587-627..

30. Maki, D., Alvarado, C., Hassemer, C., and Zilz, M., Relation of the Inanimate Hospital Environment to Endemic Nosocomial Infection. *New Engl. J. Med.,* 1982. 307(25): p. 1562-1566.

31. Streifel, A., P. Stevens, and F. Rhame, In-hospital Source of Airborne Penicillium Species Spores. *J. Clin. Microbiol.,* 1987. 25(1): p. 1-4.

32. Morey, P. and C. Williams, Porous Insulation in Buildings: A Potential Source of Microorganisms. *Proceedings Indoor Air '90 5th International Conference,* 1990. p. 1-6.

33. Fox, B., Chamberlin, L., Kulich, P., Rae, E., and Webster, L., Heavy Contamination of Operating Room Air by Penicillium Species: Identification of the Source and Attempts at Decontamination. *Am. J. Infect. Control,* 1990. 18(5): p. 300-306.

34. McJilton, C., Reynolds, S., Streifel, A.J., and Pearson, R., Bacteria and Indoor Odor Problems — Three Case Studies. *Am. Ind. Hyg. Assoc.,* 1990. 51(10): p. 545-549.

35. Kugler, J.W., Armitage, J.D., and Helms, C.M., Nosocomial Legionnaires' Disease: Occurrence in Recipients of Bone Marrow Transplants. *Am. J. Med.,* 1983. 74: p. 281-288.

36. Muraca, P.W., V.L. Yu, and A. Goetz, Disinfection of Water Distribution Systems for Legionella: A Review of Application Procedures and Methodologies. *Infect. Control Hosp. Epidemiol.,* 1990. 11(2): p. 79-88.

37. Muraca, P., J.E. Stout, and V.L. Yu, Comparative Assessment of Chlorine, Heat, Ozone, and UV Light for Killing Legionella pneumophila within a Model Plumbing System. *Appl. Environ. Microbiol.,* 1987. 53(2): p. 447-453.

38. Colville, A., Crowley, J., Dearden, D., Slack, R., and See, J.V., Outbreak of Legionnaires' Disease at University Hospital, Nottingham. *Epidemiology, Microbiology and Control. Epidemiol. Infect.,* 1993. 195.

39. Vickers, R.M., Yu, V.L., Hanna, S., and Muraca, P., Determinants of Legionella pneumophila Contamination of Water Distribution Systems, *Infect. Control,* 1987. 80: p. 78-86.

40. Personal communication, Michael Osterholm, Acute Disease Section, Minnesota Department of Health, 1995.

CHAPTER 7

Exposure Risk and Prevention of Aerial Transmission of Tuberculosis in Health Care Settings

George P. Kubica

Tuberculosis (TB); the disease dates to antiquity. The causative agent was isolated in 1882 by Koch,[1] and Riley et al.[2] demonstrated clearly that the disease could be spread to sentinel guinea pigs solely by airborne droplet nuclei, without the need for direct contact with the infectious TB patient. Yet, TB still remains an enigma. That nosocomial disease occurs is evidenced by numerous recent outbreaks (see pertinent articles in References 3 and 4), and the conjoint recognition of multiple drug-resistant TB (MDR-TB) and HIV infection or active AIDS disease[5] portends a more rapid progression of and a higher mortality from MDR-TB.

Most of our current knowledge on the aerodynamics of TB is based upon experimental studies: (1) temporal survival of artificially generated aerosols of *Mycobacterium tuberculosis* or surrogate organisms;[6,7] (2) detection of naturally generated airborne TB in the uniquely susceptible guinea pig;[2] and (3) retrospective analyses of institutional outbreaks of the disease.[2,8-10] Does this mean we are still in the dark ages; that our knowledge is so obscure, so veiled that we cannot safely discuss the mechanisms for transmission and control of airborne TB? Must we wait for solid evidence of size or configuration of airborne droplet nuclei and numbers of tubercle bacilli contained therein before we make logical decisions about control methodologies? I think not. I believe that, with currently available information coupled with some good scientific intuition and a sprinkling of oft-replicated anecdotal experiences, that one might paint a reasonably reliable picture of the exposure risk and methods to minimize (but probably not *prevent*) the aerial transmission of TB.

1-56670-144-9/96/$0.00+$.50
© 1996 by CRC Press, Inc.

Many human things "oral" — talking, singing, coughing, sneezing — generate tiny (<5 μm) droplet nuclei in numbers that range from hundreds to millions. Under such conditions we might expect a much higher incidence of TB were it not for the following observations. Studies with experimentally generated aerosols of tubercle bacilli indicate that in excess of 90% of the bacilli in droplet nuclei lose viability almost immediately, while the survivors exhibit a half-life of about six hours.[6] The survivors are rapidly distributed by convectional mixing into large volumes of air (e.g., the roughly 1000 cubic feet in an average hospital room), thereby reducing further their concentration in the prevailing "infectious air." Aerodynamically, the size of the droplet nucleus must be < 5 μm to reach the alveolus of the lung, while the tortuous path through the bronchotracheal tree to arrive at the alveolar destination depletes the starting bacterial population by about 50%.[11]

In spite of all the seemingly heartening facts, many gray, even unknown, areas make a true assessment of risk almost impossible. Why is this?

1. No one has ever cultured airborne tubercle bacilli discharged by an infectious tuberculous patient. Currently, there is no rapid method to determine numbers of bacilli discharged by the patient, thus making it very difficult (if not impossible) to make a reliable prospective risk assessment for healthcare workers (HCWs) or others exposed to the source case. Even if a rapid method did exist, its value would be limited to that finite point in time when the numbers of discharged bacilli (if any) were determined. If the patient had no other respiratory acts (cough, sneeze, etc.), he/she probably would not disseminate much TB, regardless of how positive the sputum was. In addition, HCWs undoubtedly are exposed to other, often undiagnosed TB cases[4] and this is why a good infection control program (including repeat tuberculin testing) remains a most valuable yardstick for infection risk.

2. There is a wide variation in the infectiousness of tuberculous patients.[8,12-14] Because naturally generated tubercle bacilli have never been cultured from air, the determination of numbers of bacilli discharged must be done retrospectively, using the numbers of infected persons as a measure only of the minimal numbers of infectious particles expelled by the source case. The total number of droplet nuclei generated cannot yet be determined because the necessary technology has not been developed; until such methodology is forthcoming, we can only draw inferences from studies in which infection of HCWs is detected by the documented conversion of the tuberculin skin test from negative to positive. Studies reported to date reveal that the average number of TB-containing droplet nuclei discharged by different infectious patients may vary from 1.25 to 250 per hour.[13] These numbers are a reflection not only of the presence of tubercle bacilli in the sputum (or other infectious site), but also of the existence of open cavitary disease in the lung, frequency and forcefulness of cough,[15] and willingness of the patient to "cover the cough."[3,4] All other things being equal, the chance for transmission of TB is directly proportional to the concentration of infectious droplet nuclei discharged in a given time.

Presently, retrospective determination of the probability for infection is determined using the Riley modification of the Reed-Frost equation[12,16] (commonly referred to as the Wells-Riley equation[16]): $P = 1 - e^{-Iqpt/Q}$, where P = probability for infection, I = number of infectors, q = number of infectious quanta/ hour/infector, p = pulmonary ventilation/person in cfm, t = time of exposure in hours, and Q = room ventilation in cfm. Within this formula, Iq/Q equals the so-called steady-state concentration of infectious particles. When the exponent (i.e., Iqpt/Q) is multiplied by the total number of exposed susceptibles, it gives a first approximation of the number of new cases to be expected. Exhibiting the complete Wells-Riley formula in exponential form (as above) also takes into account the possibility of not being infected. Table 1 illustrates the probability of becoming infected when exposed for varying times to two different hospitalized patients generating (on average) either 13 or 250 TB-containing droplet nuclei per hour in a hospital room of 960 cubic feet ($10 \times 12 \times 8$ feet) with hypothetical air changes per hour (ACH) varying from 6 to 100.

Table 1 Risk of Infection (%) Related to Ventilation (ACH), Time in Room (min) and Infectivity (qph)[1]

		13 qph		250 qph	
ACH	CFM	15 min	60 min	15 min	60 min
6	96	1.2	4.7	21	60
10	160	0.7	2.8	13	42
25	400	0.3	1.1	5	20
100	1600	—	—	1.4	5

[1] Data calculated using Wells-Riley equation, assuming only one infector (I) and pulmonary ventilation (p) as 0.353 cfm. All other figures available from the above table.

3. Retrospective analyses of experimental studies,[2] using guinea pigs as the sentinels for airborne droplet nuclei containing *M. tuberculosis,* indicate that a single tubercle bacillus can infect this highly susceptible experimental animal. Extrapolation of this data to human HCWs on the same ward in which the guinea pig study was performed strongly suggests[12] that it takes only a single droplet nucleus containing one or a few tubercle bacilli to infect humans. Although there is recent evidence to indicate racial differences in susceptibility to TB infection when exposed in the same environment,[33] retrospective data would imply that there is no threshold limit value (TLV) below which a risk of TB infection would be unlikely. Because (1) the droplet nuclei coughed up by a TB patient are rapidly and randomly distributed throughout the air of the room and (2) so many bacilli are lost to natural attrition, the risk for infection is relatively low, as attested to both by the hypothetical data in Table 1 and the many still tuberculin-negative pulmonary care physicians who have treated tuberculous patients. Even in the face of MDR-TB, the risk for infection is not increased (except for those with conjoint AIDS or HIV infection; see References 3 and 4). On the other hand, the chances for successful treatment of an acquired MDR-TB infection are reduced when compared to those for a fully susceptible strain of TB.

The retrospective analyses of data from well-documented nosocomial TB epidemics or experimental studies have shown that, in large, open facilities (e.g., office space), droplet nuclei containing tubercle bacilli become widely dispersed by convectional mixing so that their concentration may be as low as one droplet nucleus per 8000 to 12,000 cubic feet of institutional air.[8,13] In the confines of hospital isolation rooms, the steady-state concentration of airborne *M. tuberculosis* may be higher, resulting in an increased risk of infection for those who must be in the room for more than a few minutes. Additionally, even in the best balanced and properly maintained negative pressure isolation room, there is a recognized "belch back" of potential TB-containing air from the room of a source case into the adjoining corridor.[16,17] If this happens, then any infectious droplet nuclei that escape the confines of the patient isolation room could possibly move anywhere in the health-care setting where the now contaminated air may move.

The preceding discussion reveals how generally difficult (at least statistically) it is to become infected with TB, but recognizing the absence of a TLV for this disease, one is then faced with the practical problem of removing or significantly reducing the numbers of infectious droplet nuclei. This latter problem is further exacerbated by the fact that most of our attention at control of TB is directed to rapid identification and treatment of the source case of TB.[4] The need for better management of the known tuberculous patient and the continuation of adequate funding to reach the goal of elimination of TB have been addressed.[18-21] Less attention has been devoted to the active transmission of TB by undiagnosed, or slow-to-be-diagnosed, cases that have resulted in unexpected outbreaks in homeless shelters, prisons, and among HCWs (see Reference 21 for more details).

Three methods are currently and easily available to control airborne TB-containing droplet nuclei. These are ventilation, high efficiency particulate air (HEPA) filtration, and ultraviolet germicidal irradiation (UVGI). The relative effectiveness of the three methods has been reviewed recently by several writers.[12,13,22,23] Because of the vagaries of reproducing exactly the naturally generated TB droplet nuclei, or the ventilation system, or the population of exposed tuberculin-negative hosts, there has *never* been a satisfactorily controlled study that indicates unequivocally the protective impact for HCWs of any of the three environmental controls addressed above;[13] however, controlled studies of UVGI are now planned in shelters for the homeless (Nardell, E.A. and Riley, R.L., personal communications). Most certainly, *none* of the three interventions will provide 100% protection from infection, so it is generally desirable to use a combination of two or more of the available methods in order to minimize most effectively the numbers of airborne organisms left in the air to infect HCWs, other patients, or visitors to the institution.

Let's look briefly at these three interventions.

1. **Ventilation** — In areas at high risk for airborne diseases, federal recommendations[24] suggest 6 to 12 negative air changes (related to risk), with exhaust directly to the outside. Interestingly, the numbers of ACH recommended in different hospital/clinic areas have changed from one edition to the next in publications from both ASHRAE and Health and Human Services. Neither organization provides supportive documentation for the arbitrarily selected airflow rates.[16] Most certainly, increased numbers of ACH can become physically uncomfortable to both patients and staff, as well as economically difficult to sustain during times of energy conservation. Additionally, air turbulence generated by objects (e.g., beds, nightstands, chairs, people, etc.) in the room can

interfere both with convectional mixing of air and with desired directional airflow. Keene and Sansone[17] as well as Hermans and Streifel[16] have shown that, even in well-balanced, negative pressure directional flow rooms, there is a demonstrable spill of room air out into an adjoining corridor whenever the room door is opened. Fraser et al.[25] showed that nearly half the hospital isolation rooms they evaluated were under positive pressure to the corridor, and the presence of an anteroom seemed not to affect the correctness of airflow. My personal experience in several hospitals and clinics that have experienced mini-epidemics of TB among the staff would corroborate these observations. Failure of directional and dilutional airflow systems, for a variety of reasons, has prompted one airflow engineer and consultant to warn that most well-designed hospital systems will be dysfunctional in 5 to 10 years, and the net effect of these failures might be a sufficient reverse pressurization that could change "…infection control to infection encouragement."[26]

2. **Filtration** — This intervention has been widely used in biological laboratories where removal of airborne infectious organisms is necessary before air from the infectious laboratory work area is exhausted or recirculated. Commonly, the filters selected are HEPA filters capable of removing 99.97% of particles >0.3 μm in size. Ongoing problems of maintenance and testing, together with the greater cost to move the air through the increased resistance of such filters has often limited the use of HEPA filters to small areas (e.g., booths used for sputum induction or aerosolized medications). In recent years, several HEPA filtration products have been marketed for the purposes of reducing numbers of airborne infectious particles in the air and maintaining a negative inward flow of air from corridor to isolation room; one such unit has been described and tested.[27] Such filtration units often suffer the same deficiencies of a central HVAC supply, viz., maintenance. Filtration efficiency must be monitored, filters changed when indicated, and belts and motors subjected to periodic checks. Maintenance personnel can easily compromise the efficiency of such "roll-in" units by pushing furniture against the filter intakes during room cleaning. When attempts are made to move large volumes of room air (e.g., 25 ACH) by suction through small HEPA filters (e.g., 2×2 feet), it can be demonstrated that some of the air in the room is as efficiently mixed and retained by convection currents as it is extracted by the HEPA filter unit. Because of the small size of most HEPA units currently used, rapid extraction occurs only for that air closest to the exhaust filter fan (i.e., within a few inches). Fixtures in the room physically contribute to both increased turbulence and eddying of the air, while temperature differentials about the room encourage a thorough and rapid convectional mixing that can significantly (and negatively) impact the ability of directional flow to remove infectious airborne droplet nuclei. If randomly distributed tubercle bacilli exist in this recirculated air, the risk for infection within the patient room is still real. Even though the HEPA unit may extract large amounts of air, the most effective air movement is that in close proximity to the exhaust fan, while rapid entrainment of air from distant corners or from the mid to upper room is difficult to accomplish. This is perhaps best illustrated by filling a bathtub or large kitchen sink with water and then adding a few drops of concentrated food coloring (this can simulate the droplet nuclei) before pulling the drain plug (let this latter represent the HEPA exhaust filter). Before all the water exits the tub or sink, the food coloring is evenly distributed throughout the volume of water that has not yet gone down the drain. Air mixes in similar fashion. This latter

demonstration is described to point out that filtration, too, can have its draw-backs, but when no other system or approach is available, this may be the only alternative to provide some measure of protection for HCWs, patients, and visitors.

3. **UVGI** — This is the only one of the three interventions that has been tested against airborne tubercle bacilli (or the surrogate vaccine strain, BCG).[7,12,22] Perhaps the greatest concern about UVGI is that of safety for the occupants of the room(s) in which it may be used. Unlike the longer wavelength, and much more damaging, UV-A and UV-B in sunlight, UVGI (or UV-C, 254 nm) is readily absorbed by the stratum corneum, it does not cause cataracts, and it is highly unlikely that the dosage sufficient to cause skin cancer in intensely irradiated hairless mice would be attained from exposure of humans to properly installed and monitored upper room UVGI fixtures.[22] The National Institute of Occupational Safety and Health (NIOSH) limit of 6 mJ/cm^2 (or 0.2 µW/cm^2) for an 8-h exposure to UV-C (i.e., UVGI) irradiation in the upper room provides more than adequate protection to persons working in the lower room. Too, the newer multilouvered UVGI fixtures commonly keep the intense UV-C confined to the upper air, while levels in the lower room are only a fraction of that allowed by NIOSH standards. In contrast, exposure to the much more damaging and more harmful UV-A and UV-B found in sunlight (but not in UVGI) that one might expect after a 4-h outdoor activity on a bright, sunny day may exceed 700 mJ/cm^2, yet there is much less concern expressed by NIOSH for this universal danger. Nevertheless, when using UVGI in the upper air, the meth-odology suffers the same limitations of both ventilation and filtration — getting the target organisms into the path of the UV irradiation. Riley demonstrated that convectional mixing alone was adequate to produce a reduction of airborne BCG organisms to levels comparable to that of 10 to 30 or more ACH.[7] Permutt[28] has emphasized the need for good air mixing if such beneficial effects of UVGI are to be realized. Alas, no studies to date have successfully convinced all investigators that upper air UVGI can adequately protect against lower room exposure to airborne TB. Some very compelling results have demonstrated the potentially protective effects of upper room UVGI against measles,[29] influenza,[30] and tuberculosis.[31] Recently, we've seen results of a 3-year experience in Boston in which UVGI was the only intervention used (no central HVAC existed) to protect against transmission of MDR-TB in an 18-bed ward. After more than 200,000 man hours of exposure (to both infectious TB cases and "potentially" harmful UV), and use of older type UV fixtures that sometimes provided 1 µW/cm^2 energy at work level (i.e., >5 times NIOSH recommended dose), there have been (1) no tuberculin conversions among the HCW, (2) no erythema, and (3) no photokeratoconjunctivitis reported from either patients or HCWs.[35]

Without a TLV for TB, it is difficult to know what an acceptable annual risk for infection might be. Figures such as 5 and 2% have been suggested and the California Department of Health Services together with the University of Cali-fornia School of Public Health suggested a risk of 1 in 10,000 (0.01%).[32] To achieve such protection by ventilation alone would be an insurmountable task. Let's set up a hypothetical 1000 cubic foot hospital isolation room to house a previously described "source case" who emitted 13 infectious droplet nuclei per hour.[8] If we had the currently recommended 6 ACH,[24] the risk for infection if an

HCW spent a total of 1 h in the patient's room would be 4.5%. To reduce that risk to 2% or 0.01% (as determined by the Wells-Riley formula[12]) would require 14 or a whopping 2700 ACH, respectively. Imagine what the necessary ACH would be if (1) exposure were prolonged or (2) the number of infectious droplet nuclei generated each hour were 60 or 250 (see Table 2).

Table 2   ACH to Provide Indicated Risk Reduction
           for HCW Exposed to 13 qph for 1 or 8 h[1]

Original 6 ACH	1 h 4.5% risk	8 h 31% risk
Risk Red'n To		
10%	NA	20
2%	14	93
0.01%	2700	18,600

[1] Data calculated using Wells-Riley equation, assuming only one infector (I) and pulmonary ventilation (p) as 0.353 cfm. Room volume is 1000 cubic feet. The (cfm) for indicated ACH are: for 6 (100); 14 (233); 20 (333); 93 (1550); 2700 (45,000); and for 18,600 (310,000).

Remember, none of the foregoing three interventions is 100% effective, and the plethora of anecdotal and published data generated in the past 3 or 4 years indicates that ventilation alone is inadequate. With the explosion of TB cases in recent years,[3,4] the needs for better management of patients and adequate funding to hopefully eliminate TB,[18-21] and the dim projections for the future of this disease under current treatment and control modalities,[21] it should be abundantly clear that every available intervention for control of the disease should be invoked where and when applicable. We don't have time to resolve all the controversial issues about control of airborne TB — the shape of the droplet nucleus, the numbers of bacilli in a single droplet, the development of a method to count their numbers in a forcefully expelled cough, or the precise feasibility of all the control interventions available.

The relative costs of the three current technologies here discussed (i.e., UVGI, HEPA filtration, and ventilation) for control of airborne droplet nuclei containing tubercle bacilli are 1, 3, and 10, respectively (data from both Nardell and Kubica). Knowing this, a considerable impact can be made on the control of nosocomial airborne TB and the reduction in the risk of infection for HCWs or other persons present in the hospital environment if we utilize a symphony of these currently available technologies: ventilation for comfort and air circulation (perhaps 4 to 6 ACH), upper room UVGI (even throughout a hospital or clinic to protect the public as well as HCWs) to minimize numbers of surviving tubercle bacilli, and HEPA filtration in rooms with very low ceilings (<8 feet) or in areas where air may be recirculated. Because of the often unknown, unexpected, and potentially widespread dissemination of infectious droplet nuclei throughout a hospital or clinic by a commonly mobile, undiagnosed infectious TB case,[4] Riley[34] recently suggested whole building irradiation with UVGI. This would probably serve to reduce the numbers of any nosocomial infectious agent spread by airborne droplet nuclei (e.g., TB, measles, influenza). The present problems of increasing TB and particularly the rising numbers of MDR-TB, dictate an all out effort at control. This effort should demand no less than a total commitment of all appropriate interventions.

REFERENCES

1. Koch, R. Die Aetiologie der Tuberkulose. *Berl. Klin. Wschr.* 1882; 19:221-230.
2. Riley, RL, CC Mills, F O'Grady, LU Sultan, F Wittesstadt, DN Shivpuri. Infectiousness of air from a tuberculosis ward. *Am. Rev. Respir. Dis.* 1962; 85:511-525.
3. Centers for Disease Control and Prevention. Guidelines for preventing the transmission of tuberculosis in health-care settings, with special focus on HIV related issues. *MMWR* 1990; 39(RR-17):1-29.
4. Centers for Disease Control and Prevention. Draft guidelines for preventing the transmission of tuberculosis in health-care facilities, 2nd Ed. Notice of Comment Period. 1993;(Oct. 12). *Federal Register* 58(195): 52810-52854.
5. American Thoracic Society (Position Paper). Mycobacterioses and the acquired immunodeficiency syndrome. *Am. Rev. Respir. Dis.* 1987; 136:492-496.
6. Loudon, RG, LR Bumgarner, J Lacy, GK Coffman. Aerial transmission of mycobacteria. *Am. Rev. Respir. Dis.* 1969; 100:165-171.
7. Riley, RL, M Knight, G Middlebrook. Ultraviolet susceptibility of BCG and virulent tubercle bacilli. *Am. Rev. Respir. Dis.* 1976; 113:413-418.
8. Nardell, EA, J Keegan, SA Cheney, SE Etkind. Airborne infection. Theoretical limits of protection achievable by building ventilation. *Am. Rev. Respir. Dis.* 1991; 144:302-306.
9. Catanzaro, A. Nosocomial tuberculosis. *Am. Rev. Respir. Dis.* 1982; 125:559-562.
10. Centers for Disease Control. *Mycobacterium tuberculosis* transmission in a health clinic — Florida 1988. *MMWR* 1989; 38:256-264.
11. Hatch,TF. Behavior of microscopic particles in the air and in the respiratory system. Aerobiology, Washington, DC. *Am. Assoc. Adv. Sci.* Pub. No. 17, 1942, pp 102-105.
12. Riley, RL, EA Nardell. Controlling transmission of tuberculosis in health care facilities: ventilation, filtration and ultraviolet air disinfection. In Plant, Technology and Safety Management Series, Controlling Occupational Exposures to Tuberculosis. Oakbrook Terrace, IL. Joint Commission on Accreditation of Healthcare Organizations. 1993; pp 25-31.
13. Nardell, EA. Environmental control of tuberculosis. *Med. Clin. N. Am.* 1993, 77:1315-1334.
14. Sultan, L, W Nyka, C Mills, et al. Tuberculosis disseminators — a study of the variability of aerial infectivity of tuberculosis patients. *Am. Rev. Respir. Dis.* 1960; 82:358–369.
15. Loudon, RG, SK Spohn. Cough frequency and infectivity in patients with pulmonary tuberculosis. *Am. Rev. Respir. Dis.* 1969; 99:109-111.
16. Hermans, RD, AJ Streifel. Ventilation Design. In Biernbaum, PJ, M Lippmann, Eds. *Proceedings of the Workshop on Engineering Controls for Preventing Airborne Infections in Workers in Health Care and Related Facilities.* Cincinnati; US Dept of Health and Human Services, Public Health Service, CDC, 1994. DHHS publication no. (NIOSH) 94-106.
17. Keene, JH, EB Sansone. Airborne transfer of contaminants in ventilated spaces. *Lab Anim. Sci.* 1984; 34(5):453-457.

18. Reichman, LB. The U-shaped curve of concern. *Am. Rev. Respir. Dis.* 1991; 144:741-742.
19. Brudney, K, J Dobkin. Resurgent tuberculosis in New York City: HIV, homelessness, and the decline of TB control program. *Am. Rev. Respir. Dis.* 1991; 144:745-749.
20. American Thoracic Society. Centers for Disease Control national tuberculosis training initiative core curriculum on tuberculosis. New York: *Am. Thoracic Soc.* 1990:1-40.
21. Bloom, B, CJL Murray. Tuberculosis: commentary on a reemergent killer. *Science* 1992; 257:1055-1064.
22. Nardell, EA. Fans, filters, or rays? Pros and cons of the current environmental tuberculosis control technologies. *Infect. Control Hosp. Epidemiol.* 1993; 14:681-685.
23. Macher, JM. The use of germicidal lamps to control tuberculosis in health-care facilities. *Infect. Control Hosp. Epidemiol.* 1993; 14:723-729.
24. Health Resources and Services Administration. Guidelines for construction and equipment of hospital and medical facilities. 1983/84 Ed. U.S. DHHS, PHS, Bureau of Health Maintenance Organizations and Resources Development, Office of Health Facilities, Rockville, MD 20857, DHHS Publication No.(HRS-M-HF) 84-1. July 1984.
25. Fraser, VJ, K Johnson, J Primack, M Jones, G Medoff, WC Dunagan. Evaluation of rooms with negative pressure ventilation used for respiratory isolation in seven midwestern hospitals. *Infect. Control Hosp. Epidemiol.* 1993; 14:623-628.
26. Lindberg, P. Improving hospital ventilation systems for tuberculosis infection control. In Plant, Technology and Safety Management Series, Controlling Occupational Exposures to Tuberculosis. Oakbrook Terrace, IL. Joint Commission on Accreditation of Healthcare Organizations, 1993; pp 19-23.
27. Marier, RL, T Nelson. A ventilation-filtration unit for respiratory isolation. *Infect. Control Hosp. Epidemiol.* 1993; 14:700-705.
28. Permutt, S. The impact of ultraviolet light on transmission of infection. At American College of Chest Physicians Consensus Conference: Institutional Infection Control Measures for Tuberculosis in the Era of Multiple Drug Resistance. Nov. 13, 1993, Chicago.
29. Riley, EL, G Murphy, RL Riley. Airborne spread of measles in a suburban elementary school. *Am. J. Epidemiol.* 1978;107:421.
30. McLean, R. The effect of ultraviolet radiation upon the transmission of epidemic influenza in long-term hospital patients: International Conference on Asian Influenza, Feb 17-19,1960, Bethesda. Discussion in *Am. Rev. Respir. Dis.* 83(Suppl.):36, 1961.
31. Nardell, E, B McInnis, R Riley, S Weidhaus. Ultraviolet light air disinfection to reduce tuberculosis transmission in a shelter for the homeless. *Am. Thoracic Soc. (Abstr.)* Mar. 1988, pg 257.
32. Nicas, M, JE Sprinson, SE Royce, RJ Harrison, JM Macher. Isolation rooms for tuberculosis control. *Infect. Control Hosp. Epidemiol.* 1993; 14:619-622.
33. Stead, WW, JW Senner, WT Reddick, et al. Racial differences in susceptibility to infection by *Mycobacterium tuberculosis. New Engl. J. Med.* 1990;322:422-427.
34. Riley, RL. Ultraviolet air disinfection: rationale for whole building irradiation. *Infect. Control Hosp. Epidemiol.* 1994; 15:324-328.
35. Turner, M, K McGowan, GJ Cuchural et al. Environmental controls and PPD conversion rates in an inpatient tuberculosis unit. Abstract, Annual Meeting of American Thoracic Society, May 22-25,1994, Boston, MA.

CHAPTER 8

Indoor Air Pollution: Understanding the Mechanisms of the Effects

Rebecca Bascom, Jana Kesavanathan, and David L. Swift

INTRODUCTION

Diverse materials are present in ambient air that is inhaled into the respiratory tract at a rate of 10,000 to 20,000 l per day.[1] Pollutants are materials that can exert adverse health effects at relatively low concentrations; ozone, for example, causes lung inflammation at concentrations of 0.08 to 0.12 ppm. In the indoor environment, exposure to single pollutants is distinctly uncommon; more typically humans inhale complex mixtures. One of the essential challenges for people interested in understanding the effects of the indoor environment on human health is to develop the concepts and methods to understand how indoor pollutant exposures alter human health. It is difficult to document objectively the symptom complexes of which workers complain. A common characteristic is that traditional medical physical evaluations do not demonstrate findings despite a fair degree of consistency in the symptom complexes reported by the workers.[2]

The purpose of this chapter is to review the basic processes by which different categories of pollutants act on the respiratory tract. Studies of the nose will receive particular attention since upper respiratory complaints are typically more prevalent than lower respiratory complaints in indoor environments, and diagnostic methods for the nose lag behind those for the lung.

Agents can be classified on how they affect human tissue. Corrosives are so termed because they cause direct destruction of structural tissue. Irritants are materials that cause symptoms and inflammation, i.e., alteration of the structural tissue and influx of cells and mediators into the inflamed tissue with redness, swelling, heat, and pain. Sensitizers are materials that induce a specific immunologic response (e.g., rodent allergy) with the well-known properties of specificity, diversity, tremendous amplification, and the ability to distinguish between self and non-self.

1-56670-144-9/96/$0.00+$.50

Because the profile of pollutants in the indoor environment is so complex, we have chosen to investigate how these categories of agents affect the upper respiratory mucosa. The hope is that responses to irritants and sensitizers can be well documented in controlled human exposure studies, and that then worker populations can be examined and an understanding achieved of the nature of the process that is causing symptoms.

The investigative approach will be outlined below, with examples of the current status of our understanding. The first step is to identify biologic variability in the symptomatic response to a particular situation. The second step is to form a hypothesis as to the exposure causing the symptoms and to the possible mechanism. The third step is to perform controlled challenges in people with a variable history, to determine whether controlled exposure reproduces the symptoms of interest and confirms the presence of the biological variability. The fourth step is to seek objective measures of response, guided by hypotheses as to the mechanism of the response. The fifth step is to study particular aspects of the response, selecting subjects that display the characteristic of interest. The sixth step is to try to modify the response, both to provide therapeutics for the affected people, and to understand the mechanism. Finally, the findings of the controlled studies are tested in the field, in epidemiologic studies.

STUDIES OF IMMEDIATE HYPERSENSITIVITY

Twenty years ago, investigators began to develop an understanding of the clinical features of IgE-mediated allergic responses. Their approach was simple: subjects with and without a history of seasonal (fall) rhinitis were recruited and characterized. The first step was thus achieved; biological variability in the symptomatic response was identified. It was hypothesized that ragweed, a pollen common in the fall, was responsible for the symptoms, and an IgE-mediated, allergic response was suspected. The second step was thus achieved; a responsible exposure was hypothesized, as was a possible mechanism.

The subjects were then exposed to ragweed, the agent thought responsible for the symptoms. Individuals with a history of fall rhinitis were more likely to develop symptoms of itching, sneezing, rhinorrhea, and congestion following controlled exposure to ragweed than those without a history of fall rhinitis. The third step was thus achieved; controlled exposure reproduced the symptoms of interest and confirmed the presence of the biological variability in the response to that agent.

Subsequent studies sought objective evidence to corroborate the ragweed-induced symptoms.[3] These studies were guided by the hypothesis that an IgE-mediated response was occurring. Sneezes were counted, nasal resistance was measured, and nasal lavage was performed to determine the pattern of cells and medicators that were present at the time of symptoms. Over time, elevations in histamine and leukotriene C_4 were demonstrated in nasal lavage fluid, indicating that mast cell activation had occurred. Concentrations of albumin and kinins increased, indicating the presence of increased vascular permeability. Thus the fourth step was achieved; objective measures of response were sought, guided by hypotheses as to the mechanism of the response.

Subsequent studies focused on the "late phase response" — the symptoms of rhinorrhea and congestion that were occurring 4 to 8 h after the initial exposure. For these studies, individuals were selected that had demonstrated a late phase response to antigen challenge on a screening challenge. It was understood that this response was subselected for study, and no attempt was made to obtain a representative sample of the population. It was hoped that a study of this response would lend insight into the mechanism of allergic disease. Studies of the late phase were in fact revealing; influxes of basophils, eosinophils, neutrophils, and mononuclear cells were demonstrated[4] and a re-release of histamine and leukotriene C_4 was demonstrated.[5] The fifth step was thus achieved; to study particular aspects of the response, selecting subjects that display the characteristic of interest.

Studies to modify the response to allergen challenge have been many: antihistamines have been shown to decrease acute symptoms and oral steroids have been shown to block the late response to allergen exposure, including the cellular influx.[6] Topical steroids have been shown to block both the acute and the late response. Immunotherapy has been tested as well. Allergen avoidance has been studied as a means of reducing symptoms and signs of disease.[7] These observations achieved the sixth step: to try to modify the response, both to provide therapeutics for the affected people and to understand the mechanism.

Finally the findings and hypotheses developed in the controlled studies have been applied to clinical trials of medications, immunotherapy, and allergen avoidance. With these studies, the hypotheses of the mechanism of the allergic response have been further developed and modified.

STUDIES OF THE UPPER RESPIRATORY IRRITANT RESPONSE

Biologic variability in the symptomatic response to irritants has been described in the medical literature for four decades. Clinicians have reported variable symptoms with exposure to environmental tobacco smoke (ETS) and odors (i.e., volatile organic mixtures). The first step in this approach was thus achieved long ago; biologic variability was identified in the symptomatic response to a particular situation. For many years, this increased response to ETS was called "tobacco smoke allergy." This term highlighted the presence of clinical biological variability and implied a mechanistic hypothesis of an IgE-related mechanism.

Studies became bogged down at the second step; skin testing with tobacco leaf extracts failed to demonstrate an immediate wheal and flare in subjects with historical symptoms. The hypothesis that "environmental tobacco smoke allergy" was an IgE-mediated response had no support. However, the clinical observation of biological variability in the symptomatic response persisted.

We initiated studies of upper respiratory responses to environmental tobacco smoke in the late 1980s. Our laboratory asked 77 healthy nonsmoking subjects whether they had a history of rhinitis symptoms with ETS exposure: nearly one third reported a history of rhinorrhea, congestion or sneezing.[8] This confirmed the impression of biological variability in the history of responses to tobacco smoke.

To test the hypothesis that ETS was the source of the rhinitis symptoms and that an IgE mechanism was involved, controlled human exposure challenges were performed. There was a significant correlation between the magnitude of rhinitis symptoms on historical ETS exposure and with controlled exposure to smoke. We sought the mechanism of the response to smoke. Our initial hypothesis was that ETS rhinitis was caused by a classic IgE-mediated response. If true, then the mediators of immediate hypersensitivity should be demonstrable in nasal lavage fluid following controlled smoke exposure. The results of these studies were negative: despite the occurrence of symptoms, there was no increase in histamine (indicating an absence of mast cell activation), no increase in kinins or albumin (indicating an absence of increased vascular permeability), and no increase in markers of glandular activation.[8] *These studies indicated that ETS rhinitis is not caused by an IgE-mediated mechanism.* These studies also showed that vascular leak, a common component of respiratory irritant responses in rodent models,[9] was not occurring in the human response. Intranasal challenge of human subjects with a simpler irritant, capsaicin, also failed to result in increased vascular permeability.[10] Thus the second step was achieved; we formed a hypothesis as to the exposure causing the symptoms and to the possible mechanism. We had shown that ETS was the agent likely to be responsible for the different symptoms, but had no mechanistic explanation. At this point, however, it was unclear whether there would be objective evidence of differential responsiveness.

We next sought physiologic evidence for altered responsiveness. *Physiologic measurements of nasal resistance, using posterior rhinomanometry, demonstrated clear objective of biologic variability following controlled exposure to smoke.*[8] Increased nasal resistance occurred following exposure to a brief, high level of smoke in subjects with a history of ETS rhinitis, but not in subjects with no history of ETS rhinitis. We therefore have chosen to label this phenomenon as "ETS sensitivity," not ETS allergy.

We hypothesized that the symptoms of rhinitis were caused primarily by vascular congestion. Pretreatment with the vasoconstrictor, oxymetazoline, blocked the congestive response to smoke,[11] lending support to this hypothesis. The vasculature that alters nasal congestion is the deep cavernous plexus, and direct measurement of blood flow in that region is difficult. The laser Doppler technique, used in clinical studies in recent years, measures flow in the superficial subepithelial capillary plexus. Other studies have shown that the blood flow to these two regions is differentially regulated, therefore the laser Doppler technique cannot be applied in this case. These studies began to accomplish the fourth step which is to seek objective measures of response, guided by hypotheses as to the mechanism of the response. They also selected subjects for the congestive response through prescreening (step five) and began to try to modify the response (step six). Other pharmacologic studies showed no evidence of cholinergic reflexes involved in the differential response to smoke.

The occurrence of nasal congestion was confirmed independently using acoustic rhinomanometry, which showed a diffuse decrease in nasal volume with smoke exposure.[12,13] Exposure-response studies and time-course experiments were performed. These showed that exposure to the vapor phase of smoke caused congestion among the ETS-sensitive subjects only. Both subject groups developed congestion with prolonged exposure to moderate concentrations of smoke. The congestion resolves typically after half an hour, and there is no late-phase response 4 to 8 h later. These studies demonstrated an important difference between IgE allergics and nonallergics, while there appears to be a quantitative difference in the response to smoke among ETS sensitive and ETS nonsensitive subjects.

Additional hypotheses have begun to be pursued. One hypothesis is that subjects who demonstrate congestion with smoke exposure will have an altered response to exogenous vasodilator neuropeptides. Thus far, our studies indicate a significant nasal congestion with substance P among subjects with smoke-induced nasal congestion, but not among subjects without smoke-induced nasal congestion.[14] The possible explanations for this include an increased delivered dose of substance P to the blood vessels (possibly due to a reduction in epithelial neutral endopeptidase), an altered expression of substance P receptors on the nasal vasculature, or an altered signal transduction at the substance P receptor.

More recently, we have been interested in mucociliary clearance as a biologic marker of the respiratory tract and hypothesized that mucociliary clearance may be differentially altered by smoke exposure. In animal studies, low level irritant exposure tends to increase mucociliary clearance (teleologically understandable as helping to clear the irritants from the mucosal surface), while high level irritant exposure eventually reduces mucociliary clearance. Our initial studies have shown marked interindividual variation in the effect of smoke on mucociliary clearance; many individuals demonstrate a marked acceleration of mucociliary clearance with smoke exposure. A subset, however (all of who are historically ETS sensitive), demonstrates marked reductions in mucociliary clearance.[15] Possible explanations for this finding are that these latter individuals have a shift in their exposure-response curve, either due to an altered delivered dose of smoke or to pre-existing decreased epithelial functional reserve. We are actively pursuing an understanding of the mechanism of this differential effect.

In summary, considerable progress has been made in studying the effect of irritants on the upper respiratory tract and in beginning to understand biological variability in the symptoms that occur with smoke exposure. To reiterate the main findings: biologic variability in symptoms experienced with ETS exposure does occur. The primary response is one of vascular congestion; there is no evidence for an IgE-mediated allergic response. Differences in the objective congestive response are demonstrable under some, but not all conditions. Brief, high levels of smoke, and more prolonged exposure to the vapor fraction of the smoke elicit congestion in the ETS sensitive, but not ETS nonsensitive subjects. Substance P may be involved in the responsiveness, and mucociliary clearance may be differentially altered in some subjects.

ACKNOWLEDGMENTS

The author wishes to acknowledge gratefully the collaboration of Drs. Lawrence Lichtenstein, Robert Naclerio, David Proud and Anne Kagey-Sobotka, Joan Hudgin Leonard, and Thomas K. Fitzgerald.

REFERENCES

1. Bascom R. Air pollution. In: Mygind N, Naclerio RM, eds. Allergic and Non-allergic Rhinitis. Clinical Rhinitis. Copenhagen: Munksgaard, 1993; 32-45.

2. Kreiss K. The epidemiology of building-related complaints and illness. In: James E. Cone MJH, eds. Problem Buildings: Building-Associated Illness and the Sick Building Syndrome. Philadelphia: Hanley & Belfus, Inc., 1989:575-592.

3. Bascom R, Proud D, Togias AG, Peters SP, Norman PS, Kagey-Sobotka A, Lichtenstein LM, Naclerio RM. Nasal provocation: an approach to study the mediators of allergic and non-allergic rhinitis. XII International Congress of Allergy and Clinical Immunology, 1986. Washington D.C. The C.V. Mosby Company, St. Louis, 113-120.

4. Bascom R, Wachs M, Naclerio R, Pipkorn U, Galli S, Lichtenstein L. Basophil influx occurs after nasal antigen challenge: effects of topical corticosteroid pre-treatment. *J. Allergy Clin. Immunol.* 1988;81:580-589.

5. Naclerio RM, Proud D, Togias AG, Adkinson Jr. NF, Meyers DA, Kagey-Sobotka A, Plaut M, Norman PS, Lichtenstein LM. Inflammatory mediators in late antigen-induced rhinitis. *New Engl. J. Med.* 1985;313:65-70.

6. Bascom R, Pipkorn U, Lichtenstein LM, Naclerio RM. The influx of inflammatory cells into nasal washings during the late response to antigen challenge. Effect of systemic steroid pretreatment. *Am. Rev. Respir. Dis.* 1988;138:406-412.

7. Pope AM, Patterson R, Burge H, ed. Indoor Allergens. Assessing and Controlling Adverse Health Effects. Committee on the Health Effects of Indoor Allergens (R. Bascom, committee member), Division of Health Promotion and Disease Prevention, Institute of Medicine. ed. Washington, D.C.: National Academy Press, 1993;308.

8. Bascom R, Kulle T, Kagey-Sobotka A, Proud D. Upper respiratory tract environmental tobacco smoke sensitivity. *Am. Rev. Respir. Dis.* 1991;143:1304-1311.

9. Lundberg JM, Saria A, Martking ER. Capsaicin pretreatment abolishes cigarette smoke-induced oedema in rat tracheo-bronchial mucosa. *Eur. J. Pharmacol.* 1983;86:317-318.

10. Bascom R, Kagey-Sobotka A, Proud D. Effect of intranasal capsaicin on symptoms and mediator release. *J. Pharmacol. Exp. Ther.* 1991;259:1323-1327.

11. Bascom R, Fitzgerald TK. A vasoconstrictor partially alters the nasal response to sidestream tobacco smoke. *J. Respir. Crit. Care. Med.* 1994;149:A391.

12. Swift DL, Nadarajah J, Fitzgerald TK, Permutt T, Bascom R. Acoustic rhinometry as a tool for human inhalation toxicology studies. Proceedings of the meeting of the American Industrial Hygiene Association 1992.

13. Willes S, Fitzgerald TK, Permutt T, Sauder L, Bascom R. Respiratory effects of prolonged sidestream tobacco smoke exposure and effect of filtration. *Am. Rev. Respir. Dis.* 1991;143:A90.

14. Leonard JF, Fitzgerald TK, Bascom R. Substance P nasal challenge in environmental tobacco smoke responsive subjects. *Am. Rev. Respir. Dis.* 1992;145:A87.
15. Bascom R, Kesavanathan J, Fitzgerald TK, Cheng K-H, Swift DL. Sidestream tobacco smoke acutely alters human nasal mucociliary clearance. *Environ. Health Perspect.* 1995;103:1026-1030..

CHAPTER 9

The Effect of Ozone on House Dust Mite Allergen-Induced Nasal Inflammation in Asthmatics

Robert B. Devlin, Lisa A. Dailey, Jacqueline Carter, Lisa Cazares, Woodrow A. Setzer, and David B. Peden

ABSTRACT

Ozone may play a significant role in the exacerbation of airway disease in asthmatics, either by priming the airway mucosa such that cellular responses to allergen are enhanced or by exerting an intrinsic effect on airway inflammation. Previous investigations of nonasthmatic subjects reveal that ozone induces both nasal and bronchial inflammation, suggesting that nasal responses to ozone may be used as a surrogate marker for the effect of this pollutant on bronchial mucosal inflammation. In this study, the effect of exposure to 0.4 ppm ozone on nasal inflammation in 11 allergic asthmatics sensitive to *Dermatophygoides farinae* was examined. By employing a "split nose" design, in which only one side of the nose was challenged with allergen while the contralateral side was challenged with saline, the primary effect of ozone on nasal inflammation as well as its influence on allergen-induced inflammation was examined. The findings of this study demonstrate that ozone exerts a primary inflammatory effect on nasal tissues of asthmatics as well as augmenting allergen-induced cellular influx in these subjects.

INTRODUCTION

Exposure to ambient ozone is a recognized public health hazard. People at increased risk for the adverse effects of ozone likely include those with airway allergic diseases, such as extrinsic asthma.[1-3] Epidemiological studies reveal decreases in lung function on both asthmatics and nonasthmatics after exposure to increased levels of ambient ozone.[4,5] Molfino et al.[6] reported that allergic asthmatics had increased responses to inhaled allergen bronchoprovocation after exposure to 0.12 ppm ozone for one hour. Kriet and co-workers[7] found that both asthmatic and nonasthmatic subjects have increased respiratory symptoms, airflow obstruction, and bronchial responsiveness to methacholine following exposure to 0.4 ppm ozone. However, the effect of ozone was more severe in the asthmatic population. Thus, understanding the pathophysiology of ozone-induced changes in asthma is important so that treatments and strategies for minimizing this effect can be developed.

Airway inflammation with eosinophils is a cardinal feature of asthma. Controlled exposure of nonallergic and nonasthmatic subjects to ozone is associated with both decreased lung function and an influx of neutrophils as revealed with bronchoalveolar lavage, suggesting that ozone causes airway inflammation in humans.[8-10] Ozone also induces inflammatory changes in the upper airways of normal and allergic subjects.[11,12] Koren et al.[13] reported that mast cell tryptase was also increased in nasal lavage fluid obtained from nonallergic volunteers after ozone exposure, indicating that mast cell function may be effected by ozone. In studies of allergic rhinitics exposed to 0.5 ppm ozone for four hours, significant increases in the numbers of both neutrophils and eosinophils were noted after ozone exposure. Interestingly, cellular responses to allergen challenge were not augmented by ozone exposure. Nonetheless, these results indicate that ozone acts as a stimulus for eosinophil influx into nasal secretions.

Comparison of the effect of ozone on nasal and bronchial inflammation in normal subjects reveals that nasal inflammatory changes after ozone exposure mimic those that occur in the lower airways.[8,11,13] This correlation may be important because *in vivo* cellular and biochemical studies of the effect of ozone in allergic reactions can be more easily and safely studied in a nasal mucosa as opposed to studies of the bronchial mucosa. In this study, the effect of 0.4 ppm ozone on nasal inflammation in allergic asthmatics sensitive to *D. farinae* was examined. By employing a split nose design, in which allergen was applied to only one side of the nose while saline was applied to the contralateral side, the effect of ozone exposure on baseline and allergen-induced nasal inflammation of asthmatics was examined.

METHODS

Subjects

The protocol used in this study was reviewed and approved by both the Committee for the Protection of the Rights of Human Subjects and the General Clinical Research Center Advisory Committee of the University of North Carolina. Eleven mild asthmatics with skin test reactivity to *D. farinae* extract (Greer Laboratories) who were otherwise healthy nonsmoking males or females between the ages of 18 and 35 were recruited into this study. A battery of 18 antigens common to this area were employed, although only persons sensitive to the common house dust mite, *D. farinae,* were enrolled.

Ozone Exposure and Allergen Challenge Protocol

Each subject underwent two separate 2-h exposures to 0.4 ppm ozone and clean filtered air in a randomized double-blind fashion without exercise, followed by allergen challenge of one nostril as described below. Exposures were separated by a minimum of four weeks and performed at the U.S. EPA Clinical Studies Facility in Chapel Hill, NC. Baseline spirometry and separate nasal lavage of both the nostril to be challenged with allergen and the control nostril (lavage one) was performed prior to either clean air or ozone exposure. Immediately after exposure, nasal lavage fluid was again obtained on the right and left nares (lavage two). Administration of increasing doses of allergen (10, 100, 1000, and 10,000 AU, Greer Laboratories, Lenoir, NC) was then performed in 15-min intervals (on one side of the nose only) until nasal symptom scores exceeded 5 (based on subjective scoring of rhinorrhea, sneezing [or itching], and congestion using the following scale: no symptoms = 0, trace symptoms = 0.5, mild symptoms = 1, moderate symptoms = 2 and severe symptoms = 3, maximal score = 9) or 10,000 AU of *D. farinae* extract had been given. The control nostril was challenged with saline. Four hours after the allergen challenge, a third bilateral nasal lavage was obtained (lavage three). The next morning, subjects returned for a fourth nasal lavage. Thus, nasal lavage fluid (NLF) from both the allergen and control nostril was obtained at baseline, immediately prior to allergen challenge (which occurred immediately after the chamber exposure), 4 and 18 h after allergen challenge.

Nasal Lavage Technique

As stated above, a split nose design was employed for this study in which the nostril which was challenged with allergen and the contralateral (control) nostril were lavaged independently of one another. This was done so that the intrinsic effect of ozone on nasal inflammation (the effect of ozone exposure on nasal inflammation without allergen challenge) as well as the priming effect of ozone on allergen-induced inflammation could be examined simultaneously in each subject. Briefly, 4 ml of saline was sprayed into the nares using a hand-held nebulizer which delivers 100 μl per actuation.[15] Each lavage was comprised of eight sets of five sprays. Lavage fluid was recovered by forceful expulsion of lavage fluid into a specimen cup immediately after each set of five actuations. Samples were then transported on ice to the laboratory where slides for differential cell counts of uncentrifuged NLF were made.

Analysis of Cells in NLF

Cell viability was assessed by tryptan blue exclusion and the number of total cells was determined from unprocessed NLF using a hemocytometer. A 0.4 ml aliquot of nasal lavage fluid was cytocentrifuged onto a microscope slide and stained with a modified Wright's stain (Diff-Quik™, Difco, Inc.). Differential counts were than made on a minimum of 300 cells and the percentage of the total cell count comprised of eosinophils (Eos), neutrophils (PMNs), and epithelial cells was calculated and expressed as cell number/ml NLF.

Statistics

Comparison of the amount of allergen required to elicit nasal symptoms after clean air exposure to that after ozone exposure was carried out as a paired analysis using the Wilcoxon signed rank statistic.[16] Exact p values for the rank sum statistic were calculated rather than employing standard tables for these values. Analysis of the effect of ozone on cellular influx was performed by examining the difference between a given post challenge timepoint and baseline values for eosinophil and neutrophil numbers obtained during ozone exposure sessions vs. differences noted during the clean air session using the formulas listed below. Wilcoxon's signed rank statistic was used in a one-sided test. All quantities were $\log_{10}$ transformed prior to analysis. To determine the validity of the split nose design, the effect of allergen was examined using a similar comparison of baseline and post allergen or post saline challenge values for specific components of NLF using the formula listed below.

Formula for examination of the effect of ozone alone at a given timepoint:

$$\Sigma(X_{yc} - X_{bc})_{ozone} - \Sigma(X_{yc} - X_{bc})_{air} = 0.0 \text{ (null hypothesis)}$$

Formula for examination of the priming effect of ozone on response to allergen:

$$\Sigma(X_{ya} - X_{ba})_{ozone} - \Sigma(X_{ya} - X_{ba})_{air} = 0.0 \text{ (null hypothesis)}$$

Formula for examination of the response to allergen without ozone (allergen effect):

$$\Sigma(X_{ya} - X_{ba})_{air} - \Sigma(X_{yc} - X_{bc})_{air} = 0.0 \text{ (null hypothesis)}$$

X = given nasal lavage parameter for a given subject, b = baseline lavage (lavage 1) y = lavage for time point of interest (lavage 2, 3, or 4, see above), a = allergen nostril, c = control nostril.

RESULTS

Confirmation of the Split Nose Design

Eleven allergic asthmatics were enrolled into this study, with ten of the eleven completing both (clean air and ozone) exposure protocols. Figure 1 depicts the comparison of eosinophil levels in NLF in all four lavages obtained from the allergen and saline challenged nostrils following exposure to clean air. Eosinophil numbers are significantly enhanced by allergen both 4 and 18 h after allergen challenge (Figure 1A). These data demonstrate a clear segregation of the effect of allergen from the effect of a control challenge in the nasal airway, minimizing the likelihood that allergen induced an effect on the contralateral (saline) nostril.

The Effect of Ozone on Allergen-Induced Nasal Inflammation

As shown in Figure 2, the mean quantity of *D. farinae* extract required to induce a nasal symptom score of ≥ 5 was 493 ± 154 AU after clean air exposure vs. 154 ± 95 AU after exposure 0.4 ppm ozone for 2 h at rest. Although not reaching significance ($p = 0.09$), sensitivity to allergen appears to be somewhat increased after ozone exposure. Figure 3A depicts the number of PMNs recovered after allergen challenge preceded by both ozone and clean air exposure, demonstrating a significant increase in PMN influx following ozone exposure 18 h after allergen challenge ($p = 0.024$). Displayed in Figure 3B is the actual number of Eos/ml NLF obtained in lavages 1 through 4. Eosinophil influx to the nasal airway due to allergen challenge was somewhat increased after ozone exposure, though this increase did not reach statistical significance.

The Intrinsic Effect of Ozone on Nasal Inflammation

Figure 4 depicts the absolute cell count in NLF for both PMNs (Figure 4A) and eosinophils (Figure 4B) obtained in lavages 1 through 4 in the control nostril. Ozone exposure increased both PMN and eosinophil recovery (compared to baseline values) 4 and 18 h after exposure (lavages 3 and 4) vs. that after clean air exposure.

DISCUSSION

In this preliminary study, the effect of exposure to 0.4 ppm ozone for 2 h without exercise on both allergen-induced and baseline nasal inflammation of the nasal mucosa of asthmatics with documented allergy to a common allergen (house dust mite, *D. farinae*) was investigated. This level and duration of ozone exposure was selected because it is the same as that which has induced neutrophil influx into the nasal tissues of normal subjects. Each subject underwent ozone and clean air exposures which were followed by unilateral administration of allergen (with increasing doses of allergen ranging from 10 to 10,000 allergen units) to the nasal mucosa accompanied by saline challenge to the contralateral nostril. This approach allowed for examination of the effect of ozone on baseline nasal inflammation (in the saline nostril) as well as the effect of ozone on allergen-induced inflammation.

A

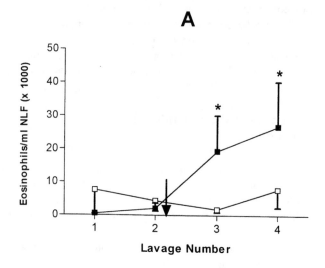

B

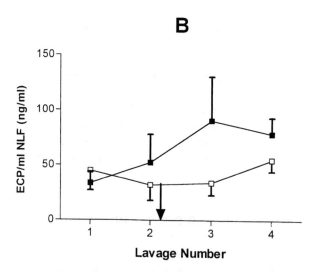

Figure 1   Demonstration of the split nose allergen challenge design. (A) Depicts the number of eosinophils (cells × 1000/ml NLF) recovered in NLF from paired lavages of the allergen challenged (open squares) and control nostrils (solid squares) of allergic asthmatics (n = 10) immediately before (lavage 1) and after (lavage 2) clean air exposure as well as 4 (lavage 3) and 18 (lavage 4) h after subsequent allergen challenge. (B) Depicts the concentration of eosiniophil cationic protein (ECP; ng/ml) of these NLF samples. Allergen challenge results in significant increases in eosinophils 4 and 18 h after allergen challenge only in the allergen-challenged nostril. Changes in ECP concentrations had a similar trend. Shown are mean values ±SEM, $p < 0.05$ is indicated by an asterisk. (From Peden, D.B. et al., *Am. J. Respir. Crit. Care Med.*, 1995, 151(5), 1336-1345. With permission.)

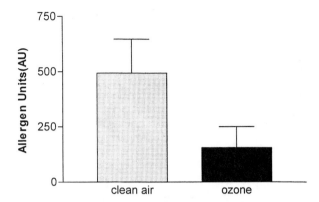

Figure 2   Effect of 0.4 ppm ozone exposure on dose of *D. farinae* allergen extract required to induce nasal symptoms. Allergen dose is expressed as allergen units (AU). The mean dose (n = 10) required after ozone exposure is depicted with an open bar and the mean dose required after clean air exposure is shown with a closed bar. Error bars represent SEM. (From Peden, D.B. et al., *Am. J. Respir. Crit. Care Med.*, 1995, 151(5), 1336-1345. With permission.)

NLF was collected from each nostril prior to exposure to establish the baseline inflammatory state on the exposure/challenge day using a hand-held nebulizer to deliver 4 ml of a gentle saline spray to each nostril. NLF was also collected immediately after the chamber (clean air or ozone) exposure, as well as 4 and 18 h after nasal allergen challenge. NLF was assayed for eosinophil and neutrophil numbers. Ten subjects completed exposures to both clean air and ozone. The mean amount of allergen required to induce symptoms after clean air exposure was 493 ± 154 AU vs. 154 ± 95 AU after exposure to ozone ($p = 0.09$). Despite not reaching significance, this trend suggests that ozone may increase airway responses to allergen, as has been reported by other investigators whom have demonstrated that ozone exposures of 0.12 at rest for 1 h[6] and 0.30 ppm with moderate exercise for 3 h[18] enhance bronchial reactivity of seasonal asthmatics to inhaled allergen.

Examination of NLF from the allergen-challenged nostril revealed that ozone exposure increases subsequent cellular responses to allergen as well. A nonsignificant trend for increased eosinophil numbers was found in NLF 4 h after allergen challenge. Ozone exposure also had an effect on PMNs recovered in NLF after allergen challenge, with a trend for increases in PMNs being noted 18 h after challenge. Thus, ozone exposure appears to enhance airway responses to allergen in allergic asthmatics. It should be noted that ozone appears to increase allergen-induced inflammation even though less allergen was given to many of the asthmatics. If equal doses of allergen had been given to each subject, it is likely that an even more vigorous priming effect of ozone would have been observed.

Ozone also had an effect on nasal inflammation which was independent of allergen. Like observations made in allergic rhinitics,[14] ozone exposure alone resulted in significant increases in eosinophil influx over baseline levels 4 h after

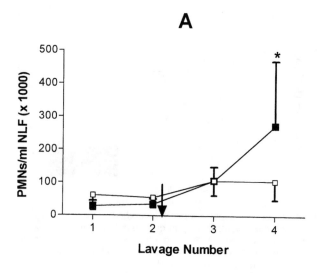

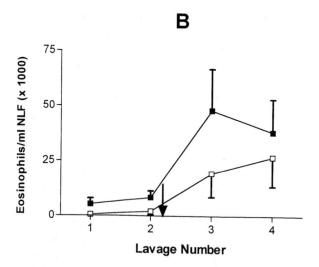

Figure 3   Effect of 0.4 ppm ozone exposure on allergen-induced inflammatory cell influx into the nasal airway of *D. farinae* sensitive asthmatics. (A) Depicts PMN recovery in NLF from the allergen nostril of allergic asthmatics (n = 10) immediately before (lavage 1) and after (lavage 2) clean air (closed squares) or ozone (open squares) exposure as well as 4 (lavage 3) and 18 (lavage 4) h after subsequent allergen challenge. (B) Depicts similar analysis of eosinophil recovery in NLF from the allergen-challenged nostrils. Values shown represent mean ±SEM. Significance (*p* <0.05) is noted by an asterisk and represents difference in cell influx with clean air vs. ozone exposure (see Methods). (From Peden, D.B. et al., *Am. J. Respir. Crit. Care Med.*, 1995, 151(5), 1336-1345. With permission.)

A

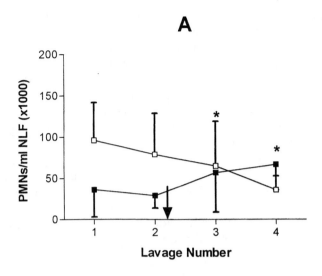

B

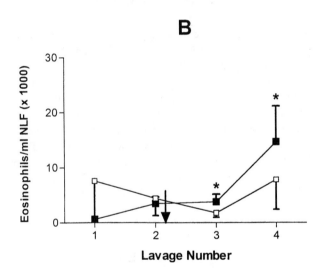

Figure 4 Intrinsic effect of 0.4 ppm ozone exposure on inflammatory cell influx into the control nostril of *D. farinae* sensitive asthmatics. (A) Depicts PMN recovery in NLF from the allergen nostril of allergic asthmatics (n = 10) immediately before (lavage 1) and after (lavage 2) clean air (closed squares) or ozone (open squares) exposure as well as 4 (lavage 3) and 18 (lavage 4) h after subsequent allergen challenge. (B) Depicts similar analysis of eosinophil recovery in NLF from the allergen-challenged nostrils. Values shown represent mean ±SEM. Significance ($p < 0.05$) is noted by an asterisk and represents difference in cell influx associated with clean air vs. ozone exposure (see Methods). (From Peden, D.B. et al., *Am. J. Respir. Crit. Care Med.*, 1995, 151(5), 1336-1345. With permission.)

ozone exposure was complete. As observed in studies of nonallergic subjects and allergic rhinitics,[11,14] ozone induces an influx of PMNs to the nasal mucosa. While this effect was not as striking in allergic asthmatics as it was in normal subjects, the asthmatics were exposed at rest whereas the normals were subjected to heavy exercise, likely increasing the amount of ozone to which the nasal mucosa was exposed. However, normal subjects did not have an ozone-induced increase in eosinophils to the nasal mucosa as do both allergic rhinitics studied by Bascom et al.[14] and the allergic asthmatics examined in this report, indicating that ozone induces an eosinophil influx to the nasal mucosa only in allergic subjects.

In summary, nasal inflammation of allergic asthmatics is augmented by ozone. Allergen-induced eosinophilic inflammation also appears to be augmented by ozone exposure. Additionally, ozone also caused an influx of both eosinophils and neutrophils into the nasal airway not due to allergen challenge. If ozone exerts similar changes in the lower airway, then chronic or repeated exposure to ozone is likely to exert a harmful effect on lung function of asthmatics.

ACKNOWLEDGMENTS

The authors would like to acknowledge the technical assistance of Debra Levin, R.N., Maryann Bassett, R.N., and Joleen Soukup of TRC Environmental Inc., Chapel Hill, NC, as well as Rhonda Rocker, R.N., of the General Clinical Research Center of the University of North Carolina. We would also like to thank Hillel Koren, Ph.D., Director of the Human Studies Division of the U.S. EPA and Philip A. Bromberg, M.D., Director of the Center for Environmental Medicine and Lung Biology, UNC School of Medicine for their expert advice.

These studies were supported by funds provided by U.S. EPA Cooperative Agreement CR817643 administered through the Center for Environmental Medicine and Lung Biology and NIH grant 5-MO1-RR00046 administered through the General Clinical Research Center of the School of Medicine of the University of North Carolina.

REFERENCES

1. Paulo M, Gong H. Respiratory effects of ozone: whom to protect, when and how? *J. Respir. Dis.* 1991; 12:482-499.
2. Lippman M. Health effects of ozone: a critical review. *J. Air Pollut. Control Assoc.* 1989; 39:672-695.
3. Weiss KB, Wagener DK. Changing Patterns of Asthma Mortality, *JAMA,* October 3, 1990, Vol. 264 No. 13.
4. Bates DV, Sizto R. Air pollution and hospital admissions in Southern Ontario: the acid summer haze effect. *Environ. Res.* 1987; 43:317-331.
5. Lioy PJ, Vollmuth TA, Lippmann M. Persistence of peak flow decrement in children following ozone exposures exceeding the national ambient air quality standard. *J. Air Pollut. Control Assoc.* 1985; 35:1068-1071.

6. Molfino NA, Wright SC, Katz I, Tarlo S, Silverman F, McClean PA, Szalai JP, Raizenne M, Slutsky AS, Zamel N. Effect of low concentrations of ozone on inhaled allergen responses in asthmatic subjects. *Lancet* 1991; 338:199-203.

7. Kreit JW, Gross KB, Moore T, Lorenzen TJ, D'Arcy J, Eschenbacher WL. Ozone-induced changes in pulmonary function and bronchial responsiveness in asthmatics. *J. Appl. Physiol.* 1989; 66:217-222.

8. Koren HS, Devlin RB, Graham DE, Mann R, McGee MP, Horstman DH, Kozumbo WJ, Becker S, House DE, McDonnell WF, Bromberg PA. Ozone-induced inflammation in the lower airways of humans. *Am. Rev. Respir. Dis.* 1989; 139:407-415.

9. Devlin RB, McDonnell WF, Mann R, Becker S, House DE, Schreinemachers D, Koren HS. Exposure of humans to ambient levels of ozone for 6.6 hours causes cellular and biochemical changes in the lung. *Am. J. Respir. Cell Mol. Biol.* 1991; 4:72-81.

10. Folinsbee LJ, McDonnell WF, Horstman DH. Pulmonary function and symptom responses after 6.6 hour exposure to 0.12 ppm ozone exposure. *J. Air Pollut. Control Assoc.* 1988; 38:28-35.

11. Graham DE, Koren HS. Biomarkers of inflammation in ozone-exposed humans. Comparison of the nasal and bronchoalveolar lavage. *Am. Rev. Respir. Dis.* 1990; 142:152-161.

12. Frischer TM, Kuehr J, Pullwitt A, Meinert R, Forster J, Studnicka M, Koren H. Ambient ozone causes upper airways inflammation in children. *Am. Rev. Respir. Dis.* 1993; 148:961-964.

13. Koren HS, Hatch GE, Graham DE. Nasal lavage as a tool in assessing acute inflammation in response to inhaled pollutants. *Toxicology* 1990; 60:15-25.

14. Bascom R, Naclerio RM, Fitzgerald TK, Kagey-Sobota A, Proud D. Effect of ozone inhalation on the response to nasal challenge with antigen of allergic subjects. *Am. Rev. Respir. Dis.* 1990; 142:594-601.

15. Raphael GD, Druce HM, Baraniuk JN, Kaliner MA. Pathophysiology of rhinitis-1. Assessment of the sources of protein in methacholine-induced nasal secretions. *Am. Rev. Respir. Dis.* 1988; 138:414-420.

16. Hettmansperger, TP. *Statistical Inference Based on Ranks.* New York: Wiley Press 1984.

17. Platts-Mills TAE, de Weck AL. Dust mite allergens and asthma — a world wide problem: report of an international workshop, Bad Kreuznach FRG. *J. Allergy Clin. Immunol.* 1988; 83:416-427.

18. Jorres R, Nowak D, Magnussen H. Effects of ozone exposure on airway responsiveness to inhaled allergens in subjects with allergic asthma or rhinitis. *Am. J. Respir. Crit. Care Med.* 1994; 149:A154 (abstract).

CHAPTER 10

Health Effects
of Biological Contaminants

Harriet A. Burge

INTRODUCTION

Since the last Oak Ridge National Laboratory symposium on indoor air, bioaerosols have become widely recognized as important components of indoor pollutant aerosols. However, with a few notable exceptions, little progress has been made in the identification of important bioaerosol-related disease agents, characterization of dose-response relationships, or the development of methods for studying bioaerosols. Most of the information on bioaerosols remains in proceedings rather than being a part of the standard peer-reviewed literature, and field investigations aimed at solving individual problems remain a significant challenge.

This brief review will focus on areas where significant progress has been made and published in the peer-reviewed journal literature, on especially important or exciting ideas from other sources, and on an approach to risk assessment that uses currently available data on the health effects of toxic bioaerosols.

DISEASES RELATED TO BIOAEROSOLS

Infections

Infectious disease remains the most important of the bioaerosol-related diseases from a public health perspective, and has recently been the subject of intense public scrutiny. Notable outbreaks of airborne infectious disease have occurred in India (pneumonic plague) (Anon., 1994), and the Southwestern U.S. (Hantavirus respiratory syndrome) (Farr, 1994). Tuberculosis has reemerged in a multiple drug-resistant form (Neville et al., 1994), and outbreaks have occurred in prisons

(Valway et al., 1994), shelters for the homeless (Nardell, 1989), and on commercial aircraft (Hickman et al., 1995). Emphasizing the current public interest in airborne infectious disease is the popular film "Outbreak" in which Dustin Hoffman plays the part of an aerobiologist.

Research studies have clearly documented that some forms of the common cold are consistently and most effectively transmitted via the airborne route (Dick et al., 1987). The utility of dilution ventilation for the prevention of upper respiratory infections has been suggested in a small epidemiologic study (Brundage et al., 1988), and the limitations on ventilation-based control of tuberculosis have been elegantly calculated (Riley and Nardell, 1993).

Hypersensitivity Diseases

The hypersensitivity diseases, particularly asthma, are currently receiving a great deal of attention, and new information is becoming available on the role of many bioaerosols in both the initial sensitization step and on symptom development. It is now clear that dust mite and cockroach allergens play a major role in the development of asthma (Platts-Mills and de Weck, 1989; Morris et al., 1986). On the other hand, evidence is increasing that sensitization to other allergens (pollen, fungal animal dander allergens) is a significant risk factor for symptom development in both adults and children (Beaumont et al., 1985; Gergen et al., 1987; Licorish et al., 1985; Newill et al., 1995).

Allergic alveolitis (hypersensitivity pneumonitis) has long been recognized as a bioaerosol-associated disease. New studies have explored the role of glucans and endotoxins in both this disease and in organic dust toxic syndrome (Fogelmark et al., 1994) and patterns of exposure causing each of these diseases (Malmberg et al., 1993).

Toxicoses

The most dramatic progress that has been made with respect to toxin exposure in indoor air has been the accumulation of evidence supporting a causal relationship between exposure to endotoxin and symptoms of the sick building syndrome (Teeuw et al., 1994). An excellent review of the health effects of endotoxin is provided in Chapter 11 of this volume.

Much attention has been focused on glucans, toxic components of fungal cell walls that may have toxic properties similar to those of endotoxin, as agents of sick building syndrome (Rylander et al., 1992), although the evidence for this association remains weak. On the other hand, as mentioned above, glucans may be involved in organic dust toxic syndrome which has been the focus of intense interest in the occupational health community (Fogelmark et al., 1994).

The role of other toxic metabolites of the fungi (mycotoxins) in indoor air remains controversial (Flannagan, 1987). These toxins are unequivocally dangerous, and the literature on their toxicity by ingestion is large (Betina, 1989). The aflatoxins probably do cause cancer by inhalation in heavily exposed occupational groups (peanut handlers [Sorenson et al., 1984], farmers harvesting and/or otherwise handling moldy corn [Sorenson et al., 1981]). The risk for such effects has been calculated by Baxter et al. (1981).

Toxins produced by *Stachybotrys atra (chartarum)* have been the focus of great controversy with respect to indoor air exposures, stimulated by the report of a case of poisoning in a Chicago home (Croft et al., 1986). Toxin content has been measured in spores (Sorenson et al., 1987), and from growth on building materials (Nikulin et al., 1994), and the effects of satratoxin on protein synthesis in rat alveolar macrophages and proliferation of mouse thymocytes has been examined (Sorenson et al., 1987). These data have been sufficient to stimulate the development of rules for controlling exposure to the spores of this fungus in New York State (Anon., 1993).

Volatile organic compounds produced by fungi and bacteria have been anecdotally linked to cases of building-related discomfort, irritation, and possibly illness, although no convincing evidence has yet been gathered to support these connections. Additional work has been done to document the kinds of volatile compounds released by the fungi (Batterman et al., 1991).

KINDS OF EVIDENCE USED TO ASSOCIATE EXPOSURE AND DISEASE

Part of the controversy that has arisen regarding the relative importance of different bioaerosols is rooted in the casual acceptance of anecdotal information as "proof." Ideally, one would like to see statistically relevant proof that exposure to a particular agent (biological or otherwise) actually causes disease in human populations. Such proof (or, for that matter, the ability to derive such proof) is generally not available for most bioaerosol-related diseases. However, several different kinds of evidence can provide strong support for agent/disease relationships. Note, by the way, that the words "no evidence" can mean either that the studies have not been done, or that the studies that have been done have not provided any evidence. In the second case, no evidence may mean either that the relationship does not exist or that the studies were inadequately designed or performed to provide evidence.

Experimental (What Can Be)

Experimental evidence demonstrates that a specific exposure *can* cause a disease under the experimental conditions. Whether or not this kind of evidence provides support for the actual exposure/disease relationships depends on the specific experimental design. For example, exposure of guinea pigs has demonstrated that aerosols of *Mycobacterium tuberculosis* can cause tuberculosis in guinea pigs, and that a single active cell can result in infection in this highly sensitive animal (Riley and Nardell, 1993). This proves that the agent can survive airborne transport and arrive in a virulent state within a susceptible animal. It does not prove that the primary means for transmission of disease to humans is through the airborne route, nor that a single agent will cause the disease. Likewise, the fact that satratoxin causes changes in macrophage protein synthesis (Sorenson et al., 1987) demonstrates that the toxin can cause such a change at the specific dose in the isolated cells, but only suggests that such effects might also occur in humans.

Therefore, unless human subjects are used, and experiments are very carefully designed, experimental evidence only demonstrates "what can be" not "what is." For example, it has been demonstrated that the hepatitis B virus will cause aerosol-transmitted infections under experimental or unusual laboratory conditions (Almeida et al., 1971), but the disease is rarely, if ever, transmitted by air under natural conditions because the disease does not naturally cause aerosols to be produced. On the other hand, using human subjects, Dick et al. (1987) elegantly demonstrated experimentally that some common cold viruses are not only able to be transmitted via air, but that air is the most likely means of transmission.

Epidemiologic (What Is)

Epidemiologists demonstrate that a specific agent is statistically likely to be causing a particular disease, connecting hypothetical or experimental exposure/disease relationships to human subjects. Thus, epidemiological evidence can be used to document that dust mite allergens play a major role in asthma (Platts-Mills and de Weck, 1989), and that endotoxin plays a prominent role in sick building syndrome (Teeuw et al., 1994).

On the other hand, epidemiologic studies have failed to document the commonly held opinion that total fungal bioburden in houses is related to either the incidence of asthma or symptoms of asthma. This absence of corroboration of "common knowledge" may mean that the studies were not designed properly, or that common knowledge is wrong.

Note also that epidemiological studies *have not been conducted* that explore connections between *S. atra* on surfaces and any respiratory disease. This doesn't mean that there is no connection, only that the studies have not been done.

Extrapolation (What Might Be)

Extrapolation is essentially the process of guessing on good evidence. With respect to aerobiology, this means the use of evidence from one part of the aerobiology equation to infer an effect at another level. When the level to be inferred is the human health impact level, this process can be called risk assessment.

The Aerobiology Equation

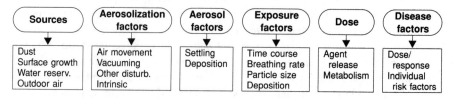

For tuberculosis, there is some data that can be used for each of the factors in this equation, and calculation of the risk of disease is not only possible, but relatively accurate (Riley and Nardell, 1993). For asthma related to dust mite allergens, the sources are relatively well characterized. However, aerosolization, aerosol, exposure, and dose factors remain hypothetical, and risk assessments make a giant leap from source strengths to the disease state, with "threshold limit values" related to content of dust, not to actual exposure or dose response (Platts-Mills and de Weck, 1989). On the other hand, epidemiologic data has confirmed this relationship, so that, although we don't know actually how much dust mite allergen must be inhaled to cause disease, we do know that there is a statistical association between the amount of allergen in dust and the odds of having the disease.

For the satratoxins, as for dust mite, we know the source, we have some idea about how much agent is present/aerosol particle (although aerosolization terms remain unknown), and we have some noninhalation human data and some animal data on dose/response relationships. Based on these pieces of data, we can develop a hypothetical risk assessment model that might tell us how many aerosol particles (e.g., spores) might be necessary to cause disease (see below). Another approach would be to perform epidemiologic studies that would relate some disease state to either an aerosol or a source measure. Neither of these approaches has yet been taken, and so we have no good evidence (other than anecdotal) that satratoxin causes human disease by inhalation.

Risk Assessment Models

Although a number of assumptions are necessary, it should be possible to develop a risk model for acute toxic effects due to exposure to *S. atra* spores. We have the following pieces of data:

S. atra spores are approximately 8 to 11×5 to 10 μm. We can then construct an average spore that is 9.5×7.5 μm. Using the formula for the volume of a prolate spheroid, $4/3(\pi ab^2)$, to approximate the volume of a spore, where a is half of the long axis and b is half of the short axis, then each spore is approximately 2.8×10^{-10} cm^3 (or g), or 2.8×10^{-7} mg/spore.

Using data from Sorenson et al. (1987) dust with 85% spores contains 9.5 ng Satratoxin H/mg of dust, or 11.5 ng satratoxin H/mg spores. Therefore,

$$\frac{2.8 \times 10^{-7} \text{ mg}}{\text{spore}} \times \frac{11.5 \text{ ng toxin}}{\text{mg}} = 3.1 \times 10^{-6} \text{ ng toxin/spore}$$

At this point, a number of assumptions must be made, since actual data is not available. The following are worst-case assumptions:

1. All inhaled spores are deposited somewhere in the respiratory tract, and each spore releases all of its toxin.
2. All of the toxin released is effective (i.e., none is metabolized to an inactive state before having its toxic effect).
3. A cumulative dose of 1 ng of toxin is of concern (this is an extremely conservative estimate).

If we assume that a person inhales about 1 m³ of air in 8 h, we can then construct the following:

Table 1   Risk Model for Acute Toxic Effects from *S. atra* Spore Exposure

If you have	You accumulate	And the time it takes to accumulate 1 ng toxin is
100 spores/m³	3.1×10^{-4} ng toxin/8 h	1100 days
1000 spores/m³	3.1×10^{-3} ng toxin/8 h	110 days
10000 spores/m³	3.1×10^{-2} ng toxin/8 h	11 days
100000 spores/m³	0.31 ng toxin/8 h	1.1 days
1000000 spores/m³	3.1 ng toxin/8 h	0.1 days

Remember that these are conservative estimates based on several worst case assumptions. On the other hand, other toxins are probably simultaneously present that are not taken into consideration. This exercise is meant only to serve as a model for one method that could be used to establish realistic guidelines for exposure to specific biological agents. The exercise also makes clear exactly the information that is missing in order to develop a truly realistic model. In this case, the nature of the inhalation exposure/dose/response relationship for toxins contained in *S. atra* spores remains unknown.

Note also that this model requires measurement of *airborne* spore concentrations. We have not established source/aerosolization/aerosol terms. To do this would require developing additional models that describe the relationship between levels of toxin-containing particles in sources (all of those in a particular environment) and levels of these particles in air. We are currently collecting data aimed at developing such models for dust-borne toxigenic fungi.

Best Guesses

The fact remains that, until we do have additional data for many bioaerosols that allow risk assessment, we must operate on best guesses based on anecdotal information. These best guesses are often based on worst-case scenarios by investigators who are concerned for the health of building occupants. However, unless one can balance all of the factors in a particular environment, it is difficult to know where to draw the line. There are nonexposure related factors (i.e., the psychological damage caused by overreacting, economic damage to owners and subsequently to occupants, etc.) that can outweigh the dangers of the actual exposures. In addition, the assumption that an obvious agent is causing disease in the absence of proof may blind the investigator to the real cause.

Taking all of these factors into consideration requires more individual knowledge than any of us have. Indoor air investigations that include bioaerosols are necessarily team efforts that require large amounts of intuition, common sense, patience, and compassion.

REFERENCES

Almedia JD, Chisholm GD, Kulatilake AE, MacGregor AB, MacKay DH, O'Donogue EPN, Shackman R, Waterson AP. 1971. Possible airborne spread of serum-hepatitis virus within a haemodialysis unit. *Lancet* 2, 849.

Anonymous. 1993. Guidelines on Assessment and Remediation of *Stachybotrys atra* in indoor environments. Based on a panel discussion, May 7, 1993, District Council 37 AFSCME, 125 Barclay Place New York.

Anonymous. 1994. Human plague — India, 1994. *MMWR* 43(38):689-91.

Batterman S, Bartoletta N, Burge H. 1991. Fungal volatiles of potential relevance to indoor air quality, Paper 91-62.9, presented at the Annual Meeting of the Air and Waste Management Association, Vancouver, Canada, June 1991.

Baxter CS, Wey HE, Burg WR. 1981. A prospective analysis of the potential risk associated with inhalation of aflatoxin-contaminated grain dusts. *Food Cosmet. Toxicol.* 19:763-769.

Beaumont F, Kauffman HF, Eluitor HJ, DeVries K. 1985. Sequential sampling of fungal air spores inside and outside the homes of mould-sensitive, asthmatic patients: a search for a relationship to obstructive reactions. *Ann. Allergy* 55(5):740-6.

Betina V. 1989. Mycotoxins: Chemical, Biological, and Environmental Aspects. Elsevier, Amsterdam.

Brundage JF, Scott RM, Lednar WM, Smith DW, Miller RN. 1988. Building associated risk of febrile acute respiratory diseases in army trainees. *JAMA* 259(14):2108.

Creasia DA, Thurman JD, Wannemacher RW, Bunner DL. 1990. Acute inhalation toxicity of T-2 mycotoxin in the rat and guinea pig. *Fund. Appl. Toxicol.* 14:54.

Croft WA, Jarvis BB, Yatawara CS. 1986. Airborne outbreak of trichothecene toxicosis. *Atmos. Environ.* 20:549-552.

Dick EC, Jennings LC, Mink KA, Wartgow CP, Inborn SL. 1987. Aerosol transmission of rhinovirus colds. *J. Infect. Dis.* 156:442.

Farr RW. 1994. Hantavirus pulmonary syndrome. *West Virginia Med. J.* 90(10):422-5.

Flannagan B. 1987. Mycotoxins in the air. *Int. Biodeterior.* 23:73-78.

Fogelmark B, Sjostrand M, Rylander R. 1994. Pulmonary inflammation induced by repeated inhalations of beta(1,3)-D-glucan and endotoxin. *Intern. J. Exp. Pathol.* 75(2):85-80.

Gergen PJ, Turkeltaub PC, Kovar MG. 1987. The prevalence of allergic skin test reactivity to eight common aeroallergens in the US population: results from the second National Health and Nutrition Examination Survey. *J. Allergy Clin. Immunol.* 80:669.

Hickman C, MacDonald KL, Osterhoolm MT, et al. 1995. Exposure of passengers and flight crew to *Mycobacterium tuberculosis* on commercial aircraft, 1992–1995. *MMWR* 44(8):137-139.

Licorish K, Novey HS, Kozak P, Fairshter RD, Wilson AF. 1985. Role of Alternaria and Penicillium spores in the pathogenesis of asthma. *J. Allergy Clin. Immunol.* 76(6):819-25.

Malmberg P, Rask–Andersen A, Rosenhall L. 1993. Exposure to microorganisms associated with allergic alveolitis and febrile reactions to mold dust in farmers. *Chest* 103(4):1202-9.

Monto AS. 1987. Influenza: quantifying morbidity and mortality. *Am. J. Med.* 829 (Suppl. 6A), 20.

Morris EC, Smith TF, Kelly LB. 1986. Cockroach is a significant allergen for inner city children (abstract), J. Allergy Clin. Immunol. 77:206.

Nardell EA. 1989. Tuberculosis in homeless, residential care facilities, prisons, nursing homes, and other close communities. *Semin. Respir. Infect.* 4:206.

Neville K, Bromberg A, Bromberg R, Bonk S, Hanna BA, Rom WN. 1994. The third epidemic — multidrug-resistant tuberculosis. *Chest* 105(1):45-8.

Newill CA, Eggleston PA, Prenger VL, Fish JE, Diamond EL, Wei Q, Evans R III. 1995. Prospective study of occupational asthma to laboratory animal allergens: stability of airway responsiveness to methacholine challenge for one year. *J. Allergy Clin. Immunol.* 95(3):707-715.

Nikulin M, Pasanen AL, Berg S, Hintikka EL. 1994. *Stachybotrys atra* growth and toxin production in some building materials and fodder under different relative humidities. *Appl. Environ. Microbiol.* 60(9):3421-3424.

Nolan CM, Elarth AM, Barr H, Saeed AM, Risser DR. 1991. An outbreak of tuberculosis in a shelter for homeless men. A description of its evolution and control. *Am. Rev. Respir. Dis.* 143(2):257-61.

Pasanen A-L, Nikulin M, Tuomainen M, Berg S, Parikka P, Hintikka E-L. 1993. Laboratory experiments on membrane filter sampling of airborne mycotoxins produced by *Stachybotrys atra* Corda. *Atmos. Environ.* 27A:9-13.

Platts-Mills TAE, de Weck AL. 1989. Dust mite allergens and asthma — A world wide problem. *J. Allergy Clin. Immunol.* 83:416.

Riley RL, Nardell EA. 1993. Controlling transmission of tuberculosis in health care facilities: ventilation, filtration, and ultraviolet air disinfection. *PTSM,* Series No 1, 25.

Rylander R, Persson K, Goto H, Yuasa K, Tanaka S. 1992. Airborne beta-1,3-glucan may be related to symptoms in sick buildings. *Indoor Environ.* 1:263-267.

Sorenson WG, Frazer DG, Jarvis BB, Simpson J, Robinson VA. 1987. Trichothecene mycotoxins in aerosolized conidia of *Stachybotrys atra. Appl. Environ. Microbiol.* 53(6):1370-5.

Sorenson WG, Jones W, Simpson J, Davidson JI. 1984. Aflatoxin in respirable airborne peanut dust. *J. Toxicol. Environ. Health* 14:525-533.

Sorenson WG, Simpson JP, Peach MJ III, Thedill TD, Olenchock SA. 1981. Aflatoxin in respirable corn dust particles. *J. Toxicol. Environ. Health* 7:669-672.

Teeuw KB, Vandenbrouckegrauls CMJE, Verhoef J. 1994. Airborne Gram-negative bacteria and endotoxin in sick building syndrome — a study in Dutch governmental office buildings. *Arch. Intern. Med.* 154(20):2339-2345.

Valway SE, Richards SB, Kovacovich, J, Greifinger RB, Crawford JT, Dooley SW. 1994. Outbreak of multi-drug resistant tuberculosis in a New York State prison. 1991. *Amer. J. Epidemiol.* 140(2):113-22.

CHAPTER 11

Bacterial Endotoxins: A Review of Health Effects and Potential Impact in the Indoor Environment

Donald K. Milton

INTRODUCTION

Endotoxin has fascinated generations of biomedical investigators and is still an active subject of research today. Most of the fascination grew out of interest in infection and septic shock. However, endotoxin is ubiquitous in the environment, and we are constantly inhaling airborne endotoxin. This paper will present an overview of endotoxin biology and health effects with a goal of understanding the potential impact of inhaled endotoxin in the indoor environment. Key to this understanding is that small doses (ng/kg) can produce significant responses. It is also important to recognize that endotoxin biology is characterized by heterogeneity at every level including chemical structure, cellular receptors, cell types bearing receptors, the patterns of cellular responses, and the associated health effects. Finally, the endotoxin response of a single animal will be heterogeneous over time depending on its recent history of endotoxin exposure. This heterogeneity is a source of the fascination and of great frustration for those who have tried to study endotoxin.

GRAM-NEGATIVE BACTERIAL ENDOTOXIN

Endotoxin is not a single molecule but rather a class of biological molecules with certain characteristic toxic effects. Chemically, endotoxins are composed of lipopolysaccharides (LPS) from the outer membrane of Gram-negative bacteria (GNB). LPS is an integral part of the outer-membrane and is released by disruption of the cell membrane and by budding of the membrane during growth.[1] The lipid portion of LPS (lipid A) is unique among biological membrane lipids and is responsible for the molecule's characteristic toxicity. We may consider LPS as a class of molecules because bacterial species differ widely in the polysaccharide portion of the molecule and to a lesser extent in the composition of their lipid A. Even within one organism, molecules of LPS are somewhat heterogeneous.[2,3]

The outer part of the polysaccharide of LPS, the O-antigen, is the primary antigenic component of the Gram-negative cell surface. O-antigens have structural heterogeneity between species, within a species by serotype, and within a single organism by the number of repeating units of the antigenic polysaccharide. LPS are also heterogeneous in their core polysaccharide and lipid A components, although these are relatively constant within species. The structure of LPS has been recently reviewed.[4] The lipid A portion of the molecule is composed, in most cases, of a diglucosamine backbone with attached hydroxylated fatty acids and phosphate groups. Nonhydroxylated fatty acids may also be attached to the hydroxy fatty acids. Important differences in toxicity are associated with structural variation in lipid A.[5-8] Bacterial species may produce lipid A with greater or lesser acute toxicity due to the length, number, and arrangement of hydroxylated and nonhydroxylated fatty acids and of phosphate groups on the diglucosamine backbone of lipid A. Some lipid A structures are almost devoid of acute toxicity while retaining important immunological effects including adjuvant activity stimulating specific immune responses.[9]

Environmental endotoxin measurement was recently reviewed[10] and will not be discussed in detail here. Because of differences in methods for sample collection, extraction, and assay, it is not always possible to compare endotoxin measurements from published studies. Standard units of measurement can be problematic. Measurements made by bioassay methods predominate because of their sensitivity even though some may lack specificity. In this review endotoxin levels from the literature will be reported in the units used in the original. Where possible these will also be converted to endotoxin units (EU) with reference to the U.S. reference standard EC5 to facilitate comparison.

CELL BIOLOGY OF ENDOTOXIN

Endotoxin may be unique among airborne contaminants because it is recognized by more than one specific high-affinity receptor.[11] These receptors are present in both soluble and cell membrane forms. The highest affinity receptor is the macrophage (MΦ) cell surface protein CD14 which binds LPS that is bound to plasma LPS binding protein (LBP).[12] Binding of LPS-LBP complexes (at pg/ml concentrations) to MΦ CD14 stimulates tumor necrosis factor (TNFα) release. CD14 is also expressed by neutrophils which become adherent and secrete TNFα after binding LPS-LPB.[13-15] A soluble form of CD14, when bound to LPS-LPB complexes, binds to and activates both epithelial and endothelial cells.[16,17] A second component of plasma, septin, can also opsinize LPS and promote recognition by CD14.[18] In addition to CD14, two other cell surface proteins expressed by a wide range of cells including lymphocytes, neutrophils, and endothelial cells are known to specifically bind endotoxin[19,20] and may be important in responses to, and degradation of endotoxin.

Endotoxin has a variety of important cellular effects. The key cell responding to endotoxin appears to be the MΦ. Macrophages exposed to pg/ml concentrations of endotoxin produce TNFα, interleukins 1, 6, 8, 12, and platelet activating factor.[21-23] These potent mediators activate other immune effector cells. Structural differences among lipid A are now known to affect the pattern of cytokines elicited from macrophages by LPS and are associated with the differing levels of acute toxicity.[9,24] At birth circulating monocytes are already capable of responding to endotoxin in the characteristic manner, and endotoxin exposure enhances the ability of immature monocytes to express MHC II proteins and present antigen to T lymphocytes.[26,27] Later in life, endotoxins are potent adjuvants both for humoral and cellular immunity.[7,27] The ability of LPS to stimulate cell-mediated immunity is in part due to the ability of LPS to stimulate IL-12 production by macrophages resulting in IFN-γ production by NK cells and differentiation of T_H naive cells to committed T_H1 cells.[28,29] Endotoxin also promotes B-lymphocyte isotype switching to IgE production in the presence of IL-4.[30] Neutrophils and basophils exposed to endotoxin are primed for increased inflammatory responses.[31,32] Thus, endotoxin has a range of biological activities that may play important roles in lung health when it is inhaled.

A characteristic aspect of endotoxin biology is the phenomenon of endotoxin tolerance. Tolerance to endotoxin occurs approximately 24 h after exposure and lasts for about 72 h.[33] It is not an antibody-mediated phenomenon. Rather, it appears to be a change in the ability of cells bearing CD14 receptors to produce cytokines in response to further endotoxin exposure.[34,35] It was this phenomenon that made early investigators consider endotoxin a possible cause of the Monday symptoms in cotton mill workers.[36]

While the focus of this review is on adverse health effects, not all of endotoxin's effects are deleterious. Endotoxin's ability to reverse tolerance to polysaccharide antigens may prove useful in the design of new vaccines.[9,37] The antitumor effects of endotoxin lead to recent trials of endotoxin as a cancer chemotheraputic agent.[38] The epidemiologic evidence suggests that inhalation of endotoxin reduces the mortality from lung cancer in heavily exposed populations.[39-41] The biological plausibility of this finding was supported by an animal model of endotoxin inhalation and lung cancer.[42]

ENVIRONMENTAL EXPOSURE

Airborne endotoxin is ubiquitous in the outdoor environment, most likely arising from aerosolization of the normal GNB flora of leaves.[43,44] Outdoor levels are usually less than 0.2 ng/m^3 (2 EU/m^3); (Milton, unpublished observations). Outdoor levels of airborne endotoxin were measured as part of various studies in addition to the data shown here. These include data from Colorado, Massachusetts, Ohio, and Washington State at various times of year. The samples generally fell in the range 0.05 to 0.5 ng/m^3. The lowest values outdoors were measured in winter after snowfall, and the highest were during the growing season or down wind of a known source. High levels are found in a variety of environments, including cotton mills, agricultural workplaces, machining operations, offices, laboratories, homes, and swimming pools.[45-50] The highest reported levels (2 to 27 μg/m^3) are associated with processing vegetable fibers (e.g., cotton mills) or fecal materials in agriculture, with waste treatment and recirculating industrial washwater spray, and also with mist from ultrasonic and other "cool-mist" humidifiers. Thus, a very large fraction of the population has potential for significant exposure and health effects.

AIRFLOW OBSTRUCTION

Endotoxin is associated with acute and chronic airflow limitation in occupational studies. Inhaled GNB were first shown to be a cause of mill fever, and implicated in the pathogenesis of byssinosis by an investigation of illness among mattress makers.[51] Experimental human exposure to cotton dust found a linear log-dose-response relationship between acute airflow limitation and airborne endotoxin concentration, but no correlation with dust exposure.[52,53] The threshold for this effect was approximately 9 ng/m^3 (90 EU/m^3) in sensitive subjects and 100 ng/m^3 (1000 EU/m^3) in cotton mill workers.

Workplace studies found that across the workshift drops in FEV$_1$ were associated with exposure in excess of 1 μg/m^3 in poultry workers,[54] and with exposure above 180 ng/m^3 in swine confinement workers.[55] However, not all studies found significant acute effects.[56-58] These negative results may have derived from endotoxin measurement error,[59-62] use of area samples, or small numbers of subjects. In contrast with the negative results, a study of workers exposed to humidifier droplet aerosols found significant acute airflow obstruction associated with very low endotoxin levels (geometric mean 64 pg/m^3) as compared with work in areas with yet lower endotoxin levels (geometric mean 18 to 19 pg/m^3).[63] Two recent workplace studies found that across shift change in lung function correlated with endotoxin exposure in ranges similar to those in the experimental cotton dust exposures. Smid et al.,[64] using personal sampling to define homogenous endotoxin exposure groups, found across shift drops in maximal mid-expiratory flow and an across week decline in FEV$_1$ among workers with group mean endotoxin exposure between 30 and 40 ng/m^3 (300 to 400 EU/m^3). A direct association between acute airflow obstruction and simultaneously measured personal breathing zone endotoxin exposure was recently described among workers in a fiberglass manufacturing plant.[65] Exposures between 4.3 and 15 ng/m^3 (43 to 150 EU/m^3, 8 h time-weighted average) were associated with a 5% or greater decline in peak expiratory flow across the workshift, and with a 5% or greater decline in peak expiratory flow between the start of shift and arising the following morning, consistent with an inflammatory response to endotoxin inhalation. The previous studies examined endotoxin exposure-response among workers exposed to complex agricultural aerosols including cotton dust. There is some evidence that the presence of other constituents in the agricultural dusts besides endotoxin make important contributions to the inflammatory effects of exposure.[66] Therefore, the study in fiberglass workers, which was controlled for other known respiratory irritants and did not contain agricultural dust, adds weight to the concept that endotoxin itself can produce the acute effects observed. Thus, on balance the epidemiologic evidence supports a causal effect of endotoxin in acute airflow obstruction and airway inflammation following inhalation of endotoxin in a variety of environments.

A number of animal models support the epidemiologic findings of airflow obstruction and airway inflammation following endotoxin inhalation. The species used include rabbits,[36,67,68] hamsters,[69] rats,[70] mice,[71] and guinea pigs.[72-75] While the dose, time-course of exposure, and endpoint as well as species varied among these studies, the conclusions are similar. Inhaled endotoxin is a potent inflammatory stimulus causing moderate airflow obstruction and transient bronchial hyperresponsiveness after acute exposure.

In addition to the acute effects, chronic airflow obstruction has been associated with endotoxin exposure in cross-sectional epidemiologic studies. A study of cotton mill workers in Shanghai[56] found that an estimate of cumulative endotoxin exposure was associated with bronchitic symptoms and with low baseline FEV_1. Cotton dust exposure was not correlated with lung function or symptoms. In the animal feed industry, cumulative endotoxin exposure based on personal sampling and homogenous exposure groupings was associated with significantly reduced FEV_1 and expiratory flow rates. The association with cumulative inhalable dust was weaker than that for endotoxin.[76] In the fiberglass industry, the number of years since starting frequent work in a high endotoxin exposed area was associated with low baseline FEV_1 and high peak flow amplitude, a measure of bronchial hyperreactivity.[65] Only one longitudinal study of lung function and endotoxin exposure has been reported. Based on area samples in a cotton mill, no exposure related effects were observed at 5 years follow-up.[77]

PARENCHYMAL LUNG DISEASE

Because of its potent and protean inflammatory effects, and because experimental inhalation of enterobacterial LPS produces systemic symptoms in humans, endotoxin is accepted as a cause of humidifier fever.[78] Such inhalation fevers are generally thought not to have chronic effects.[79,80] However, animal models of endotoxin inhalation and intratracheal instillation have shown that acute inflammatory responses to endotoxin result in changes in the airway epithelium and mucus production and also produce centrilobular emphysema.[81-83] A recent report[84a] suggests that workers intermittently exposed to very high levels of endotoxin in air developed an inhalation fever associated with persistently depressed pulmonary diffusion capacity consistent with emphysema.

Endotoxin may also play a role in some cases of granulomatous lung disease with lymphocytic alveolitis. First, airborne endotoxin was associated with two outbreaks of hypersensitivity pneumonitis (HP).[48,84] In one of these outbreaks, no other bioaerosol could be identified. Second, recent advances in understanding the pathogenesis of lymphocytic alveolitis and granuloma suggest that natural or nonspecific immunity may play an important role. Instillation of either IL-1 or TNFα was reported to produce granulomata in guinea pigs, and the effect was inhibited by antibody to each of the cytokines.[85] Denis and colleagues[86] found that whole cells and extracts of an organism associated with farmer's lung, a form of hypersensitivity pneumonitis, can directly stimulate macrophage production of IL-1 and TNFα without specific recognition by antibody or T-cell receptors. A recent animal model of chronic endotoxin inhalation found that after 5 weeks, guinea pigs developed increased eosinophils in the lung, and that when insoluble β(1-3)-D-glucan was inhaled with the endotoxin the inflammatory reaction was increased and early granulomata were found.[87] Glucan by itself had little effect. Thus, because LPS is among the most potent inducers of key cytokines, it is biologically plausible that endotoxin exposure plays a role in granulomatous pneumonitis through this nonspecific immune mechanism, possibly in combination with other agents or insoluble particles. Alternatively it is possible that endotoxin plays a role because LPS is an adjuvant for cell-mediated immunity, as described above, and because cell-mediated immunity is thought to play an important role in HP.[88] Whether endotoxin's occasional association with parenchymal lung disease is causal remains to be proven.

HUMAN EXPERIMENTAL STUDIES WITH ISOLATED LPS

Early studies of experimental LPS inhalation demonstrated that endotoxin's effects by this route include fever, cough, diffuse aches, nausea, malaise, shortness of breath, and acute air flow obstruction.[36,67] Normal subjects appeared to have a threshold of approximately 80 μg (LPS delivered as an aerosol over a few seconds to minutes) for acute decline in FEV_1. Rylander reported minimal or no change in FEV_1 among normal volunteers exposed to 20 to 30 μg of LPS from *Enterobacter agglomerans* but obtained a 5 to 8% decline at levels of 200 to 300 μg delivered as short boluses.[89] Cavagna and colleagues found that FEV_1 dropped 25% after inhalation of 40 μg of LPS in subjects with pre-existing airway disease.[67] Similarly, subjects with chronic nonspecific lung disease and bronchial hyperreactivity (BHR) experienced an 18% decline in FEV_1 beginning at 30 min and lasting for 4 h.[90] A late reaction was also observed between the seventh and ninth hour after inhalation of *Hamophilus influenzae* LPS in these subjects.

Endotoxin inhaled by adult asthmatic subjects as a single 20 μg puff of *Escherichia coli* LPS produced a decrease in FEV_1 of 6.7% starting at 15 min and lasting for >5 h.[91] The PD_{50} sGaw and PD_{20} FEV_1 responses to methacholine challenge were significantly reduced 5 h after LPS inhalation. In 5 of 8 asthmatic subjects, PD_{20} FEV_1 remained low at 24 and 48 h after LPS.[92] Nonasthmatics did not show a change in FEV_1 or bronchial responsiveness to histamine following LPS inhalation.[91]

As indicated by the prolonged bronchial reactivity in some subjects, experimental LPS inhalation in humans results in airway and parenchymal inflammation.

In studies of asthmatics, a systemic inflammatory response was evidenced by a significant rise in serum TNFα at 60 min, of PMNs at 6 h, and of C-reactive protein at 24 and 48 h after inhalation of 20 µg of LPS. The increase in circulating PMNs at 6 h was linearly related to the acute decline in FEV_1 between 15 min and 6 h.[92] In normal subjects, inhalation of 25 µg of LPS resulted in increased PMNs and lymphocytes in bronchoalveolar lavage (BAL) performed 3 h after LPS challenge.[93] In addition, LPS inhalation has been shown to produced a transient diffusion abnormality apparently due to acute pneumonitis.[94]

A major drawback of the experimental studies with isolated LPS has been the high dose rates used. By contrast with dose rates of 20 to 300 µg LPS over a few minutes, environmental exposures are often at lower levels for longer periods of time, except for some cases of humidifier fever. Experimental exposure to cotton dust described above[52,53] found thresholds for airway effects that corresponded to doses of 2 to 8 ng, assuming 50% deposition, over 4 to 6 h. Thus, either other materials associated with environmental endotoxin or the dose rate play an important role in determining the response to endotoxin.

ENDOTOXIN IN THE INDOOR ENVIRONMENT

The data on endotoxin in nonindustrial indoor environments comes mostly from residential studies. The presence of endotoxin in house dust was suspected because extracts of house dust used for allergy testing were known to contain endotoxin.[95-97] Confirmation that endotoxin is present in house dust itself, however, was only recently reported.[98] Because endotoxin potentiated histamine release produced by dust mite extracts,[31,99] an investigation was undertaken to determine whether house dust endotoxin affected the clinical severity of asthma.[100] Twenty-eight asthmatic subjects, 20 with allergy to dust mite, were examined on three occasions during a period of 3 months. Dyspnea scores, medication usage, and lung function were measured. A pooled house dust sample from the mattress and bedroom floor was collected and analyzed on two occasions for each subject. Endotoxin content (measured by a chromogenic Limulus assay) and dust mite antigen (estimated by quantitative guanine measurement using HPLC) were not correlated. The median house dust endotoxin concentration was 1.1 ng/mg and median guanine concentration was 0.07 mg/100 mg. Subjects with house dust endotoxin above the median had significantly higher dyspnea scores, more β-agonist and oral steroid usage, and lower FEV_1/FVC ratios. There were no differences in symptom scores or medication usage between subjects above and below the median for dust mite (guanine) content of house dust. Most of the subjects homes contained *Der p I* levels, based on guanine levels, below the threshold of 10 µg/g associated with asthmatic exacerbations. Thus, susceptible populations may have significant health effects from endotoxin exposure at the levels found in homes.

There are few reports of endotoxin measurements outside industrial and agricultural settings. Avaliable data show that airborne endotoxin in residential settings can reach levels at which acute bronchoconstriction was observed in experimental and workplace studies. We measured airborne endotoxin levels measured in homes of workers participating in an occupational study. The highest level 7.6 ng/m^3 (76 EU/m^3) was in the range (4.2 to 15 ng/m^3) associated with acute effects in the same study.[65] A recent study of "sick buildings" measured airborne endotoxin in 22 Swedish homes.[101] The levels ranged between undetected and 18 ng/m.^{3}

In a recent pilot study of endotoxin levels in house dust, we examined dust from 12 homes. Samples were obtained from the mattress, a bedroom rug, the kitchen, and the living room or TV room sofa and rug. Each sample was sieved and weighed, and a 25 to 50 mg aliquot of each sample was subjected to immediate buffer extraction and assayed for endotoxin using standard methods. The endotoxin content of dust ranged from 16 to 440 EU/mg. To study the reproducibility of the assay and extraction technique for endotoxin in house dust, certain samples were divided into separate aliquots and replicate extractions were made. Correlation between the repeated extraction and assay were high for both simultaneous (r = 0.95) and sequentially extracted (r = 0.90) samples. The endotoxin levels in house dust were similar to those reported in Belgium.[100]

A second pilot study was undertaken to examine the relationship between settled dust and airborne dust in homes. A total of 12 homes were sampled for house dust and airborne endotoxin in the living room. Aerosol samples were collected for 24 h at 4 lpm using standard methods.[10] Overall, the airborne and settled dust concentrations were correlated (r = 0.88). Airborne endotoxin and total particulate concentration measured on the same filter were not correlated (r = 0.01, data not shown). Household characteristics, such as pets and wall-to-wall carpet, may influence the relationship of airborne to settled dust. The highest levels in air and dust were from an apartment with wall-to-wall carpet damaged by cat urine.

There are few published data on the health effects of airborne endotoxin in sick building syndrome. Rylander and colleagues studied 22 "flats," 18 where symptoms of sick buildings had been reported, and 4 without.[101] Bedroom air samples were collected for 2 to 4 h for a total of 200 to 400 l. Efforts were made to disturb the rooms every 15 min during sampling by walking around and rearranging the pillows, etc. The endotoxin levels obtained ranged from 0 to 18 ng/m^3. By dividing flats into three groups according to endotoxin level, a dose response for increasing reports of cough, breathing difficulty, itchy eyes, nausea, and fatigue were observed with increasing endotoxin category. It is striking that the symptom with the steepest exposure-response across the exposure categories was fatigue, reported by 10% of those in homes with <0.1 ng/m^3 and by 91% of those in homes with >0.2 ng/m^3. The authors also measured β(1-3)-D-glucan and found it associated with nasal irritation and hoarseness. Glucan was also associated with fatigue, but not as strongly as was endotoxin. Unfortunately, no information on the types of structures, heating, or ventilation system were provided, and outdoor levels of endotoxin and season of sampling were not reported. The levels of endotoxin found were generally low and the highest category included values in the range of common outdoor levels. However, because experimental endotoxin inhalation is associated with general malaise, this result is biologically plausible. Because fatigue is one of the frequent components of sick building syndrome, this report of an association between endotoxin and fatigue at low exposure levels is worthy of follow-up investigation.

Rylander also reports on four sick buildings identified from newspaper reports: two schools, a post office, and a day-care center.[102] An office building without known history of complaints was used as a control. No data was provided regarding the construction materials, size, ages, or ventilation systems in the buildings. Questionnaire data suggest that the office building (405 respondents) was much larger than the complaint buildings (6 to 14 respondents). Questionnaire responses of the occupants were compared with airborne levels of endotoxin and $\beta(1\text{-}3)\text{-D-glucan}$. Air samples were collected over 2 to 4 h while using a fan to create an artificial disturbance. Endotoxin in the office building was reported as low (0.06 ng/m^3, presumably arithmetic mean) and in the other buildings ranged from 0.19 to 0.43 ng/m^3. The proportion of individuals reporting symptoms was analyzed for correlation with levels of endotoxin and glucan. The size of the correlations was not shown, but p values reported suggested a correlation of endotoxin with skin rash and of glucan with cough and itching. The result appears to be driven by a high prevalence of rashes in the day-care center; the building with the highest endotoxin. Cough and itching were the primary complaints in the schools, which were the buildings with high glucan levels. Because it is not clear that these structures or their occupants are comparable, it is difficult to draw any conclusions from this study.

The largest study of endotoxin indoors was a study of 19 randomly selected government office buildings in The Netherlands.[103] Building-related symptom reports were gathered from 1355 employees and the study also collected data on job satisfaction, building characteristics, and physical and chemical parameters as well as airborne microorganisms and endotoxin. The primary analysis was made after dichotomizing mechanically ventilated buildings on the basis of over-all mean symptom prevalence: <15% (n = 4), and ≥15% (n = 7). They found indoor endotoxin levels ranging from 28 to 43 ng/m^3 in naturally ventilated buildings and from 27 to 67 ng/m^3 in buildings with mechanical ventilation and low symptom prevalence. However, buildings with high symptom prevalence had endotoxin levels ranging from 100 to 408 ng/m^3. The correlation of symptoms and endotoxin appeared to be driven by mucosal irritation symptoms. In a logistic regression analysis of the data stratified according to endotoxin level,[104] the odds of mucosal irritation given endotoxin >200 ng/m^3 were 2.16 compared with exposure below 100 ng/m^3 when controlled for gender, age, job category, smoking, job satisfaction, number of occupants in the respondents room, and years of work in the building. No other parameter clearly distinguished the "sick" and "healthy" buildings, except that endotoxin was higher in buildings with sealed windows.

While this study seems well designed and otherwise convincing, concern is raised by the endotoxin levels reported. The levels indoors in the naturally ventilated buildings are much higher than those observed in residential and recreational environments without clear sources such as contaminated humidifiers.[105] The levels in the sick buildings are close to the levels frequently reported from swine barns and cotton mills. Although the authors stated that background-outdoor levels were measured none were reported. The study used a type of Limulus assay that can respond to glucans resulting in overestimation of endotoxin levels. However, the authors were able to remove activity in the Limulus assay by adding polymyxin, after heating samples to 100°C, suggesting that they actually measured endotoxin. The control standard endotoxin used was not described so that the results cannot be converted to standard units. It seems unlikely however, that a low potency standard could account for overestimation of background endotoxin levels by a factor of 100.[106] It may be that the levels are internally consistent. Yet, given the uncertainty about the endotoxin measurements, it is hard to give much weight to this study.

A recent follow up to the Danish Town Hall Study suggests an important role for GNB in sick building syndrome.[107] Twelve buildings were selected based on the results of the earlier study to give a range of complaint rates. Questionnaires were administered to 870 building occupants and dust was collected in each building. Analyses of dust included counts of culturable bacteria and fungi with some amount of speciation, endotoxin and mite allergen measurement, histamine releasing capacity, specific surface area, and volatile organic emissions testing. Of these, the number of culturable GNB was the most closely associated with the complaints of fatigue, heavy headedness, headache, and with hoarseness/sore throat symptoms in an analysis of building symptom frequency. In addition, an exposure response was evident in the data when percent GNB were plotted against the frequency of hoarseness/sore throat symptoms in each building. The data were also analyzed on the basis of individual responses by logistic regression. Significant odds ratios were found for all four symptoms and these persisted when controlled for history of eczema, hay fever, asthma, and for gender.

The authors suggested that the new town hall findings may be due to airborne endotoxin, but did not find a correlation between symptoms and endotoxin in settled dust. This implies that airborne endotoxin was correlated with culturable GNB but not with endotoxin in settled dust. The relationship between levels of endotoxin and culturable and countable bacteria are complex[106,108,109] and depend on the environment. It may be worth noting that one of the first studies of endotoxin and GNB in cotton mills found an association of symptom frequency with airborne GNB but not with airborne endotoxin.[110] A possible explanation for the town hall findings, in keeping with endotoxin being the primary toxic agent in GNB, is inhibition of the Limulus assay. The endotoxin levels reported in the dust samples from the town halls are very low (median 20.8 EU/g) as compared with house dust levels (16 to 440 EU/mg, see above). The Limulus method used is not sensitive to glucans,[111] but is susceptible to undetected interference.[112] In our laboratory, concentrated house dust extracts tend to inhibit the Limulus assay and must be properly diluted for assay. Other possible explanations are that the extraction method, not described, may have failed to release endotoxin from the dust matrix or that samples lost endotoxin potency during storage prior to assay.

The two papers just described are the only reported large scale investigations of endotoxin and sick building syndrome — one measured airborne endotoxin and found unexpectedly high levels, while the other measured settled dust and found surprisingly low levels. Both found associations suggesting that endotoxin may be an important factor in some of the common nonspecific and irritant symptoms in sick buildings. The results are certainly provocative and the relationships to fatigue and mucosal symptoms are biologically plausible. It is clear that future studies must make extensive efforts to validate endotoxin measurements in the indoor environment.

The source of endotoxin indoors is not known. A recent study of small room humidifiers found that very high endotoxin levels can be generated when the reservoir contains endotoxin.[105] In that study, ultrasonic and cool mist type humidifiers were spiked with endotoxin 1 µg of *Pseudomonas aeruginsoa* LPS to simulate levels of endotoxin reported in the literature.[78] The resulting levels of endotoxin in room air ranged between 360 and 1100 ng/m^3. Evaporative "warm mist" and steam humidifiers did not emit appreciable amounts of endotoxin into room air. Because large scale cool mist humidifiers are sometimes used in buildings and can be contaminated with GNB,[113] significant endotoxin exposure could occur through the ventilation systems. This has been suspected but never proven.[114] In the absence of humidification, however, little is known about sources. A recent study found unusual lipid A structures in some indoor samples.[115] This may suggest that responsible sources indoors are not the same as those found in occupational environments.

In summary, inhaled endotoxin causes systemic symptoms at high exposure levels and important effects on pulmonary function at levels found in a wide range of industrial and some nonindustrial settings (>4.5 ng/m^3). Indoor environments can have very high levels of endotoxin in the presence of poorly maintained humidifiers and can, under other as yet poorly defined circumstances, have moderately high levels that are associated with effects on pulmonary function. There is also data to suggest that the upper end of normal indoor endotoxin levels are associated with increased severity of asthma and with sick building syndrome. Thus, there may be effects of building-related, low-level endotoxin exposure on susceptible occupants. Given the small number of studies and the paucity of data on indoor exposure and response, we can only conclude that this is an interesting area for further investigation.

REFERENCES

1. Helander I, Lounatmaa K. Cotton bacterial endotoxin assessed by electron microscopy. *Br. J. Indust. Med.* 1981;38:394-96.
2. Nowotny A. Heterogeneity of endotoxin. In: Reitschel ET ed. Handbook of Endotoxin Volume I: Chemistry of Endotoxin. New York: Elsevier, 1984;308-38.
3. Rietschel ET, Brade H. Bacterial Endotoxins. *Sci. Am.* 1992;267:55-61.
4. Sonesson HRA, Zähringer U, Grimmecke HD, Westphal O, Rietschel ET. Bacterial Endotoxin: chemical structure and biological activity. In: Brigham K ed. Endotoxin and the Lungs. New York: Marcel Dekker, Inc., 1994;1-20.
5. Grabarek J, Her GR, Reinhold VN, Hawiger J. Endotoxic lipid A interaction with human platelets. Structure-function analysis of lipid A homologs obtained from Salmonella minnesota Re595 lipopolysaccharide. *J. Biol. Chem.* 1990;265:8117-21.

6. Baker PJ, Taylor CE, Stashak PW, et al. Inactivation of suppressor T cell activity by the nontoxic lipopolysaccharide of *Rhodopseudomonas sphaeroides*. *Infect. Immun.* 1990;58:2862-8.

7. Ukei S, Iida J, Shiba T, Kusumoto S, Azuma I. Adjuvant and antitumor activities of synthetic lipid A analogues. *Vaccine* 1986;4:21-24.

8. Takayama K, Olsen M, Datta P, Hunter RL. Adjuvant activity of non-ionic block copolymers. V. Modulation of antibody isotype by lipopolysaccharides, lipid A and precursors. *Vaccine* 1991;9:257-65.

9. Baker PJ, Hraba T, Taylor CE, et al. Structural features that influence the ability of lipid A and its analogs to abolish expression of suppressor T cell activity. *Infect. Immun.* 1992;60:2694-701.

10. Milton DK. Endotoxin. In: Burge HA ed. Bioaerosols. Boca Raton, FL: Lewis Publishers, 1995; 77-86.

11. Wright SD. Multiple receptors for endotoxin. *Curr. Opin. Immunol.* 1991;3:83-90.

12. Wright SD, Ramos RA, Tobias PS, Ulevitch RJ, Mathison JC. CD14, a receptor for complexes of lipopolysaccharide (LPS) and LPS binding protein. *Science* 1990;249:1431-3.

13. Haziot A, Tsuberi BZ, Goyert SM. Neutrophil CD14: biochemical properties and role in the secretion of tumor necrosis factor-alpha in response to lipopolysaccharide. *J. Immunol.* 1993;150:5556-65.

14. Worthen GS, Avdi N, Vukajlovich S, Tobias PS. Neutrophil adherence induced by lipopolysaccharide *in vitro*. Role of plasma component interaction with lipopolysaccharide. *J. Clin. Invest.* 1992;90:2526-35.

15. Wright SD, Ramos RA, Hermanowski-Vosatka A, Rockwell P, Detmers PA. Activation of the adhesive capacity of CR3 on neutrophils by endotoxin: dependence on lipopolysaccharide binding protein and CD14. *J. Exp. Med.* 1991;173:1281-6.

16. Haziot A, Rong GW, Silver J, Goyert SM. Recombinant soluble CD14 mediates the activation of endothelial cells by lipopolysaccharide. *J. Immunol.* 1993;151:1500-7.

17. Pugin J, Schurer-Maly CC, Leturcq D, Moriarty A, Ulevitch RJ, Tobias PS. Lipopolysaccharide activation of human endothelial and epithelial cells is mediated by lipopolysaccharide-binding protein and soluble CD14. *Proc. Natl. Acad. Sci. U.S.A.* 1993;90:2744-8.

18. Wright SD, Ramos RA, Patel M, Miller DS. Septin: a factor in plasma that opsonizes lipopolysaccharide-bearing particles for recognition by CD14 on phagocytes. *J. Exp. Med.* 1992;176:719-27.

19. Morrison DC, Silverstein R, Bright SW, Chen TY, Flebbe LM, Lei MG. Monoclonal antibody to mouse lipopolysaccharide receptor protects mice against the lethal effects of endotoxin. *J. Infect. Dis.* 1990;162:1063- 8.

20. Ulevitch RJ. Recognition of bacterial endotoxins by receptor-dependent mechanisms. *Adv. Immunol.* 1993;53:267- 89.

21. Rylander R, Beijer L. Inhalation of endotoxin stimulates alveolar macrophage production of platelet-activating factor. *Am. Rev. Resp. Dis.* 1987;135:83-6.

22. LeMay DR, LeMay LG, Kluger MJ, D'Alecy LG. Plasma profiles of IL-6 and TNF with fever-inducing doses of lipopolysaccharide in dogs. *Am. J. Physiol.* 1990;259:R126- 32.

23. Schindler R, Mancilla J, Endres S, Ghorbani R, Clark SC, Dinarello CA. Correlations and interactions in the production of interleukin-6 (IL-6), IL-1, and tumor necrosis factor (TNF) in human blood mononuclear cells: IL-6 suppresses IL-1 and TNF. *Blood* 1990;75:40-7.

24. Henricson BE, Benjamin WR, Vogel SN. Differential cytokine induction by doses of lipopolysaccharide and monophosphoryl lipid A that result in equivalent early endotoxin tolerance. *Infect. Immun.* 1990;58:2429-37.

25. Weatherstone KB, Rich EA. Tumor necrosis factor/cachectin and interleukin-1 secretion by cord blood monocytes from premature and term neonates. *Pediatr. Res.* 1989;25:342-6.

26. Santiago-Schwarz F, McHugh DM, Fleit HB. Functional analysis of monocyte-macrophages derived from nonadherent cord blood progenitor cells: correlation with the ontogeny of cell surface proteins. *J. Leukoc. Biol.* 1989;46:230-8.

27. Alving CR, Richards RL. Liposomes containing lipid A: a potent nontoxic adjuvant for a human malaria sporozoite vaccine. *Immunol. Lett.* 1990;25:275-9.

28. Scott P. IL-12: initiation cytokine for cell-mediated immunity. *Science* 1993;260:496-7.

29. Gazzinelli RT, Hieny S, Wynn S, Wolf S, Sher A. IL-12 is required for T-independent induction of IFN-γ by *Toxoplasma gondii* and mediates resistance against the parasite in *scid* mice. *J. Immunol.* 1993;150:86A.

30. Snapper CM, Pecanha LM, Levine AD, Mond JJ. IgE class switching is critically dependent upon the nature of the B cell activator, in addition to the presence of IL-4. *J. Immunol.* 1991;147:1163-70.

31. Clementsen P, Milman N, Kilian M, Fomsgaard A, Baek L, Norn S. Endotoxin from *Haemophilus influenzae* enhances IgE-mediated and non-immunological histamine release. *Allergy* 1990;45:10-7.

32. Vosbeck K, Tobias P, Mueller H, et al. Priming of polymorphonuclear granulocytes by lipopolysaccharides and its complexes with lipopolysaccharide binding protein and high density lipoprotein. *J. Leukoc. Biol.* 1990;47:97-104.

33. Beeson PB. Tolerance to bacterial pyrogens I. factors influencing its development. *J. Exp. Med.* 1947;86:29-38.

34. Henricson BE, Neta R, Vogel SN. An interleukin-1 receptor antagonist blocks lipopolysaccharide-induced colony-stimulating factor production and early endotoxin tolerance. *Infect. Immun.* 1991;59:1188-91.

35. Henricson BE, Perera PY, Qureshi N, Takayama K, Vogel SN. *Rhodopseudomonas sphaeroides* lipid A derivatives block *in vitro* induction of tumor necrosis factor and endotoxin tolerance by smooth lipopolysaccharide and monophosphoryl lipid A. *Infect. Immun.* 1992;60:4285-90.

36. Pernis B, Vigliani EC, Cavagna C, Finulli M. The role of bacterial endotoxins in occupational diseases caused by inhaling vegetable dusts. *Br. J. Indust. Med.* 1961;18:120-9.

37. Richards RL, Swartz GJ, Schultz C, et al. Immunogenicity of liposomal malaria sporozoite antigen in monkeys: adjuvant effects of aluminium hydroxide and nonpyrogenic liposomal lipid A. *Vaccine* 1989;7:506-12.

38. Engelhardt R, Mackensen A, Galanos C. Phase I trial of intravenously administered endotoxin *(Salmonella abortus equi)* in cancer patients. *Cancer Res.* 1991;51:2524- 30.

39. Enterline PE, Sykora JL, Keleti G, Lange JH. Endotoxins, cotton dust, and cancer. *Lancet* 1985;2:934- 5.

40. Hodgson JT, Jones RD. Mortality of workers in the British cotton industry in 1968-1984. *Scand. J. Work Environ. Health.* 1990;16:113-20.

41. Rylander R. Environmental exposures with decreased risks for lung cancer? *Int. J. Epidemiol.* 1990;19:S67-72.

42. Lange JH. Anticancer properties of inhaled cotton dust: a pilot experimental investigation. *J. Environ. Sci. Health* 1992;A27(2):505-14.

43. Edmonds RL. Aerobiology, The Ecological Systems Approach. Stroudsburg: Dowden, Hutchinson & Ross, 1979.

44. Andrews JH, Hirano SS. Microbial Ecology of Leaves. New York: Springer-Verlag, 1992.

45. Mattsby-Baltzer I, Sandin M, Ahlstrîm B, et al. Microbial growth and accumulation in industrial metal-working fluids. *Appl. Environ. Microbiol.* 1989;55:2681-9.

46. Milton DK. Endotoxin in Metal Working Fluids: Report to United Auto Workers — General Motors Joint National Committee on Occupational Health and Safety. Harvard School of Public Health, 1992.

47. Jacobs RR. Airborne endotoxins: an association with occupational lung disease. *Appl. Ind. Hyg.* 1989;4:50-6.

48. Rose CS, Newman LS, Martyny JW, et al. Outbreak of hypersensitivity pneumonitis in an indoor swimming pool: clinical, pathophysiologic, radiographic, pathologic, lavage and environmental findings. *Am. Rev. Resp. Dis.* 1990;141:A315.

49. Rylander R, Morey P. Airborne endotoxin in industries processing vegetable fibers. *Am. Ind. Hyg. Assoc. J.* 1982;43:811-2.

50. Rylander R, Vesterlund J. Airborne endotoxins in various occupational environments. *Prog. Clin. Biol. Res.* 1982;93:399-409.

51. Neal PA, Schneiter R, Caminita BH. Report on acute illness among rural mattress makers using low grade stained cotton. *J. Am. Med. Assoc.* 1942;119:1074-82.

52. Rylander R, Haglind P, M. L. Endotoxin in cotton dust and respiratory function decrement among cotton workers in an experimental cardroom. *Am. Rev. Respir. Dis.* 1985;131:209-13.

53. Castellan RM, Olenchock SA, Kinsley KB, Hankinson JL. Inhaled endotoxin and decreased spirometric values, an exposure-response relation for cotton dust. *N. Engl. J. Med.* 1987;317:605-10.

54. Thelin A, Tegler Ö, Rylander R. Lung reactions during poultry handling related to dust and bacterial endotoxin levels. *Eur. J. Respir. Dis.* 1984;65:266-71.

55. Donham K, Haglind P, Peterson Y, Rylander R, Belin L. Environmental and health studies of farm workers in Swedish swine confinement buildings. *Br. J. Ind. Med.* 1989;46:31-7.

56. Kennedy SM, Christiani DC, Eisen EA, et al. Cotton dust and endotoxin exposure-response relationships in cotton textile workers. *Am. Rev. Respir. Dis.* 1987;135:194-200.

57. Rylander R, Haglind P, Butcher BT. Reactions during work shift among cotton mill workers. *Chest* 1983;84:403- 7.

58. Hagmar L, Schütz A, Hallberg T, Sjöholm A. Health effects of exposure to endotoxins and organic dust in poultry slaughter-house workers. *Int. Arch. Occup. Environ. Health* 1990;62:159-64.

59. Milton DK, Gere RJ, Feldman HA, Greaves IA. Endotoxin measurement: aerosol sampling and application of a new Limulus method. *Am. Indust. Hyg. Assoc. J.* 1990;51:331-7.

60. Olenchock SA, Lewis DM, Mull JC. Effects of different extraction protocols on endotoxin analysis of airborne grain dusts. *Scand. J. Work Environ. Health.* 1989;15:430-5.

61. Gordon T, Galdanes K, Brosseau L. Comparison of sampling media for endotoxin-containing aerosols. *Appl. Occup. Environ. Hyg.* 1992;7:472-7.

62. Sonesson A, Larsson L, Schutz A, Hagmar L, Hallberg T. Comparison of the Limulus amebocyte lysate test and gas chromatography-mass spectromety for measuring lipopolysaccharides (endotoxins) in airborne dust from poultry processing industries. *Appl. Environ. Microbiol.* 1990;56:1271-8.

63. Kateman E, Heederik D, Pal TM, Smeets M, Smid T, Spitteler M. Relationship of airborne microorganisms with the lung function and leucocyte levels of workers with a history of humidifier fever. *Scand. J. Work Environ. Health.* 1990;16:428-33.

64. Smid T, Heederik D, Houba R, Quanjer PH. Dust- and endotoxin-related acute lung function changes and work-related symptoms in workers in the animal feed industry. *Am. J. Ind. Med.* 1994;25:877-88.

65. Milton DK, Kriebel D, Wypij D, Walters M, Hammond SK, Evans J. Endotoxin exposure-response in fiberglass manufacturing. *Am. J. Ind. Med.* 1996; in press.

66. Clapp WD, Thorne PS, Frees KL, Zhang X, Lux CR, Schwartz DA. The effects of inhalation of grain dust extract and endotoxin on upper and lower airways. *Chest* 1993;104:825-30.

67. Cavagna G, Foa V, Vigliani EC. Effects in man and rabbits of inhalation of cotton dust or extracts and purified endotoxins. *Br. J. Indust. Med.* 1969;26:314-21.

68. Edwards JH, Cockcroft A. Inhalation challenge in humidifier fever. *Clin. Allergy* 1981;11:227-35.

69. Hudson AR, Kilburn KH, Halprin GM, McKenzie WN. Granulocyte recruitment to airways exposed to endotoxin aerosols. *Am. Rev. Resp. Dis.* 1977;115:89-95.

70. Pauwels RA, Kips JC, Peleman RA, Van Der Straeten ME. The effect of endotoxin inhalation on airway responsiveness and cellular influx in rats. *Am. Rev. Resp. Dis.* 1990;141:540-545.

71. Schwartz DA, Thorne PS, Jagielo PJ, White GE, Bleuer SA, Frees KL. Endotoxin responsiveness and grain dust-induced inflammation in the lower respiratory tract. *Am. J. Physiol.* 1994;267:L609-17.

72. Gordon T. Acute respiratory effects of endotoxin-contaminated machining fluid aerosols in guinea pigs. *Fund. Appl. Toxicol.* 1992;19:117-123.

73. Gordon T, Balmes J, Fine J, Sheppard D. Airway oedema and obstruction in guinea pigs exposed to inhaled endotoxin. *Br. J. Ind. Med.* 1991;48:629-635.

74. Ryan LK, Karol MH. Acute respiratory response of guinea pigs to lipopolysaccharide, lipid A, and monophosphoryl lipid A from *Salmonella minnesota*. *Am. Rev. Respir. Dis.* 1989;140:1429-35.

75. Gordon T. Dose-dependent pulmonary effects of inhaled endotoxin in guinea pigs. *Environ. Res.* 1992;59:416-426.

76. Smid T, Heederik D, Houba R, Quanjer PH. Dust- and endotoxin-related respiratory effects in the animal feed industry. *Am. Rev. Respir. Dis.* 1992;146:1474-9.

77. Christiani DC, Ye TT, Wegman DH, Eisen EA, Dai HL, Lu PL. Cotton dust exposure, across-shift drop in FEV_1, and five-year change in lung function. *Am. J. Respir. Crit. Care Med.* 1994;150:1250-5.

78. Rylander R, Haglind P. Airborne endotoxins and humidifier disease. *Clin. Allergy* 1984;14:109-12.

79. Von Essen S, Robbins RA, Thompson AB, Rennard SI. Organic dust toxic syndrome: an acute febrile reaction to organic dust exposure distinct from hypersensitivity pneumonitis. *J. Toxicol. Clin. Toxicol.* 1990;28:389-420.

80. do Pico GA. Report on diseases. *Am. J. Ind. Med.* 1986;10:261-5.

81. Folkerts G, Henricks PAJ, Slootweg PJ, Nijkamp FP. Endotoxin-induced inflammation and injury of the guinea pig respiratory airways cause bronchial hyporeactivity. *Am. Rev. Respir. Dis.* 1988;137:1441-8.

82. Snider GL, Ciccolella DE, Morris SM, Stone PJ, Lucey EC. Putative role of neutrophil elastase in the pathogenesis of emphysema. *Ann. N.Y. Acad. Sci.* 1991;624:45- 59.

83. Milton DK, Godleski JJ, Feldman HA, Greaves IA. Toxicity of intratracheally instilled cotton dust, cellulose, and endotoxin. *Am. Rev. Respir. Dis.* 1990;142:184-92.

84. Flaherty DK, Deck FH, Hood MA, et al. A Cytophaga species endotoxin as a putative agent of occupation-related lung disease. *Infect. Immun.* 1984;43:213-6.

84a. Milton DK, Amsel J, Reed CE, Enright PL, Brown LR, Aughenbaugh GL, Morey, PR. Cross-sectional follow-up of a flu-like respiratory illness among fiberglass manufacturing employees: endotoxin exposure associated with two distinct sequelae. *Am. J. Ind. Med.* 1995;28:469-88.

85. Mori Y, Kojima Y, Sugawara K, Konishi K, Tamura M, Tomichi N. Experimental granuloma formation by intratracheal injection of interleukin 1 (IL-1) and tumor necrosis factor-alpha (TNF-alpha). *Am. Rev. Resp. Dis.* 1993;147:A81.

86. Denis M, Cormier Y, Tardif J, Ghadirian E, Laviolette M. Hypersensitivity pneumonitis: whole *Micropolyspora faeni* or antigens thereof stimulate the release of proinflammatory cytokines from macrophages. *Am. J. Respir. Cell Mol. Biol.* 1991;5:198-203.

87. Fogelmark B, Sjostrand M, Rylander R. Pulmonary inflammation induced by repeated inhalations of beta(1,3)-D-glucan and endotoxin. *Int. J. Exp. Pathol.* 1994;75:85-90.

88. Salvaggio JE. Hypersensitivity pneumonitis. *J. Allergy Clin. Immunol.* 1987;79:558-70.

89. Rylander R, Bake B, Fischer JJ, Helander IM. Pulmonary function and symptoms after inhalation of endotoxin. *Am. Rev. Respir. Dis.* 1989;140:981-6.

90. Van der Zwan JC, Orie NG, Kauffman HF, Wiers PW, de Vries K. Bronchial obstructive reactions after inhalation with endotoxin and precipitinogens of Haemophilus influenzae in patients with chronic non-specific lung disease. *Clin. Allergy* 1982;12:547-59.

91. Michel O, Duchateau J, Sergysels R. Effect of inhaled endotoxin on bronchial reactivity in asthmatic and normal subjects. *J. Appl. Physiol.* 1989;66:1059-64.

92. Michel O, Ginanni R, Le Bon B, Content J, Duchateau J, Sergysels R. Inflammatory response to acute inhalation of endotoxin in asthmatic patients. *Am. Rev. Respir. Dis.* 1992;146:352-7.

93. Sandstrîm T, Bjermer L, Rylander R. Lipopolysaccharide (LPS) inhalation in healthy subjects increases neutrophils, lymphocytes and fibronectin levels in bronchoalveolar lavage fluid. *Eur. Respir. J.* 1992;5:992- 6.

94. Herbert A, Carvalheiro M, Rubenowitz E, Bake B, Rylander R. Reduction of alveolar-capillary diffusion after inhalation of endotoxin in normal subjects. *Chest* 1992;102:1095-8.

95. Peterson RA, Wicklund PE, Good RA. Endotoxin activity of a house dust extract. *J. Allergy* 1964;35:134-42.

96. Siraganian RP, Baer H, Hochstein HD, May JC. Allergenic and biologic activity of commercial preparations of house dust extract. *J. Allergy Clin. Immunol.* 1979;64:526- 33.

97. Berrens L. The allergens in house dust. *Prog. Allergy* 1970;14:259-339.

98. Michel O, Le Bon B, Duchateau J, Vertongen F, Sergysels R. Domestic endotoxin exposure in asthma. *Eur. Resp. J.* 1989;2:1019-20.

99. Michel O, Ginanni R, Le BB, Duchateau J. Effect of endotoxin contamination on the antigenic skin test response. *Ann. Allergy* 1991;66:39-42.

100. Michel O, Ginanni R, Duchateau J, Vertongen F, Le Bon B, Sergysels R. Domestic endotoxin exposure and clinical severity of asthma. *Clin. Exp. Allergy* 1991;21:441-8.

101. Rylander R, Sîrensen S, Goto H, Yuasa K, Tanaka S. The importance of endotoxin and glucan for symptoms in sick buildings. In: Bieva CJ, Courtois Y, Govaerts M ed. Present and Future of Indoor Air Quality in Proceedings of the Brussels Conference, Excerpta Medica, New York, 1989;219-26.

102. Rylander R, Persson K, Goto H, Tanaka S. Airborne beta-1,3-glucan may be related to symptoms in sick buildings. *Indoor Environ.* 1992;1:263-7.

103. Teeuw KB, Vandenbroucke-Grauls CM, Verhoef J. Airborne gram-negative bacteria and endotoxin in sick building syndrome. A study in Dutch governmental office buildings. *Arch. Intern. Med.* 1994;154:2339-45.

104. Teeuw B. Sick building syndrom: the role of airborne microorganisms and endotoxin. Doctoral Dissertation, Universitet Utrecht, Faculteit Geneeskunde, 1992.

105. Tyndall RL, Bowman EK, Ironside KS, Milton DK, Barbaree J, Lehman E. Aerosolization of microorganisms and endotoxin from home humidifiers. *Indoor Air.* 1995;5:171-8.

106. Reynolds S, Milton DK. Comparison of methods for analysis of airborne endotoxin. *Appl. Occup. Environ. Hyg.* 1993;8:761-67.

107. Gyntelberg F, Suadicani P, Nielsen JW, et al. Dust and the sick building syndrome. *Indoor Air* 1994;4:223-38.

108. Laitinen S, Nevalainen A, Kotimaa M, Liesivuori J, Martikainen PJ. Relationship between bacterial counts and endotoxin concentration in the air of wastewater treatment plants. *Appl. Environ. Microbiol.* 1992;58:3774-3776.

109. Walters M, Milton DK, Larsson L, Ford T. Airborne environmental endotoxin: a cross-validation of sampling and analysis techniques. *Appl. Environ. Microbiol.* 1994;60:996-1005.

110. Cinkotai FF, Lockwood MG, Rylander R. Airborne micro-organisms and prevalence of byssinotic symptoms in cotton mills. *Am. Ind. Hyg. Assoc. J.* 1977;38:554-559.

111. Roslansky PF, Novitsky TJ. Sensitivity of *Limulus* amebocyte lysate (LAL) to LAL-reactive glucans. *J. Clin. Microbiol.* 1991;29:2477-2483.

112. Milton DK, Feldman HA, Neuberg DS, Bruckner RJ, Greaves IA. Environmental endotoxin measurement: the kinetic limulus assay with resistant-parallel-line estimation. *Environ. Res.* 1992;57:212-30.

113. Burkhart JE, Stanevich R, Kovak B. Microorganism contamination of HVAC humidification systems: case study. *Appl. Occup. Environ. Hyg.* 1993;8:1010-4.

114. Burge S, Hedge A, Wilson S, Bass JH, Robertson A. Sick building syndrome: a study of 4979 office workers. *Ann. Occup. Hyg.* 1987;31:493-504.

115. Gradowska W, Larsson L. Determination of absolute configurations of 2- and 3-hydroxy fatty acids in organic dust by gas chromatography — mass spectrometry. *J. Microbiol. Methods* 1994;20:55-67.

CHAPTER **12**

Estimation of Allergen Concentration in Indoor Environments: Prediction of Health-Related Effects

Thomas A. E. Platts-Mills

INTRODUCTION

Estimates of human behavior in the U.S. suggest that 23 hours per day are spent indoors or in an enclosed form of transport. This means that we breathe more air indoors than outdoors and that the quality of indoor air is almost certainly of more relevance to health than that of outdoor air.[1] Several major groups of constituents contribute to the quality of indoor air; these include gases (e.g., SO_2, NO_2, O_3, and CO); particulates, notably those derived from passive smoke; volatile organic compounds (VOCs) and the biologicals, i.e., anything that can produce foreign proteins. The biologicals have several special properties that make them difficult to evaluate. First, there are many different sources each of which produces multiple different proteins. Second, these antigens are associated with a wide variety of different airborne particles. In general, these particles cannot be identified microscopically (some fungi can be) so that identification and quantification depends on immunoassay. In many cases the bulk of airborne protein is carried on "large" particles, i.e., ≥ 10 μm diameter, which fall rapidly in still air.[2-4] This makes it very difficult to make realistic estimates of the quantity inhaled. Thus, there are two major issues with inhalation of biologicals: (1) identification of the relevant proteins and development of techniques for measuring them (Table 1) and (2) the validation of techniques to sample dust or air in houses.[2,3,5,6,7]

The health impact of indoor biologicals is predominantly, but not exclusively related to the respiratory tract. Perennial rhinitis, sinusitis, asthma, and atopic dermatitis have each been related to exposure to indoor allergens. However, the

1-56670-144-9/96/$0.00+$.50

Table 1   Structural and Functional Properties of Indoor Allergens

Source	Allergen[1]	MW	Function	Sequence[2]
House dust mite:				
Dermatophagoides spp.	Group 1	25 kDa	Cysteine protease	cDNA
	Group 2	14 kDa	Unknown	cDNA
	Group 3	~30 kDa	Serine protease	Protein
	Der p 4	~60 kDa	Amylase	Protein
	Der p 5	14 kDa	Unknown	cDNA
	Der p 6	25 kDa	Chymotrypsin	Protein
	Der p 7	22–28 kDa	Unknown	cDNA
Euroglyphus maynei	*Eur m* 1	25 kDa	Cysteine protease	PCR
Blomia tropicalis	*Blo t* 5	14 kDa	Unknown	cDNA
Lepidoglyphus destructor	*Lep d* 1	14 kDa	Unknown	None
Mammals:				
Felis domesticus	*Fel d* 1	36 kDa	(Uteroglobin)	PCR
Canis familiaris	*Can f* 1	25 kDa	Unknown	cDNA
Mus musculus	*Mus m* 1	19 kDa	Calycins, pheromone	cDNA
Rattus norvegicus	*Rat n* 1	19 kDa	Binding proteins	cDNA
Cockroach:				
Blattella germanica	*Bla g* 1	20–25 kDa	Unknown	None
	Bla g 2	36 kDa	Aspartic protease	cDNA
	Bla g 4	21 kDa	Calycin	cDNA
	Bla g 5	36 kDa	Glutathione transferase	cDNA
Periplaneta americana	*Per a* 1	20–25 kDa	Unknown	None
	Per a 3	72–78 kDa	Unknown	None
Fungi:				
Aspergillus fumigatus	*Asp f* 1	18 kDa	Cytotoxin (mitogillin)	cDNA

[1] New nomenclature proposed by the WHO/IUIS subcommittee.
[2] Method given for full sequence determination, where available. However, protein sequences are incomplete, usually N terminal or internal peptide sequences have been determined.

major impact today as judged by doctor visits, school days lost, hospital admissions, and mortality is caused by asthma. Furthermore, the epidemiological evidence for the role of indoor allergy to proteins (or allergens) is much clearer for asthma than any other conditions. Thus, there is overwhelming evidence for the association between sensitization to indoor allergens and asthma. Similarly, it is clear that sensitization is dependent on exposure. Indeed, it is possible to define thresholds for allergens that will predict the risk of sensitization to a given allergen.[5,7,8-12] Although it is clear that most symptomatic sensitized individuals are exposed to allergen, there is not a good quantitative correlation between current exposure and symptoms.[13-15] This suggests that many different factors contribute to symptoms. Indeed, recent results have suggested that air pollution, beta-2-agonists and in particular, viral infections upregulate the response to allergen exposure. Thus, it now seems likely that the major role of indoor allergen exposure is to produce chronic inflammation of the respiratory tract (both lungs and nose), and that on this background, acute episodes can be precipitated by viral infections, increased allergen exposure, and other intercurrent events. Thus, it is clear that measurements of indoor exposure even if absolute should only be expected to predict a part of the health risk. This situation is further compounded because both the initial response to allergens (i.e., immune) and the subsequent response are influenced by factors including genetics of the immune response and the general health of the subject as well as intercurrent other exposures.

IMMUNOASSAYS

The quantity of the major indoor allergens in the environment is measured in µg/g of dust and ng/m^3 of air, thus assay is almost entirely dependent on sensitive immunoassays. In turn, these depend on purification of the proteins. Attempts to purify proteins from house dust have generally failed, thus immunochemistry had to follow identification of the major sources (Table 1). The major cat allergen, *Fel d* 1 was purified by Ohman in 1974 and assays based on polyclonal antibodies were available during the 1970s.[16] Purification of dust mite allergens had to await the development of techniques to grow large quantities of dust mites.[17,18] By contrast, there is no difficulty growing cockroaches, but it was not until relatively recently that the importance of cockroaches in asthma became clear.[19-24] In each case the purification of allergens has been based on identifying those proteins to which IgE antibodies are directed.[16,18,25-27] Thus, we refer to "major allergens" and it is important to understand the meaning of this term. A major allergen is generally a protein to which a majority (i.e., ≥60%) of the individuals who are allergic to that source will have IgE antibodies or positive skin tests. An alternative definition considers how much IgE antibody is directed to this protein. However, in no case is a single protein the only important allergen from a given source; neither can a single protein define which individuals are allergic to a given source. Thus, measurements of major allergens are used as a marker or index of exposure, and the issue is whether measurements of a particular protein represent a *valid index of exposure,* or not.[5,8]

Since 1984 the production of monoclonal antibodies to the major allergens has become increasingly important, both in defining the molecules and in developing assays (see Table 1). Initially inhibition radioimmunoassays were used to measure exposure, but these have been replaced by two site monoclonal antibody assays using ELISA technology.[16,18,28,29] In many cases these assays are sufficiently sensitive to measure concentrations down to 0.5 ng/ml and thus can be used to measure airborne allergens.[3,6]

SAMPLING TECHNIQUES

For inhaled gases all monitoring is carried out on air samples; it would be logical to measure inhaled allergens in airborne samples. However, for at least two and probably all the major indoor allergens, the quantity of protein airborne is dependent on physical disturbance. This disturbance is typically modeled as vacuum cleaning or making beds, however, for cat allergen domestic fans or even the outflow from a room air cleaner may be sufficient to increase airborne allergen.[3,6,30] Because the quantities of allergen airborne are low, it is necessary to sample reasonable volumes of air, e.g., 18 l/min for 30 mins (i.e., 540 l). Using much higher sampling rates, the movement of air created by the pump can become a disturbance itself;[6,31] however, the general conclusion has been that sampling of airborne allergen is primarily useful for research and is not practical for epidemiological surveys.[6,31]

The alternative to sampling the air is to obtain dust samples from reservoirs in the house. This is done routinely with a hand-held vacuum cleaner adapted to collect samples onto a filter or piece of linen.[2] There are a wide range of possible approaches to collecting samples and expressing the results. There are two widely used approaches; the first collects dust samples, assays the allergen(s) in an extract made from 100 mg of dust, and expresses the results in µg/g dust. The results for cockroach allergen are currently given in units per gram, because the absolute values are not well enough defined. The second approach is to try to collect all the dust from a reservoir, assay the allergen, and express the results as the quantity obtained from a site or an area of carpet. Despite the attractions of the second approach it has many problems; the foremost of which is that the quantity obtained depends critically on the recent domestic behavior in the house. Thus, vacuum cleaning the house just prior to a visit will dramatically reduce the total quantity of allergen recovered, but not the concentration of allergen in the dust. Another problem is that small samples will have progressively higher percentage losses during sampling. A final problem is that it is extremely difficult to standardize vacuum cleaner or cleaner operator performance. For these and other reasons two international workshops concluded that the concentration of major allergen per gram of sieved dust was the most reliable guide to indoor exposure.[5,8] From the results of a series of experiments on the relationship of dust mite allergen exposure to sensitization, it was also concluded that the concentration of Group I mite allergen per gram of dust was a valid index of exposure (see below).[8] There are a variety of situations where other methods of evaluating exposure may be preferable including:

- Evaluation of an intervention measure or intervention protocol designed to control the total dust in a reservoir[32]
- When a carpet is treated with a product, e.g., a powder such as Acarosan®, Host®, or Capture®, which will add to the weight of dust recovered; this will dilute the allergen and thus artificially decrease the concentration[33,34]
- Evaluation of particle sizes carrying allergen has only been successfully achieved on airborne samples, by using cascade impactors or sized sieves[2,6,31]

Samples need to be obtained from several different sites in a house because the highest level can be found in different sites. For dust mite the highest level may be found in bedding, carpets, or sofas.[2,7,8] No system for allotting different significance to allergen present in different parts of the house has been developed. It is possible that mite allergen in bedding or pillows has greater significance. However, once the family starts to take simple avoidance measures the bed may no longer be an important site of exposure. It has also been suggested that the mean level of allergen in a house is a sensible measure, but in many studies the bedding has much the highest level of mite allergen so that additional samples will simply dilute the result. Currently, the most widely used policy is to report the highest concentration found in a house. Cat (*Fel d* 1) or dog (*Can f* 1) allergens are found all over a house but the highest level is generally in the floor of the living room or TV room.[14,26,27,35] The highest level of cockroach allergen is most often found in the kitchen, and here the usual plan has been to obtain samples from the available sites, i.e., floor and cabinets. Again, with cockroach allergen, domestic activity in the kitchen can dramatically decrease the quantity of dust obtained, and it is necessary to "search" for adequate samples.[24,25]

CONCLUSIONS REGARDING SAMPLING

There is no simple answer to measuring exposure to indoor allergens. In the absence (at present) of a consistently successful method for measuring the allergen inhaled, whichever technique is used should be seen as an *index of,* rather than a measurement of, exposure. At present, the best validated approach is to obtain dust samples from at least three and usually four sites (bedding, bedroom floor, living room, and kitchen); assay the content of the relevant major allergens in an aliquot, and express the result as the concentration of allergen per gram of dust. In some cases other methods of monitoring will be relevant or revealing, but it is best to express results as concentration of allergen as well as the alternative technique. The important objective is to make it possible to compare results between studies.

DISEASE RELATIONSHIP

The present discussion will be restricted to consideration of asthma: first, because of its medical and economic importance and second, because it is possible to make objective measurements of the disease. Asthma is the most common chronic medical cause for children missing school, the leading cause of hospital admission among school age children, and there are more children on chronic medication for asthma than for any other disease. The evidence for a relationship between indoor allergens and asthma has to be considered in three phases: (1) the evidence for an association between immunological sensitization and asthma; (2) the evidence that a given level of exposure will predict sensitization; (3) the evidence that current exposure is related to symptoms or severity of asthma.

Sensitization and Asthma

Sensitization to the common indoor allergens can be established by skin tests for wheal and flare responses (using a 4-mm wheal diameter following an epicutaneous or prick skin test as positive) or by assays of serum IgE antibodies using the radioallergosorbent test (RAST). There is very strong evidence for an association between sensitization to indoor allergens and asthma. Clinic, emergency room, school, and prospective studies from many parts of the world have provided consistent and strong results with odds ratios between 3 and 20.[9-12,23,24,36,37] The striking feature of these results is that they are not restricted to one antigen. Thus, in many humid areas dust mite is the dominant allergen. However, in areas that are dry either because of the northern latitude or because of high altitude, the dominant allergens are those derived from domestic animals.[14,17,35,38,39] In many cities in the U.S. the dominant sensitization associated with asthma is to allergens derived from the German cockroach *Blatella germanica*.[19-24] Thus, in all studies where sensitization to indoor allergens has been appropriately assessed, the association between indoor allergens and asthma has been very strong.

Exposure to Indoor Allergens and Sensitization

That sensitization to allergens is related to exposure is obvious since no immune response can occur without exposure to the relevant antigen. The first suggestion that a specific level of exposure would increase the risk of asthma came from the original work of Voorhorst and his colleagues in The Netherlands. They suggested that the presence of 100 mites per gram of dust increased the risk and that 500 mites per gram represented a higher risk.[40] These levels were supported by subsequent studies but identifying and counting mites in dust is time consuming and (worse still) requires skill.[9] Once assays for dust mite allergens were established, studies were carried out to establish the approximate relationship between mite counts and allergen as well as to define the levels associated with disease. In 1987 the value of 2 µg of Group I dust mite allergen per gram of dust was proposed[7] and then supported by the First International Workshop on Mite Allergens and Asthma.[5] Subsequently, a series of studies from Germany, U.K., Australia, and France provided very strong evidence that 2 µg/g was a useful threshold. However, critical analysis of those studies suggests that this is a threshold for sensitization of genetically at risk individuals. Thus, the study by Lau and her colleagues in Germany was on asthmatic children.[10] Our study in England was on children with one allergic parent.[36] Recently Kueher and his colleagues in Germany have carried out an elegant study which shows clearly that the level of mite allergen necessary to sensitize "nonatopic" individuals is much higher; ~60 µg Group I mite allergen per gram.[41] However, that study confirmed the relevance of a threshold level of 2 µg/g for children who were atopic as judged by already having a positive skin test to some other allergen.[41] The threshold level can also be approached by examining those environments where mites do not flourish. Several studies have now shown that in dry areas the sensitization to mites is unusual and is not associated with asthma.[14,35] Taken together, the studies show that in a community where the houses have levels of dust mite allergen >2 µg/g dust the majority of genetically at risk children will become sensitized to dust mite allergens and this sensitization will be associated with asthma.

For allergens other than dust mite quantitative studies of the risk associated with exposure are far less well established. However, there is sufficient data to make preliminary suggestions about these relationships for cat and cockroach allergens. There are major socioeconomic differences in housing in the U.S. and these are associated with different patterns of allergen exposure. In many city projects domestic animals are not allowed and/or the families do not like indoor cats. In areas like this where less than 5% of the families own cats, the level of cat allergens in the houses is generally very low. For example, in inner city Atlanta we found very low levels of cat allergen, low prevalence of sensitization and in keeping with that, no association between cat sensitization and asthma.[24] This suggests that the levels found in houses in Atlanta are sufficiently low that they do not carry a risk of sensitization. By contrast, in a suburban area in New Mexico, 65% of the houses contained domestic animals, sensitization to cat (and dog) allergens was common and was strongly associated with asthma.[14,35] Taking the two studies together, it is possible to propose a level of between 1 and 8 µg *Fel d*1 per gram of dust as a threshold level for cat allergens. However, there are real questions about whether a threshold for cat allergens should be compared with those for mite since the cat allergen behaves so differently when airborne.[3,30,31] Also, previous studies suggest that exposure to cat allergen levels in the community are the relevant factor. This could either be because children (even if they do not have a cat in their own house) can become sensitized as a result of visiting the houses of friends who have a cat, from exposure at school or day-care centers, or from cat allergen transferred to their houses by visitors.[42] It appears that transfer on clothing is a particular problem with cat allergens because the particles of dander are sticky. Certainly, it is common to find cat allergen on walls and other surfaces. Mite allergen does not appear to stick in this way and cockroach allergen has not been studied in detail.

Exposure to cockroach allergen is measured in units per gram of dust and preliminary estimates suggest that 2 units per gram may be a useful cut-off value. In suburban areas where significant cockroach infestation is unusual, it is rare to find levels higher than 2 units per gram of *Bla g* 2 and very few children become sensitized. By contrast, in areas where many of the children become sensitized dust from 25 to 70% of the houses contains ≥2 units *Bla g* 2 per gram. Taken together there is very good evidence that there is a dose-response relationship between exposure to common indoor allergens and sensitization. Put another way this means that genetically at risk children will become allergic to the allergen (or allergens) that is present in their houses. In turn from our previous discussion it is clear that this sensitization is an important risk for nonspecific bronchial hyperreactivity (BHR) and symptomatic asthma.

Current Exposure and Symptoms of/or Severity of Asthma

The factors influencing symptoms of asthma are very complex. Taking the current model which is that exposure to allergens gives rise to "inflammation" characterized by eosinophils, that this inflammation is a major cause of BHR, and that chronic BHR underlies much of the symptoms of asthma, it is clear that there are many levels of complexity that could blur simple dose response between current exposure and symptom severity (Figure 1). Simple examples include: the inaccuracy of measurement of exposure of the lungs; individual differences in the inflammatory response or the rate of healing of the response; and absence of a simple relationship between inflammation and BHR. From many studies it is clear that even careful studies of inflammation or quantitative studies of BHR do not provide a close relationship with symptoms. Thus, it is perhaps not surprising that estimates of current exposure, measured as µg/g of dust do not provide an accurate guide to the symptoms of asthma even if the analysis is restricted to allergic individuals.[13-15,23,36] It is important to remember that sensitization and current exposure to other allergens would obscure any simple dose-response relationships. From some of the results it could be asked whether current exposure plays an important role in symptoms. There are, however, a series of avoidance experiments that strongly support the importance of current exposure and may give some guide to the quantitative relationship.[43-46] In all studies where allergic individuals have been removed from their houses, there has been a highly significant decrease in symptoms and nonspecific BHR.[47-50] However, the decrease in allergen exposure (e.g., to dust mites) when a patient moves to a hospital or a sanatorium is often >95%, e.g., from 13.4 µg/g (mean) down to ≤0.2 µg/g.[50] Thus the implication is that the level of allergen necessary to maintain symptoms is too variable to see dose-response relationships within a cross-sectional study.[13] Nonetheless, some studies do show relationships of this kind and even clear seasonal changes where symptoms increase at the time of year where exposure is highest.[51-54]

PREDICTION

From our preceding discussion it is clear that indoor allergen exposure is strongly related to sensitization and that sensitization to a relevant indoor allergen is one of the strongest known risk factors for asthma. The question then arises whether the level of allergen or allergens present within a house will provide a useful predictive guide to the prevalence or severity of asthma within a community. This question is very important because the answer should provide a guide to whether changes in houses and house management are the causes of increases in asthma prevalence. There are two kinds of studies that would provide important direct evidence: first, intervention studies from birth demonstrating that the prevalence of both sensitization and symptoms can be decreased by reducing overall exposure and second, epidemiological studies demonstrating that children raised

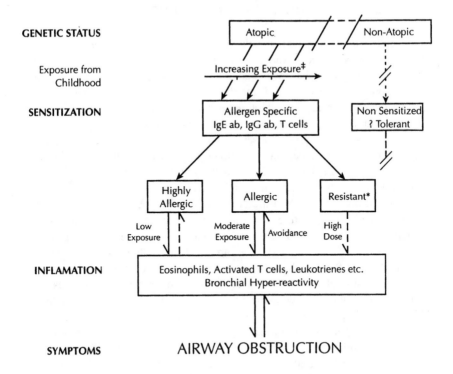

Figure 1   Dose response of the relationship between allergen exposure and asthma.

*   Under normal circumstances these individuals will have mild symptoms or no symptoms.

‡   With increasing exposure a larger % of the population will become sensitized, in addition high exposure will accelerate the process.

in houses without significant exposure to indoor allergens have a lower prevalence or severity of asthma.[55] Currently, there are no fully convincing studies of this kind. On the other hand, there is a mass of evidence that asthma is less common and indeed rare among communities living in huts or in the bush. Thus, it seems most likely that the increase in asthma is related to the change to a Western-life style but whether it is changes in housing per se that are the cause is less clear. Our current view is that many different changes have occurred that could have increased the concentration of allergen in houses, but that equally there are many changes that could alter the immune response to allergens. Thus, although it still seems very plausible that changing houses, reducing furnishings, decreasing humidity and temperature, increasing ventilation, and removing pets would reduce the prevalence of asthma, this cannot be assumed on the basis of presently available results.

CONCLUSION AND DISCUSSION

Since it is now accepted that asthma is an "inflammatory" disease of the lungs (for which the best model is inhalation of allergens by an allergic individual), it is tempting to believe that measurements of indoor allergens would provide an accurate prediction of the risk of the disease. Furthermore, there have been very striking increases in asthma which appear to correlate with changes in housing that have increased or would be expected to increase exposure to indoor allergens. Increases in asthma have been observed with "urbanization" of rural populations (e.g., Papua, New Guinea, several studies in Africa, and the Pacific Islanders migrating to New Zealand), and also with *improvements* in housing in Western society. When we try to pin down the specific changes responsible for the increase in asthma, it becomes more difficult. In some areas increases in mite allergens due to increased temperature, humidity, and furnishing could explain much of the increase in asthma. However, the increase has been equally impressive in northern Sweden[38] and Finland,[39] i.e., areas where the houses are dry and mites generally do not flourish. In these areas sensitization to animal dander has been found to be strongly associated with asthma, and it has been suggested that decreased ventilation following the *energy crisis* triggered the increase in asthma. This model is important to our present discussion for two reasons: first, decreased ventilation could change airborne allergen and other particles or gases and second, changes in ventilation rate can increase the quantity of airborne cat allergen without a change in the allergen concentration in reservoir dust.[3,30] This clearly undermines the reliability of assays of reservoir dust as an index of exposure. In some areas of the world it appears that asthma is a disease of affluence. By contrast, over the last 20 years there has been a progressive increase in asthma among the inhabitants of inner cities in North America.[56] Among this population both dust mite and cockroach allergen appear to be important.[24] In some areas of the central northern U.S. it appears that cockroach is the dominant allergen. This result supports two conclusions: that the dominant allergens associated with asthma since 1970 have been those found indoors and that increases in asthma have occurred in areas where dust mite is not relevant. Taken together, the results suggest that accurate measurements of exposure to the major indoor allergens would be important in predicting the risk of asthma in a population.

The measurement of exposure that is established is immunoassay of representative major allergens in an aliquot of reservoir dust. These assays are accurate, well standardized, and give results in absolute units (i.e., μg/g of dust). The errors involved in sampling are significant but have been extensively studied so they are well known. Studies of allergen exposure have provided consistent results for the relationship to sensitization. However, measurements of allergen on the floor are a long way from accurate measurements of allergen entering the lungs. Thus the rather poor relationship between measurements of current "exposure" and symptoms of asthma may simply reflect the indirect nature of the measurements. On the other hand, it is clear that many factors contribute to symptoms. These include "triggers" such as cold air, exercise, passive smoke, and histamine which are not thought to contribute to inflammation or BHR. There are even more complex factors such as viruses, beta-2-agonists, and diesel fumes, for each of which there is evidence for a secondary effect increasing BHR and/or inflammation.[57-59]

In conclusion, we believe that measurement of representative major allergens in dust can provide a reliable prediction of the prevalence of sensitization in a community. In turn, it is possible to predict the relationship between sensitization and asthma in a community. In some studies it appears that high levels of mite allergen can also predict the prevalence and severity of symptoms in a community. However, it is clear that asthma prevalence up to ~5% can occur in areas where dust mite allergens are not relevant. Over the last 30 years the indoor environment has changed dramatically, including major decreases in ventilation encouraged by the Department of the Environment. However, it is still not clear which aspects of these changes are most important to the increases in asthma. Thus, at present proposals to change houses should be seen as experiments designed to test a hypothesis rather than as health policy based on valid predictions.

REFERENCES

1. Pope A.M., Patterson R., Burge H., [editors]. Indoor Allergens: Assessing and Controlling Adverse Health Effects. Washington, DC: Institute of Medicine, 1993.
2. Tovey E.R., Chapman M.D., Wells C.W., Platts-Mills T.A.E. The distribution of dust mite allergen in the houses of patients with asthma. *Am. Rev. Resp. Dis.* 1981;124:630-635.
3. Luczynska C.M., Li Y., Chapman M.D., Platts-Mills T.A.E. Airborne concentrations and particle size distribution of allergen derived from domestic cats *(Felis domesticus):* Measurements using cascade impactor, liquid impinger and a two site monoclonal antibody assay for *Fel d* I. *Am. Rev. Resp. Dis.* 1990;141:361-367.
4. Platts-Mills T.A.E., Mitchell E.B., Tovey E.R., Chapman M.D., Wilkins S.R. Airborne allergen exposure, allergen avoidance and bronchial hyperreactivity. In: Kay A.B., Austen K.F., Lichtenstein L.M., eds. *Asthma: Physiology, Immunopharmacology and Treatment, Third International Symposium.* London: Academic Press, Inc., 1984;297-314.
5. Platts-Mills T.A.E., De Weck A. Dust mite allergens and asthma — A world wide problem. *Bull. WHO* 1989;66:769-780.
6. Platts-Mills T.A.E., Chapman M.D., Heymann P.W., Luczynska C.M. Measurements of airborne allergen using immunoassays. In: Solomon W., Saunders W.B., eds. *Immunology and Allergy Clinics of North America* Philadelphia: W.B. Saunders, 1989;269-283.
7. Platts-Mills T.A.E., Hayden M.L., Chapman M.D., Wilkins S.R. Seasonal variation in dust mite and grass-pollen allergens in dust from the houses of patients with asthma. *J. Allergy Clin. Immunol.* 1987;79:781-791.
8. Platts-Mills T.A.E., Thomas W.R., Aalberse R.C., Vervloet D., Chapman M.D., Co-Chairmen. Dust mite allergens and asthma: Report of a 2nd international workshop. *J. Allergy Clin. Immunol.* 1992;89:1046-1060.
9. Korsgaard J. Mite asthma and residency. A case-control study on the impact of exposure to house-dust mites in dwellings. *Am. Rev. Resp. Dis.* 1983;128:231-235.

10. Lau S., Falkenhorst G., Weber A., Werthmann I., Lind P., Buettner-Goetz P., Wahn U. High mite-allergen exposure increases the risk of sensitization in atopic children and young adults. *J. Allergy Clin. Immunol.* 1989;84:718-725.

11. Charpin D., Birnbaum J., Haddi E., Genard G., Lanteaume A., Toumi M., Faraj F., van der Brempt X., Vervloet D. Altitude and allergy to house dust mites: A paradigm of the influence of environmental exposure on allergic sensitization. *Am. Rev. Resp. Dis.* 1991;143:983-986.

12. Peat J.K., Salome C.M., Woolcock A.J. Longitudinal changes in atopy during a 4 year period: relation to bronchial hyperresponsiveness and respiratory symptoms in a population sample of Australian schoolchildren. *J. Allergy Clin. Immunol.* 1990;85:65-74.

13. Platts-Mills T.A.E., Sporik R.B., Wheatley L.M., Heymann P.W. Editorial: Is there a dose response relationship between exposure to indoor allergens and symptoms of asthma? *J. Allergy Clin. Immunol.* 1995;96:435-440.

14. Ingram J.M., Sporik R., Rose G., Honsinger R., Chapman M.D., Platts-Mills T.A.E. Quantitative assessment of exposure to dog (*Can f* I) and cat (*Fel d* I) allergens: Relationship to sensitization and asthma among children living in Los Alamos, NM. *J. Allergy Clin. Immunol.* 1995;96:449-456.

15. Marks, G.B., Tovey, E.R., Toelle, B.G., *et al.* Mite allergen (*Der p* I) concentration in houses and its relation to the presence and severity of asthma in a population of Sydney school children. *J. Allergy Clin. Immunol.* 1995;96:441-448.

16. Ohman J.L., Lowell F.C., Bloch K.J. Allergens of mammalian origin. III. Properties of a major feline allergen. *J. Immunol.* 1974;113:1668-1677.

17. Chapman M.D., Platts-Mills T.A.E. Measurement of IgG, IgA and IgE antibodies to *Dermatophagoides pteronyssinus* by antigen-binding assay, using a partially purified fraction of mite extraction (F4P1). *Clin. Exp. Immunol.* 1978;34:126-136.

18. Chapman M.D., Platts-Mills T.A.E. Purification and characterization of the major allergen from *Dermatophagoides pteronyssinus*-antigen P1. *J. Immunol.* 1980;125:587-592.

19. Hulett A.C., Dockhorn R.J. House dust mite *(D. farinae)* and cockroach allergy in a midwestern population. *Ann. Allergy* 1979;42:160-165.

20. Kang B., Vellody D., Homburger H., Yunginger J.W. Cockroach cause of allergic asthma. Its specificity and immunologic profile. *J. Allergy Clin. Immunol.* 1979;63:80-86.

21. Twarog F.J., Picone F.J., Strunk R.S., So J., Colten H.R. Immediate hypersensitivity to cockroach: isolation and purification of the major antigens. *J. Allergy Clin. Immunol.* 1976;59:154-160.

22. Pollart S.M., Chapman M.D., Fiocco G.P., Rose G., Platts-Mills T.A.E. Epidemiology of acute asthma: IgE antibodies to common inhalant allergens as a risk factor for emergency room visits. *J. Allergy Clin. Immunol.* 1989;83:875-882.

23. Gelber L.E., Seltzer L.H., Bouzoukis J.K., Pollart S.M., Chapman M.D., Platts-Mills T.A.E. Sensitization and exposure to indoor allergens as risk factors for asthma among patients presenting to hospital. *Am. Rev. Resp. Dis.* 1993;147:573-578.

24. Call R.S., Smith T.F., Morris E., Chapman M.D., Platts-Mills T.A.E. Risk factors for asthma in inner city children. *J. Pediatr.* 1992;121:862-866.

25. Pollart S.M., Mullins D.E., Vailes L.D., Hayden M.L., Platts-Mills T.A.E., Sutherland W.M., Chapman M.D. Identification, quantitation and purification of cockroach allergens using monoclonal antibodies. *J. Allergy Clin. Immunol.* 1991;87:511-521.

26. Schou C., Hansen G., Linter T., *et al*. Assay for the major dog allergen, *Can f* I: Investigation of house dust samples and commercial dog extracts. *J. Allergy Clin. Immunol.* 1991;88:847.

27. De Groot H., Goel K., Van Swicter P., *et al*. Affinity purification of a major and minor allergen from dog extract. *J. Allergy Clin. Immunol.* 1991;87:1056-61.

28. Chapman M.D., Heymann P.W., Wilkins S.R., Brown M.J., Platts-Mills T.A.E. Monoclonal immunoassays for the major dust mite *(Dermatophagoides)* allergens, *Der p* I and *Der f* I, and quantitative analysis of the allergen content of mite and house dust extracts. *J. Allergy Clin. Immunol.* 1987;80:184-194.

29. Luczynska C.M., Arruda L.K., Platts-Mills T.A.E., Miller J.D., Lopez M., Chapman M.D. A two-site monoclonal antibody ELISA for the quantitation of the major *Dermatophagoides* spp. allergens, *Der p* I and *Der f* I. *J. Immunol. Methods* 1989;118:227-235.

30. De Blay F., Chapman M.D., Platts-Mills T.A.E. Airborne cat allergen (*Fel d* I): Environmental control with the cat *in situ*. *Am. Rev. Respir. Dis.* 1991;143:1334-1339.

31. Swanson M.A., Agarwal M.K., Reed C.E. An immunochemical approach to indoor aeroallergen quantitation with a new volumetric air sampler: studies with mite, roach, cat, mouse and guinea-pig antigens. *J. Allergy Clin. Immunol.* 1987;76:724-729.

32. Carswell F. Asthma and altitude. *Clin. Exp. Allergy* 1993;23:973-975.

33. Hayden M.L., Rose G., Diduch K.B., Domson P., Chapman M.D., Heymann P.W., Platts-Mills T.A.E. Benzyl benzoate moist powder: Investigation of acarical activity in cultures and reduction of dust mite allergens in carpets. *J. Allergy Clin. Immunol.* 1992;89:536-545.

34. Woodfolk J.A., Hayden M.L., Miller J.D., Rose G., Chapman M.D., Platts-Mills T.A.E. Chemical treatment of carpets to reduce allergen: A detailed study of the effects of tannic acid on indoor allergens. *J. Allergy Clin. Immunol.* 1994;94:19-26.

35. Sporik R., Ingram J.M., Price W., Sussman J.H., Honsinger R.W., Platts-Mills T.A.E. Association of asthma with serum IgE and skin-test reactivity to allergens among children living at high altitude. Tickling the dragon's breath. *Am. J. Res. Crit. Care Med.* May 1995.

36. Sporik R., Holgate S.T., Platts-Mills T.A.E., Cogswell J. Exposure to house dust mite allergen (*Der p* I) and the development of asthma in childhood: A prospective study. *N. Eng. J. Med.* 1990;323:502-507.

37. Sears M.R., Hervison G.P., Holdaway M.D., Hewitt C.J., Flannery E.M., Silva P.A. The relative risks of sensitivity to grass pollen, house dust mite, and cat dander in the development of childhood asthma. *Clin. Exp. Allergy* 1989;19:419-424.

38. Wickman M., Nordvall L., Pershagen G., Korsgaard J., Johannsen N. Sensitization to domestic mites in a cold temperate region. *Am. Revs. Resp. Dis.* 1995;148:58-62.

39. Haahtela T., Lindholm H., Bjorksten F., Koskenvuo K., Laitenen L.A. Prevalence of asthma in Finnish young men. *Br. Med. J.* 1990;301:266-268.

40. Voorhorst R., Spieksma F.Th.M., Varekamp N. House dust mite atopy and the house dust mite *Dermatophagoides pteronyssinus* (Troussart, 1897). Leiden: Stafleu's Scientific Publishing Co., 1969.

41. Kueher J., Frischer T., Meinert R., *et al*. Mite exposure is a risk for the incidence of specific sensitization. *J. Allergy Clin. Immunol.* 1994;94:44-52.

42. Munir A.K.M., Einarsson R., Schou, C. *et al*. Allergens in school dust. *J. Allergy Clin. Immunol.* 1995;91:1067-1074.

43. Murray A.B., Ferguson A.C. Dust-free bedrooms in the treatment of asthmatic children with house dust or house dust mite allergy: a controlled trial. *Pediatrics* 1983;71:418-422.

44. Walshaw M.J., Evans C.C. Allergen avoidance in house dust mite sensitive adult asthma. *Q. J. Med.* 1986;58:199-215.

45. Colloff M.J., Lever R.S., McSharry C. A controlled trial of house dust mite eradication using natamycin in homes of patients with atopic dermatitis: Effect on clinical status and mite population. *Br. J. Dermatol.* 1989;121:199-208.

46. Ehnert B., Lau-Schadendorf S., Weber A., Buettner P., Schou C., Wahn U. Reducing domestic exposure to dust mite allergen reduces bronchial hypersensitivity in sensitive children with asthma. *J. Allergy Clin. Immunol.* 1992;90:135-138.

47. Kerrebijn K.F. Endogenous factors in childhood CNSLD: methodological aspects in population studies. In: Orie N.G.M., van der Lende R., eds. Bronchitis III. The Netherlands: Royal Vangorcum Assen, 1970:38-48.

48. Vervloet D., Penaud A., Razzouk H., Senft M., Arnaud A., Boutin C., Charpin J. Altitude and house dust mites. *J. Allergy Clin. Immunol.* 1982;69:290-296.

49. Boner A.L., Niero E., Antolini I., Valletta E.A., Gaburro D. Pulmonary function and bronchial hyperreactivity in asthmatic children with house dust mite allergy during prolonged stay in the Italian Alps (Misurina 1756m). *Ann. Allergy* 1985;54:42-45.

50. Platts-Mills T.A.E., Tovey E.R., Mitchell E.B., Moszoro H., Nock P., Wilkins S.R. Reduction of bronchial hyperreactivity during prolonged allergen avoidance. *Lancet* 1982;2:675-678.

51. Chan-Yeung M., Manfreda, J., Dimicu-Ward, H., Lam, J., Ferguson, A., Warren, P., Platts-Mills, T., Becker, A., Mite and cat allergens in homes and severity of asthma. *Am. J. Resp. Crit. Care Med.* 1995;152:1805-1811.

52. Sporik R.B., Platts-Mills T.A.E., Cogswell J.J. Exposure to house dust mite allergen of children admitted to hospital. *Clin. Exp. Allergy* 1993;23:740-746.

53. Vervloet D., Charpin D., Haddi E., N'Guyen A., Birnbaum J., Soler M., van der Brempt X. Medication requirements and house dust exposure in mite sensitive asthmatics. *Allergy* 1991;46:554-558.

54. O'Hallaren M.T., Yunginger J., Offord K.P., Somers M.J., O'Connell E.J., Ballard D.J., Sach M.I. Exposure to an aeroallergen as a possible precipitating factor in respiratory arrest in young patients with asthma. *N. Eng. J. Med.* 1991;324:359-363.

55. Arshad S.H., Mathews S., Grant C., Hide N.W. Effect of allergen avoidance on the development of allergic diseases of childhood. *Lancet* 1992;339:1493-1496.

56. Weiss K.B., Gergen P.J., Crain E.F. Inner-city asthma: the epidemiology of an emergency US public Health concern. *Chest* 1992;101:362s-367s.

57. Calhoun W.J., Dick E.C., Schwartz L.B., Busse W.W. A common cold virus, rhinovirus 16, potentiates airway inflammation after segmental antigen broncho-provocation in allergic subjects. *J. Clin. Invest.* 1994;94:2200-2208.

58. Cockcroft D.W., McParland C.P., Britto S.A., Swystun V.A., Rutherford B.C. Regular inhaled salbuterol and airway responsiveness to allergen. *Lancet* 1993;342:833-837.

59. Diaz Sanchez D., Dotson A.R., Takenaka H., Saxon A. Diesel exhaust particles induce local IgE production *in vivo* and alter the pattern of IgE messenger RNA isoforms. *J. Clin. Invest.* 1994;94:1417-1425.

CHAPTER 13

Building-Related Hypersensitivity Diseases: Sentinel Event Management and Evaluation of Building Occupants

Cecile Rose

BACKGROUND

The spectrum of building-associated illnesses are conventionally divided into two major categories: building-related illnesses (BRI) and sick building syndrome (SBS). Sick building syndrome is defined as an excess of nonspecific symptoms, including eye, nose, and throat irritation, headache, lethargy, dizziness, mental fatigue, and skin irritation, associated with building occupancy.[1] Building-related illnesses, distinguished from SBS by objective clinical findings and organ specific diagnoses, include infections, toxic reactions to specific exposures, long latency illnesses such as cancer from indoor radon or asbestos exposure, and hypersensitivity diseases. As more sensitive diagnostic techniques are applied in individuals with typical SBS symptoms, it is likely that many of these symptoms will be ascribable to subtle inflammatory effects from indoor exposures, and thus recategorized as BRI. For example, when nasal lavage is performed on building occupants with symptoms of mucous membrane irritation from exposure to environmental tobacco smoke (ETS), excess inflammatory cells are detected, probably reflecting an irritant response (see Chapter 8). Among the BRIs, hypersensitivity diseases are probably common but are difficult to recognize and causally relate to indoor exposures.

BUILDING-RELATED HYPERSENSITIVITY DISEASES

Hypersensitivity diseases arise following variable latency periods during which immunologic sensitization occurs in the exposed host. While few data exist

1-56670-144-9/96/$0.00+$.50

on latencies between exposure and onset of symptoms, latency periods for building-related hypersensitivity symptoms probably vary from a few weeks or months to many years of building occupancy. This variability may reflect differences in antigen concentration or potency, factors which are poorly understood. The spectrum of building-related hypersensitivity diseases includes allergic rhinosinusitis, asthma, and hypersensitivity pneumonitis (HP). Clinical expression depends in part on the location of antigen particle deposition in the respiratory tract and on host factors.

Rhinosinusitis related to allergen exposures in office building environments is probably common but is not well studied, in part due to variability in clinical case definitions and diagnostic criteria. Allergic rhinosinusitis is diagnosed by a combination of symptom history, physical examination, eosinophils on nasal smear, elevated serum IgE antibody levels, and immediate skin test reactivity to specific aeroallergens. Large antigenic particles (>10 to 15 μ diameter) which deposit in the nose are associated with symptoms of sneezing, rhinorrhea, and itchy, watery eyes commonly reported by those who suffer with seasonal hayfever. The frequency of such allergenic exposures in buildings and the prevalence of associated upper airway symptoms is unknown.

For asthma, the immunologic reaction is IgE mediated and involves a spectrum of inflammatory cells and mediators which cause airway inflammation, basement membrane thickening, and smooth muscle hypertrophy. The symptoms of building-related asthma include chest tightness, wheezing, coughing, and shortness of breath, occurring either immediately following building occupancy or 4 to 12 h later, or both. Diagnosis depends on symptom history, physical examination, peak air flow variability, airflow limitation on spirometry (often with a significant reaction to an inhaled bronchodilator), and/or a positive reaction to laboratory challenge with methacholine. Reductions in peak air flow related to building occupancy are helpful diagnostic clues for building-related asthma. There are a limited number of studies of building-related asthma[2-4] but only a few contain objective criteria (peak flow or spirometric data) to confirm the diagnosis of asthma.

In HP (allergic alveolitis), an antigen-induced cell-mediated immune reaction causes a T lymphocyte-predominant inflammatory effect on both the airway and the pulmonary interstitium where gas exchange occurs. Over 50 epidemiologic investigations and case reports of building-related HP have been published, and it will be the focus of further discussion in this chapter.

Most of the building antigens associated with hypersensitivity diseases are of fungal or bacterial origin and are characterized as bioaerosols. Bioaerosols are defined as airborne particles, large molecules, or volatile compounds that are living or released from a living organism.[5] The necessary conditions and events required to produce aerosolization of an organism or its parts are the presence of a reservoir, amplification (increase in numbers or concentration), and dissemination. The microbial antigens that can cause hypersensitivity diseases proliferate in common environmental niches, particularly in areas where water damage creates circumstances in which these contaminants are amplified and disseminated. Contaminated air handling units, water spray humidification systems, and water-damaged furnishings are some of the reported sources of antigen exposure causing building-related hypersensitivity diseases.[6-8]

Once an exposed host develops immune sensitization of the respiratory tract from an inhaled antigen, continued antigen exposure may lead to permanent and, in some cases, progressive disease. When the diagnosis of a hypersensitivity disease is suspected, appropriate environmental and clinical assessment must be undertaken in order to prevent permanent lung damage.

HYPERSENSITIVITY PNEUMONITIS

Hypersensitivity pneumonitis characterizes a spectrum of granulomatous, interstitial, and alveolar-filling lung diseases resulting from repeated inhalation of and sensitization to a wide variety of organic dusts and low molecular weight chemical antigens.[9] The antigens associated with HP include microbial contaminants (mainly fungi and bacteria), avian proteins, and certain low molecular weight sensitizing chemicals.

Historically, HP was considered rare and associated only with specific occupations and avocations such as farming, wood working, and bird breeding. The likelihood that unrecognized HP is actually quite common is based on several factors. The signs and symptoms of the illness are nonspecific and mimic other diseases such as influenza, viral pneumonia, and asthma. These nonspecific clinical findings are further masked by incomplete occupational and environmental history taking which results in a poor understanding of exposure-related risk factors and often leads to inadequate disease recognition. Early disease is often accompanied by a normal chest radiograph and pulmonary function, rendering standard diagnostic tools insensitive. Precipitating antibodies, once considered the gold standard for diagnosis, can be found in exposed individuals without disease,[10] and may be negative due to the wrong choice of antigen, the limitations of commercial antigen preparations, and resolution of serum abnormalities after exposure ceases.

Classical descriptions of HP often invoke the acute form of disease, with symptoms of fever, chills, myalgias, dyspnea, and cough occurring 8 to 12 h after exposure. These acute symptoms typically resolve within 24 to 48 h after onset, recurring only with repeated exposure. Insidious onset of dyspnea, cough, chest tightness, malaise, weight loss, and myalgias characterize the subacute and chronic forms of HP. Occupational organic dust exposure from activities such as baling hay contaminated with fungi or bacteria is classically associated with the acute form of illness and is known as farmer's lung.[11] The subacute and chronic forms of HP may be more common in circumstances of long-term, lower level antigen exposure as might be expected in contaminated building environments.

Traditional diagnostic criteria for HP include a history of antigen exposure, symptoms appearing several hours later, basilar crackles on exam, an abnormal chest radiograph, and restrictive changes and/or decreased diffusing capacity for carbon monoxide on pulmonary function testing.[12] However, stringent diagnostic criteria probably underestimate milder and more subacute cases of HP where the chest radiograph is normal or when symptoms are subtle and insidious in onset. Hodgson et al.[13] described a decline in the sensitivity of chest radiographs for the diagnosis of HP over the years 1950 to 1980. This meta-analysis found that chest X-rays were also less likely to be abnormal when a population-based approach to diagnosis was undertaken.

Oral corticosteroids may be useful in the management of symptoms of acute HP and in the setting of chronic or progressive lung fibrosis. However, the most important approach to therapy for HP is removal from further exposure to the inhaled sensitizer. Therefore, early recognition of circumstances in which building-related HP can occur is critical for patient management and disease prevention.

CASE STUDIES: HP OUTBREAK INVESTIGATIONS

We have identified several outbreaks of building-related HP due to repeated exposure and subsequent sensitization to microbial bioaerosols. These outbreaks were discovered following recognition of an index case in our Occupational/Environmental Medicine Clinic, and illustrate the importance of maintaining a high index of suspicion and obtaining a careful occupational and environmental history to assure early disease recognition and prevention.

In one case, a 24-year-old nonsmoking lifeguard presented with symptoms of fatigue, malaise, chest tightness, cough, and progressive dyspnea on exertion. His symptoms worsened toward the end of his work shift at an indoor swimming pool. The swimming pool was characterized by extensive water spray features which generated a respirable aerosol during use. Diagnostic evaluation of the symptomatic lifeguard showed normal resting pulmonary function but marked gas exchange abnormalities with exercise. Fiberoptic bronchoscopy showed a marked bronchoalveolar lavage lymphocytosis and multiple noncaseating granulomas on transbronchial biopsy, suggestive for HP. With removal from the pool environment, the patient's clinical abnormalities slowly improved over the next two years.

Due to the work-related pattern of his symptoms and reports of similar symptoms among co-workers, we conducted an epidemiologic investigation of current and former pool employees. Those reporting two or more work-related respiratory and systemic symptoms underwent diagnostic evaluation, including history, physical examination, rest and exercise pulmonary function studies, chest radiograph, and fiberoptic bronchoscopy. A case of HP was considered confirmed based on stringent histologic criteria including the finding of macrophage aggregates with peribronchovascular lymphoplasmacytic infiltrates with or without granulomas. We found work-related HP in 18 of 67 (27%) lifeguards, including former lifeguards who had been away from the swimming pool for prolonged time periods. Given the likelihood that transbronchial biopsy does not detect all parenchymal abnormalities due to sampling error, our finding of a 27% attack rate was surprisingly high.

Despite intensive efforts to identify and eliminate sensitizing exposures in the pool environment, subsequent use of the pool led to recurrence of respiratory and systemic symptoms among lifeguards. Questionnaire assessment of all pool employees before and after employment commenced, and medical diagnostic evaluation of those with work-related symptoms, led to recognition of granulomatous lung disease in 15 of 23 (65%).

We identified another outbreak of building-related HP among maintenance workers in an office building at a correctional facility. The index case presented with progressive weight loss, fatigue, nonproductive cough, chest tightness, and exertional dyspnea. These symptoms were worse toward the end of his work shift and at the end of his work week, initially improving over weekends and during periods away from work. Physical examination showed basilar lung crackles but was otherwise normal. His chest radiograph had diffuse, predominantly upper lobe nodules. Resting pulmonary function including lung volumes and diffusing capacity for carbon monoxide was normal. We found noncaseating granulomas on transbronchial biopsy, suggesting the diagnosis of work-related HP.

The patient described similar symptoms among co-workers occupying the same office. Questionnaire evaluation of his co-workers showed a high symptom prevalence, with eight of nine reporting two or more work-related respiratory and systemic symptoms. Clinical investigation showed histologic findings of HP in 5 of 10 exposed employees (including the index case), for a 50% attack rate. Routine diagnostic studies including chest radiograph and resting pulmonary function tests were normal or only nonspecifically abnormal except in the index case. Worksite evaluation and quantitative sampling for bioaerosols found the fungus *Penicillium* at elevated concentrations in the office building compared to outdoors. Removal of all employees from the contaminated environment resulted in resolution of clinical abnormalities in all but the index case.

APPROACH TO THE INDIVIDUAL WITH
BUILDING-ASSOCIATED RESPIRATORY SYMPTOMS

The first step in assessing an individual with building-associated symptoms is to identify the organ system affected. Reports of recurrent upper or lower respiratory symptoms (including rhinorrhea, congestion, cough, sputum production, chest tightness, wheezing, and shortness of breath) should prompt further diagnostic evaluation. The combination of recurrent building-related respiratory and systemic symptoms (the latter including fever, chills, profound malaise, myalgias, and weight loss) raises the possibility of HP.

Medical evaluation of an individual for building-related hypersensitivity disease should include several diagnostic studies. Complete medical and social histories should elicit a previous diagnosis of asthma or recurrent bronchitis, other respiratory illnesses, medication use, allergies, family history of atopy, and smoking history. Comprehensive occupational and environmental exposure histories, with attention to the temporal relation between symptoms and building occupancy, are critical in deciding whether to pursue further diagnostic evaluation including more expensive or invasive procedures. Historical clues that may influence the assessment of whether symptoms are work related include the presence of similar symptoms in co-workers and specific reports of poor indoor air quality (such as odors, problems with the heating, ventilation, and air conditioning system, a history of water intrusion into the occupied space, or obvious mold or mildew contamination) which suggest bioaerosol exposure. Physical examination, tests of lung function, and chest radiographs or CT scans are helpful when positive but do not rule out building-related respiratory illness when negative.[14] The symptom of dyspnea on exertion should prompt consideration for exercise toler-ance testing with assessment of gas exchange parameters. Methacholine challenge is useful assessing the presence and severity of nonspecific bronchial hyperre-sponsiveness, which occurs in asthma and often in HP.[15] Peak flow results recorded by the patient at least four times per day for at least two weeks of regular work, may be helpful in the diagnosis of both building-related asthma and HP. Skin prick testing to common environmental allergens helps determine the indi-vidual's atopic status and the presence of specific IgE antibodies to some indoor allergens, a useful finding in those with symptoms of building-related asthma or rhinosinusitis.

The mainstay of therapy in an individual with confirmed building-related hypersensitivity illness is removal from further exposure to the offending antigen. Identification and remediation of the building exposure is often a difficult task. Whether to allow building reoccupancy of a sensitized individual once clean-up efforts have been completed is a difficult decision, as little information exists on the adequacy of remediation efforts in eliminating antigen exposure sufficient to prevent symptom recurrence.

SENTINEL HEALTH EVENT FOLLOWUP: ASSESSMENT OF THE BUILDING AND ITS OCCUPANTS

Confirmation of the diagnosis of building-related hypersensitivity disease in an individual constitutes a sentinel health event.[16] That is, recognition of disease in one individual implies that others sharing the same contaminated work or home environment are at risk for illness. In this circumstance, appropriate intervention extends beyond the individual patient to encompass the at-risk building and its occupants, with the goals of identifying other cases of disease and eliminating the shared exposure risks.

It is essential to approach investigation with a team of experts, including physicians, epidemiologists, and industrial hygienists with an understanding of bioaerosol reservoirs, amplifiers, and disseminators in buildings. Building man-agers, maintenance personnel, engineers, and building occupants themselves should be involved to assure that risk factors are carefully sought and accurately identified. An initial walk-through survey of the building is often the first step to assess layout, look for obvious contaminant sources, and assess the need for further sampling for the presence of unusual microbial concentrations or species.

There are a number of methods for measuring bioaerosols, but these methods are not well standardized and results must be interpreted cautiously.[17] Bioaerosol sampling may be helpful if positive, but negative results are usually inconclusive and do not disprove antigen exposure. Furthermore, a positive result may not necessarily identify the relevant antigenic exposure. There is no information on bioaerosol dose-response levels, and few standards exist for defining an acceptable level. An unusual exposure may exist when bioaerosol levels indoors are at least an order of magnitude higher compared to outdoors or when the bioaerosol species detected differ markedly from indoors to outdoors. Quantitative bioaerosol analysis is expensive, as samples should be obtained at least in duplicate and usually from multiple sites in the building. For sample analysis, it is important to use a laboratory experienced in environmental microbial speciation and identification.

An important component of investigation of a building-related hypersensitivity disease sentinel event is active case finding via an epidemiologic survey of building occupants. Standardized respiratory questions[18] supplemented by additional questions about exposure, other symptoms, and timing of symptom onset are the most sensitive screening tool. A well-designed epidemiologic survey should include a clear case definition, assessment of possible confounding factors (such as smoking and pre-existing illness), selection of appropriate controls, and a carefully designed questionnaire.[19] Plans should be made at the outset for appropriate referral of potential cases for diagnostic evaluation.

CONCLUSIONS

Our investigations of building-related respiratory and systemic symptoms show that HP is not rare and that, in building environments with significant bioaerosol contamination, attack rates can be high. Symptoms reported by building occupants were often subtle, subacute, and nonspecific, and affected individuals were often treated for months for more common respiratory conditions such as asthma, bronchitis, and viral upper respiratory infections. Routine clinical diagnostic studies, including chest radiographs and tests of pulmonary function, were often normal or nonspecifically abnormal, and serum precipitating antibody analysis to standard commercial antigens was negative.

When confronted with a patient with building-related chest and/or systemic symptoms, it is imperative that the treating physician maintain a high index of suspicion for possible hypersensitivity disease and elicit a careful occupational and environmental exposure history. Early bronchosocopy may be indicated to rule out HP if the exposure and symptom histories are compelling, even in the face of nonspecific clinical findings. Any newly identified case of HP from a potentially contaminated indoor environment constitutes a sentinel health event with significant public health implications, and plans should be made for further environmental investigation.

When employers are confronted with building-related complaints, it is often useful to obtain physician consultation. An epidemiologic survey may be helpful to establish the nature of complaints, to point to etiology, and to guide corrective measures. Once these measures are completed, epidemiologic surveys can also be used to demonstrate improvement or detect recurrence of the problem. A worksite walkthrough is useful to identify obvious sources of exposure or problems with maintenance and to plan for sampling if necessary. Most importantly, corrective efforts to reduce or eliminate bioaerosols and other antigen sources are the key to prevention.

REFERENCES

1. Indoor air pollutants: exposure and health effects. Copenhagen, Denmark: World Health Organization, 1983.
2. Hoffman RE, Wood RC, Kreiss K. Building-related asthma in Denver office workers. *Am. J. Public Health* 1993; 83:89-93.
3. Finnegan MJ, Pickering CAC. Building-related illness. *Clin. Allergy* 1986; 16:389-405.
4. Burge PS, Finnegan MJ, Horsfield N, Emery D, Austwick P, Davies PS, Pickering CAC. Occupational asthma in a factory with a contaminated humidifier. *Thorax* 1985; 40:248-254.
5. Burge HA. Introduction: Guidelines for the Assessment of Bioaerosols in the Indoor Environment. American Conference of Governmental Industrial Hygienists, Cincinnati, OH, 1989, p. 1.
6. Arnow PM, Fink JN, Schlueter DP, et al. Early detection of hypersensitivity pneumonitis in office workers. *Am. J. Med.* 1978; 64:236-242.
7. Hodgson MJ, Morey PR, Simon JS, Waters TD, Fink JN. An outbreak of recurrent acute and chronic hypersensitivity pneumonitis in office workers. *Am. J. Epidemiol.* 1987; 125:631-638.
8. Bernstein RS, Sorenson WG, Garabrant D, Reaux C, Treitman RD. Exposures to respirable, airborne Penicillium from a contaminated ventilation system: clinical, environmental and epidemiologic aspects. *Am. Ind. Hyg. Assoc. J.* 1983; 44(3):161-169.
9. Rose, CS, Newman LS. Hypersensitivity pneumonitis and chronic beryllium disease. In: Schwarz MI, King TE Jr., eds. *Interstitial Lung Diseases* St. Louis, MO: Mosby Yearbook 1992, Second Edition, 231-253.
10. Burrell R, Rylander R. A critical review of the role of precipitins in hypersensitivity pneumonitis. *Eur. J. Respir. Dis.* 1981; 62:332-343.
11. Dickie HA, Rankin J. Farmer's lung: An acute granulomatous interstitial pneumonitis occurring in agricultural workers. *JAMA* 1958; 167:1069-1076.
12. Terho EO. Diagnostic criteria for farmer's lung disease. *Am. J. Ind. Med.* 1986; 10:329.
13. Hodgson MJ, Parkinson DK, Karpf M. Chest x-rays in hypersensitivity pneumonitis: a metaanalysis of secular trend. *Am. J. Ind. Med.* 1989; 16:45-53.
14. Lynch DA, Rose CS, Way D, King TE Jr. Hypersensitivity pneumonitis: Sensitivity of high resolution CT in a population-based study. *Am. J. Roentgenol.* 1992; 159:469-472.
15. Freedman PM, Ault B. Bronchial hyperreactivity to methacholine in farmer's lung disease. *J. Allergy Clin. Immunol.* 1981; 67:59-63.
16. Rutstein DD, Mullan RJ, Frazier TM, et al. Sentinel health events (occupational): a basis for physician recognition and public health surveillance. *Am. J. Public Health* 1983; 73:1054-1062.

17. Morey PR, Hodgson MJ, Sorenson WG, et al. Environmental studies in moldy office buildings: biological agents, sources and preventive measures. *Ann. Am. Conf. Govt. Ind. Hyg.* 1984; 10:21-35.
18. Ferris BG. *Epidemiology Standardization Project.* Bethesda, MD: American Thoracic Society and Division of Lung Diseases of The National Heart, Lung, and Blood Institute, 1978.
19. Kreiss K. The epidemiology of building-related complaints and illness. In: Cone JE, Hodgson MJ, eds. *Problem Buildings: Building-Associated Illness and the Sick Building Syndrome.* Philadelphia, PA: Hanley and Belfus, Inc. 1989; 44:575-592.

Part III
Neurotoxicity

Overview: Neurotoxicity of Indoor Air Pollutants: What to Measure and How to Measure It

Hugh A. Tilson

Indoor exposure to some chemicals has been linked to a number of subjective signs and symptoms characterized by a sense of discomfort and annoyance. That people may be exposed to potential neurotoxic chemicals, such as volatile organic solvents, carbon monoxide, and pesticides, has been established, but the prevalence of neurotoxic effects as a result of indoor air exposures is not clear. Because many of the chemical agents found in the indoor air setting can be smelled or irritate mucosal membranes, sensory effects can be detected and quantified using a variety of well-established psychophysical techniques suitable for both humans and laboratory animals. Indoor air exposures, however, can produce other effects that directly or indirectly involve nervous system function. For example, it has been suggested that discomfort or annoyance from indoor air exposures may reduce productivity or serve as a low level stress. In spite of the availability of procedures to measure such changes in other experimental settings, they have not been widely utilized in indoor air research.

Dr. Vernon Benignus, in Chapter 15, reviews procedures for assessing vigilance in humans as a measure of task performance. Vigilance requires a rapid detection of infrequently presented events or signals over relatively long periods of time. Acoustical noise, cold environments, and chemical stimulation are environmental conditions that are known to affect task performance in humans. Experimental results indicate that a testing strategy for chemical irritancy should begin with either a wide-range, dose-effects study or a high-level, single-exposure study, contain an explicit *a priori* hypothesis based on theory or findings of other related work, be conducted using double-blind procedures, use tasks selected on results of experiments in related environmental areas, and be designed to measure effects of irritants themselves.

Dr. Newland, in Chapter 14, reviews available procedures of assessing vigilance in animal models. One promising model concerns the use of a concurrent schedule of reinforcement in which two or more response alternatives are presented. Only one alternative can be performed at a time and the subject can switch between the two alternatives. Vigilance, defined as the detection of an unusual event or subtle change in the environment, the availability of alternative activities and an opportunity to persevere, can be studied using this behavioral technology. To use such procedures to detect neurobehavioral effects of indoor air, dependable measures should be validated with known chemicals and should be functionally similar to that of concern with humans.

Dr. Donald Lysle reviewed the literature concerning conditioned immune alterations in animals. Exposure to aversive environmental events can lead to adverse health consequences mediated through the immune system. Experimentally, it can be shown that presentation of aversive stimuli can induce pronounced alterations in several indicators of immune function. The literature clearly indicates that conditioning or learning can play an integral role in neural and endocrine responses to environmental events and that these responses can affect immunological status.

In Chapter 17 Dr. Bruce Rabin reviewed the literature concerning the responses of humans to stressful environmental changes. Events perceived as being stressful can activate a chain of events leading to the release of neurotransmitters and endocrine hormones. The activation of the hypothalamic-pituitary-adrenal axis can alter several measures of immune function. Altered immune function can have significant health consequences, including decreased resistance to disease and increased susceptibility to autoimmune disease. Conditioning or learning may play an important role in stressor-induced immune alterations in humans.

The presentations in this section underscore the possibility that exposure to indoor air pollutants can have effects on the nervous system other than an alteration in sensory function. Two frequent indications of chemical exposure include a disruption of task performance possibly through an alteration in vigilance or attention and unlearned and learned physiological responses to stress. Laboratory procedures to evaluate vigilance and stress-induced physiological changes have been developed for both humans and animal species and could be used to identify and characterize adverse health effects associated with indoor air pollution.

CHAPTER 14

Studying the Neurotoxicity of Indoor Air with Nonhuman Species: Issues in Experimental Design and Interpretation

M. Christopher Newland and Steven K. Shapiro

INTRODUCTION

The problems posed by the behavioral manifestations sometimes called the sick-building syndrome differ from those proposed by other toxicants such as lead or radon, in which there is a known toxicant and a known endpoint. With respect to sick building syndrome, it is not even known for certain that we are dealing with a neurotoxicant, and if so, which one and how many. To resolve these issues, experiments with human and nonhuman species are called for, and the uncertainties involved require that the strategies and tactics of developing experimental protocols must be given careful attention.

The basic considerations in experimental design draw from principles of both toxicology and behavior. It is necessary that dose-effect relationships be obtainable, and this means that reliable and quantitative measures of behavior must be used. It also means that a known quantity of a toxicant, or a mixture of toxicants, must be imposed. Dependent measures should be validated by use of an intervention known to replicate the disruption to be characterized. Maybe most important, the behavior of animals should be functionally similar to that of concern with humans. This does not mean that it should be topographically identical, or that the behavior should literally look the same, but rather the interactions among behavior, stimuli, and consequences should play out in the same way in human and nonhuman studies.

For example, organic solvent abuse in humans can be studied in nonhuman species by arranging for an animal to press a lever to produce a concentration of toluene.[1] It is not necessary to mimic human abuse literally by training animals

1-56670-144-9/96/$0.00+$.50
© 1996 by CRC Press, Inc.

to cover their heads with plastic bags. The self-administration paradigm is functionally similar to human solvent abuse, even if the form that it takes is very different. In considering tasks of learning, remembering, vigilance, or subjective responses to room atmospheres it will be very helpful if procedures used with human and nonhuman species resemble each other functionally. If all these criteria are in place then meaningful conclusions can be identified, and perhaps mechanism of action can be explored.

It is not entirely clear what the nature of behavioral effects of indoor air pollution is, so approaches to designing laboratory-based tests must draw from clues derived from investigations of how the problem has been reported. Two sets of clues are available: those derived from the classes of chemicals present in the atmospheres and those derived from the types of effects reported. Two of the more frequently reported classes of compounds reported in office buildings are volatile organic compounds and pesticides,[2] but if consideration is extended to other workplaces then the number of classes expands considerably.[3] Within each of these classes of compounds are subclasses with very different mechanisms of action and patterns of behavioral effects.

An array of behavioral and psychosocial effects of symptom complexes variously termed as multiple chemical sensitivities, environmental illness, and sick-building syndromes has been identified. Mølhave[4] described the sick-building syndrome as a collection of symptoms that includes (1) sensory irritation of eyes, nose, and throat; (2) skin irritation; (3) odor and taste complaints; (4) runny eyes or nose (5) asthma-like symptoms; and (6) neurotoxic symptoms such as fatigue, reduced memory or concentration, lethargy, headache, dizziness, and nausea. Other reports have described depressed mood, anxiety, restlessness, loss of interest in work, changes in libido, general apathy, confusion, sleep disturbances, irritability, and weakness.[5] These effects can be associated with a variety of potential toxicants, including volatile organic compounds, combustion products, and pesticides found in indoor air, but they also can reflect nonchemical causes including the design of the building, the number of windows, the density of workers, and job factors that contribute to satisfaction.[6,7]

The characteristics listed above are based upon performance on various tests and questionnaires. Such performance is influenced by motivation, medical and psychiatric conditions, and factors unrelated to exposure such as age, sex, intellectual ability, education, socioeconomic status, drugs, and even sleep deprivation.[5,8] In addition, adequate norms for specific reference groups do not exist. Therefore, due to the nonspecific nature of both the neurotoxic symptoms and the assessment procedures used to evaluate adverse effects of chemicals, the ability to differentiate among neurotoxic exposure, neurological disease, psychiatric disturbances, social influences, or malingering must be a multidisciplinary process.[5,8]

IDENTIFYING HAZARD

When a consistent collection of effects is reported, it can guide the design of animal studies of neurobehavioral effects of indoor air pollutants, but in interpreting the clues it is important to keep the interpretation functional, not topographical. For example, an animal model of subjective responses such as depression or malaise is not required (or available) but some of the component activities, such as a general reduction in the rate of behavior, may be. A more general approach would be to recognize the aversiveness of situations that produce such responses. In the case of indoor air, then, the aversive or reinforcing properties of the atmosphere itself can be studied. So instead of being overly literal in interpreting malaise, it might be more helpful to recognize that such a condition is unpleasant and, given a choice, an animal (human or otherwise) will behave in such a way as to avoid it. Another set of complaints associated with indoor air overlap with some seen in attention deficit disorders and here we have better defined phenomena that have been studied in both human and nonhuman species. Unfortunately, many studies of humans include tasks that would be difficult to reproduce with nonhuman situations. The reverse is sometimes true for generalizing from animal studies to human situations. The chasm between studies of human and nonhuman species raises problems in interpretations that can be avoided. The desire to avoid them at the outset lies behind much of the discussion below.

Some agents found in indoor air, such as carbon monoxide and volatile organic compounds, can produce deficits in a class of activities called vigilance.[9,10] As Laties and Merigan[9] point out, vigilance can best be described as a collection of apparently related activities, which may be trapped more or less effectively by several different procedures. There is no single procedure that can be called a "vigilance" procedure, and properly so since it may not be a unitary phenomenon but a constellation of component phenomena. Definitions of vigilance include such diverse components as changes in the EEG in the direction of arousal, sustained attention to a stimulus that changes subtly and infrequently,[11] or estimations of temporal intervals.[12-14] A search of a database like Medline for animal studies uncovers such diverse procedures as radial arm mazes and trial-based detection tasks. In most, a relatively infrequent dispersion of events in time is the most common component. Events may be scheduled to occur at infrequent and unpredictable times or a response must be made at regular and narrowly defined temporal intervals. There are opportunities for two types of errors. Intrusions or false alarms appear as premature responses or falsely reporting the presence of a stimulus. Misses appear as failing to respond at the appropriate time, responding later than would be most efficient, or failing to detect a stimulus that is present.

These characteristics of vigilance suggest a variety of approaches in designing animal studies. Precisely defined, discrete trial vigilance tasks could readily be designed but such a highly focused approach would be overly risky. The procedure might be extremely sensitive, but the greater possibility exists that a very specific task would fail to detect a hazard, either because vigilance isn't the real problem or a methodological variable in a relatively untried protocol would undermine the conclusion. A suitable compromise would be to apply procedures known to detect a variety of neurotoxicants and that have characteristics that are present in vigilance tasks. Three procedures recommend themselves and two have a long history of use in neurotoxicity. The fixed-interval (FI) and the differential reinforcement of

low rate (DRL) schedules have been used in many protocols and are even recommended in federal guidelines. A third procedure, the concurrent schedule, has
many desirable characteristics but has not yet been used much by behavioral
toxicologists.

The following discussion will contain suggestions of how to approach the
characterization of the neurotoxicity of indoor atmospheres. Several approaches
that might be taken to characterize the syndrome with nonhuman species will be
described. First, the subjective effects will be examined. Then some different
approaches to tapping some of the other behavioral actions of chemicals found
in indoor air will be presented. In a separate section, the problem of individual
susceptibility will be addressed.

Subjective Effects

Questionnaires asking about the adverse effects of indoor air and the sick-
building syndrome sometimes contain items requesting subjective responses to
the inside atmosphere. Responses on these items include vague complaints such
as feelings of malaise, fatigue, nausea, headaches, or lethargy.[4,7] While these data
are helpful, they provide little support for inferences about cause and effect since
they are inherently correlational. The subjective complaints could be related to
the atmosphere but they are so general that they could also reflect dissatisfaction
with the employer, and apparently do.[15]

These considerations can be bypassed when atmospheres are controlled experimentally. In a series of studies, volunteers were exposed to clean air or a mixture
of volatile organic solvents.[17,18] During exposure to the volatile organic compounds (VOC) the volunteers reported feeling discomfort, drowsy, uncomfortable,
and that they would like to improve the ventilation. In sum, they found the
atmosphere aversive and, given an opportunity, they would prefer clean air.
Meanwhile, there were, at most, marginal effects on conventional measures of
human performance such as digit span, finger tapping, symbol digit substitution,
and tests of attention. Dick and colleagues,[20] in another experimental exposure
also reported subjective responses as among the more sensitive indices of exposure to acetone (see also Dick and Johnson[19]). Confounders such as job dissatisfaction and workspace variables are unlikely to play a role in these experimental
setting, and the subjects are not involved in lawsuits, so subjective effects reported
can be taken seriously. It is of some interest that the computerized testing protocols did not detect effects at the lower concentrations. Such protocols are com-
promises among sensitivity, ease of use, and rapidity of testing and as such can
be considered as screening batteries. A failure to identify an effect on such a
battery raises the hoary dictum that one cannot declare a compound as safe, only
as not hazardous on the tests employed.

The subjective measures reported may be a statement about the aversiveness
of an atmosphere and may even be among the more sensitive measures of exposure. It is not possible to arrange animal studies to study the subjective measures
on an inventory such as a Profile of Moods State, but the more general question
of whether the atmosphere is aversive can be examined with well-understood

procedures. Even the stimulus characteristics of an atmosphere can be characterized against other known stimuli, at least in principle.

An irritant can be identified behaviorally by determining whether an animal (or human) will respond in such a way as to remove the stimulus. Basically, this is an avoidance paradigm and just as a rat will press a lever, or move to another room, in order to terminate an electric shock, so will it respond to terminate an aversive atmosphere.[1] Wood[21,22] exploited this insight to study the aversive properties of various concentrations of ammonia or ozone. Mice were exposed to different concentrations of these irritants in an enclosed chamber and they could respond so as to terminate the irritant by inserting their nose into a conical shaped protrusion and breaking a photobeam. After a certain number of photobeam breaks, clean humidified air was delivered to this protrusion. This is a form of a concurrent schedule, a sort of choice procedure, in which one response is maintained by the avoidance of ammonia and the other by the presentation of clean air. The tactic of presenting the subject with a choice greatly improved the sensitivity of the procedure by steepening the concentration-effect curve and reducing variability.

The importance that procedural considerations play in shaping the conclusions that can be drawn about hazard is illustrated in Figure 1.[22] The number of responses required to remove the irritant and initiate clean, humidified air into a nose-sized, conical opening was changed from 1 to 5. If only one response was required to produce the clean air, then half of the atmosphere deliveries were terminated, even when there was no ammonia in the atmosphere. Such a high rate of terminations would raise serious questions about the validity of the preparation, except that a simple increase in the response requirement to 5 reduced the number of terminations of unadulterated air to 5 to 10% and resulted in a well-defined concentration-effect curve relating the number of ammonia deliveries terminated to the concentration of ammonia. The general observation that the aversiveness of an atmosphere can be identified in nonhuman species was also replicated with ozone. The specific observation, that procedural variables like response cost profoundly influence the outcome, also needs to be noted.

Just as some atmospheres can be aversive, some can also be reinforcing, and many organic solvents can support behavior that produces the atmosphere. The reinforcing properties of toluene and other organic solvents have been well established by examining whether an animal will respond so as to produce an atmosphere.[1] It is not inconsistent to say that an atmosphere can be both reinforcing and support some aversion. Even powerfully reinforcing drugs such as cocaine can, if paired with a novel flavor, produce a flavor aversion.

Saying that an animal will behave to escape or avoid a stimulus is a first step toward identifying that compound as an irritant, but more refined descriptions of the stimulus properties can be made. Even discriminations based upon subjective states, basically what is required on a Profile of Moods Scale, can be established in nonhuman species.[23] Many examples could be provided, but an experiment by Rees et al.[24] illustrates the point. They trained mice to discriminate ethanol from saline by delivering a reinforcer following a response on one lever if ethanol administration preceded the session and on the other lever if saline injection

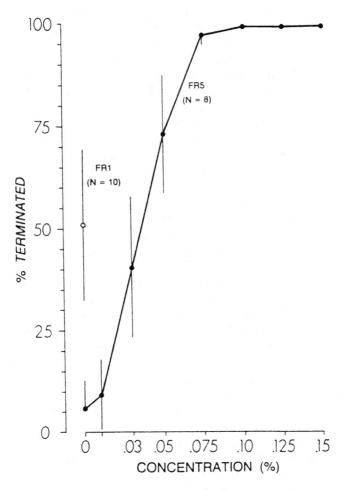

Figure 1   The percentage of ammonia deliveries terminated for different concentrations of
ammonia under a fixed-ratio 1 (FR 1) schedule (one nose poke required to
terminate exposure) or a FR 5 (five nose pokes required). (From Wood, R.W.,
Toxicol. Appl. Pharmacol., 61: 260-268 (1981). With permission.)

preceded the session. After discrimination training was complete, the mice were
exposed to atmospheres containing TCE, toluene, or halothane, and the mice
responded on the ethanol-appropriate lever during exposure to these inhalants, in
a concentration-related fashion, indicative of shared subjective stimuli among
these different agents.

This basic approach has been used to isolate very specific dimensions of the
stimulus properties of different behaviorally active compounds. The conclusions
correspond closely to other known neural effects of compounds.[23] For example,
compounds that act as dopamine agonists support generalization to other dopamine agonists but not to compounds that act on, say, acetylcholine neurotransmitter
systems. Benzodiazepines support generalization to other benzodiazepines as well

as to other compounds that act on the GABA/chloride ionophore, but not to major tranquilizers or dopamine agonists. It is even possible to distinguish receptor subtypes using drug discrimination procedures. Drugs that act on dopamine D1 receptor subtypes can be discriminated by a rat from those that act on D2 dopamine receptor subtypes.[25]

The drug-discrimination procedure is a refined and powerful experimental procedure that can be exploited to identify subjective effects of compounds with human or nonhuman species. With it an investigator can then determine private, subjective effects of atmosphere exposures and perhaps get a better descriptor of some of the subjective effects described by humans. These effects need not be limited to receptor-based action. In principle, it should be possible to identify "solvent-like effects" or even "ozone-like irritation" sensations.

The Fixed Interval (FI) Schedule

The FI schedule of reinforcement can be described simply: a reinforcer is delivered for the first response to occur after a specified amount of time has passed. For example, under a FI 5' schedule, the reinforcer follows the first response to occur after 5 in. Only one response is required, but typically many responses occur. This schedule shares with vigilance procedures an important role of time in defining the contingencies of reinforcement. The pattern of responses in FI schedule performance occurs in many species and this, coupled with its sensitivity to a variety of drugs and toxicants is one reason that the procedure is frequently used.[26] A low rate of responding appears early in the interval and response rates increase as the interval elapses. This pattern is strongly influenced by the temporal nature of the schedule and toxicant-induced disruptions in temporal discrimination are manifested in this schedule. Of course, it does not follow that all disruptions in the pattern reflect disruptions in temporal discrimination, or vigilance, since there could be other causes. For example, while the subject is *not* responding the subject *is* doing something else. Therefore, rate increases early in the interval could reflect disruptions in temporal discrimination, elevations in the rate of other activities, or nonspecific stimulant effects. The sensitivity of this schedule to several influences is part of what recommends its use in toxicity testing, especially if one is unsure as to the effects expected.

The FI schedule has been an important component in considerations of the neurotoxicity of many compounds. For example, a consistent effect of lead exposure in nonhuman species has been biphasic changes in rates of responding maintained under FI schedules of reinforcement. In different species, different laboratories, under a variety of settings, an inverted U-shaped curve has described the relationship between lead exposure and overall response rate, and the levels at which lead is behaviorally active in nonhuman animals corresponds closely to those that impair academic performance in humans.[27] It's interesting that a laboratory procedure that has no analog in the natural world does so well at identifying subtle neurotoxic effects. If one set out to derive an animal analog of IQ, it is unlikely that the FI schedule would leap to mind. This is not to say that the FI schedule is an IQ test. More likely, it presents a scheduling of events around which behavior organizes in complex ways. The correspondence between chemical effects on this schedule in animals and their sometimes nonspecific effects in humans speaks for the value of deriving tests from the primary literature according to what effects are reproducible and sensitive and should give pause to attempts to model human behavior according to whether performance bears a physical resemblance to human activity.

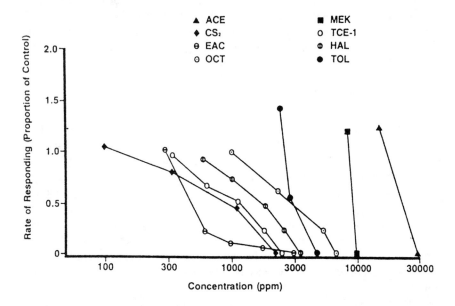

Figure 2   Average rates of responding under an FI 60-s schedule for varying concentrations of volatile organic compounds. Data are expressed as a proportion of control values. The compounds used are acetone (ACE), carbon disulfide (CS₂), ethyl acetate (EAC), n-octane (OCT), methyl-ethyl ketone (MEK), trichloroethane (TCE-1), halothane (HAL), and toluene (TOL). (From Glowa, J.R., *Behavioral Pharmacology: The Current Status,* Sieden, L.S. and R.L. Balster, Eds., (New York: Alan R. Liss, 1985). Reprinted by permission of John Wiley & Sons.)

The FI schedule can reveal dose-related effects of VOCs and can even characterize some as having stimulant-like properties at low levels of exposure. For example, Glowa[28] reported effects of eight organic solvents on mice working under an FI schedule of reinforcement (Figure 2). The pattern of effects is quite interesting, revealing the complications involved in arriving at a simple, unitary measure of the concentration of a mixture. There is about a 100-fold range of potencies described just in these eight compounds. In addition, the slope of the dose-effect curves varies widely such that the curve from one compound (carbon disulfide) spans nearly a 10- to 30-fold range of concentrations while another (methyl-ethyl ketone) goes from a rate-increasing effect to complete elimination of performance before the concentration is even doubled. Moreover, some compounds (toluene, methyl-ethyl ketone, acetone) produce rate-increasing effects, suggestive of stimulant effects, at the low end of their dose effect curves.

Differential Reinforcement of Low Rate (DRL) Schedule

The DRL schedule, like the FI schedule, is a time-based schedule of reinforcement. A reinforcer follows the occurrence of two responses separated by a specified period of time and responses during the interval result in the restarting of the timing of the interval. Thus, under a DRL 20, only interresponse times of 20 s or longer are reinforced. The number of responses per interval and number of reinforces obtained typically serve as a dependent measure.

A major advantage of the DRL schedule is that similar performance is seen in human and nonhuman species and, like the FI schedule, it has been examined with a variety of drugs and toxicants. Dose-related changes in performance have been described with a variety of compounds including decreased response rates with ethanol[29] and increased response rates occur with low doses of stimulants and anxiolytics.[30] There is also evidence that DRL schedules may be useful in distinguishing between the effects of anxiolytics and other drugs, such as stimulants.[31]

The DRL schedule has been used in humans as a measure of impulsivity (i.e., the failure to inhibit responding if required by contingencies or situational demands) among children diagnosed with attention deficit/hyperactive disorder (ADHD). Research with human and nonhuman species has often found that subjects will engage in collateral behavior as a means of regulating temporal discrimination, both during acquisition and steady state.[32-34] Furthermore, children considered as hyperactive according to varying criteria have been found to engage in more collateral behavior and emit a higher number of nonreinforced responses, than their nonhyperactive counterparts.[34-36]

Clinical research has often shown that impulsivity and hyperactivity factorially load together to form a single behavioral dimension.[37] Impulsivity appears to be the hallmark of ADHD — a disorder of response inhibition of delayed responding. Inattention, however, is a hallmark of ADHD, but is not specific to this disorder. Our understanding of ADHD in adulthood is less clear and, certainly, differential diagnosis is essential.

Various studies with nonhuman species have used the DRL paradigm to evaluate behavioral effects of toxicants. Rice and Gilbert[38] exposed newborn monkeys to either 0, 50, or 100 µg/kg/day of lead followed by steady-state concentrations until 3 years of age. Treated monkeys acquired DRL performance more slowly than controls during initial sessions, exhibited greater between session variability during terminal sessions, responded with shorter interresponse times, and had a greater ratio of responses per reinforcer under the DRL schedule. It was noted that performance differences were more subtle than those previously found on an FI schedule.[26] Schrot and Thomas[39] used a MULT FR 30 DRL 18 schedule to test the behavioral effects of six concentrations of carbon monoxide on rats. In general, fixed ratio (FR) and DRL response rates declined as a function

of increasing concentration. However, response patterning, measured by interresponse times (IRT) conditional probability distributions, was not systematically
affected by CO exposure. Moser, Coggeshall, and Balster[40] exposed rats to individual isomers and a commercial grade of xylene followed by behavioral sessions
using a DRL 10 schedule. Individual and mixed isomers produced similar biphasic
effects on response rate and concentration-dependent decreases in reinforcement
rates. Colatla et al.[41] also reported the effects of a variety of solvents on behavior
under this schedule.

The Concurrent Schedule of Reinforcement

When an animal or human performs on any task, the reinforcers provided by
that task compete with other events for the subject's behavior. At the very least,
"not performing" is an option and it is not too speculative to suggest that scratch-
ing, grooming, exploring, or (in humans) daydreaming or napping is another
activity for which there may be reinforcing consequences. Usually the alternate
reinforcers are not controlled explicitly by the experimenter, who is most concerned with getting enough behavior to analyze by making the task so reinforcing,
or failing to perform the task so aversive, that the subject performs. In a concurrent
schedule, the alternate activities are explicitly included as part of the schedule
definition. This schedule is an enormously flexible one that can be adapted to
many uses. Wood[21] arranged the availability of clean air or ammonia concurrently.
Schedule-typical patterns of responding also appear in humans when schedules
are explicitly arranged concurrently.[42] Concurrent schedules can be arranged as
a choice task, a signal detection task, a foraging task, a way of arranging more
conventional reinforcement schedules, or a vigilance task, depending upon the
particulars of the application.

The constructs of vigilance and divided attention can be understood and
described behaviorally, without regard to species, by viewing them as behavior
under concurrent schedules of reinforcement.[43] Vigilance tasks involve control
over behavior by events that are infrequent in settings that are utterly unstimu-
lating, so the availability of any alternate activity plays an important role. Questions about neurobehavioral toxicity can be framed as degree to which chemical
exposure modifies behavior under the control of very infrequently presented
stimuli or reinforcers. Divided attention tasks entail presenting a subject with two
or more concurrently available stimuli competing for the subject's attention, but
only one is relevant to the programmed consequences. Such a conceptualization
has contributed to an understanding of age effects on divided attention.[43] Maybe
more important, such a conceptualization is free of language- or cognitive-based
constructs that often are advanced to account for human behavior. Framing these
complex issues functionally in ways that pertain to the study of both human and
nonhuman behavior is an essential step that must be taken before human and
animal studies can be linked meaningfully.

There is a large body of empirical research on concurrent schedules and
among the simpler things that we know are these.[44] If there is a discrepancy

between the reinforcement rates on the two levers, then most behavior will appear on the lever producing the richer rate of reinforcement, but some responding also occurs on the leaner lever. The discrepancy in response rates on the two levers can be amplified by imposing a cost on changing levers. If the location of the richer lever changes then behavior tracks the change such that most responses shift to the richer lever, but the dynamics of this change are poorly understood.

Newland et al.[45] examined the consequences of prenatal exposure to lead or methylmercury on behavioral allocation between two schedules arranged concurrently during steady state and during transition states. Although this task was not designed as a vigilance task, its use might be instructive. Monkeys were exposed during gestation to lead or methylmercury and tested as juveniles. During 30-min experimental sessions, a monkey sat in a chair in a sound-deadened chamber and faced a panel containing two levers, a left one and a right one, and a pellet dispenser midway between them. Lever pressing was maintained according to a Random-Interval (RI) schedule of reinforcement operating independently on each lever. For example, under a RI60" schedule, a pellet was delivered following a lever press at unpredictable times, but on the average every 60 s. In the Concurrent RI60" RI60" schedule, separate and independent RI schedules operated concurrently on each lever. While responding on the left lever, the timer on the right lever continued operating. A changeover delay was applied to amplify control by the different schedules operating on the two levers. This is a refractory period that follows a change of levers and during which reinforcers are not delivered.

A monkey's behavior was first maintained under a Concurrent RI30" RI30" schedule of reinforcement. Then the reinforcement rates on the two levers were changed such that, first, the right was rich, then the left, and so forth. Both steady-state behavior and behavior during transitional states were examined as a function of exposure. To accomplish this, it was necessary to quantify the effect in such a way that dose-effect relationships can be described. Quantification of steady-state behavior was fairly straightforward, so it will be described first. Then behavior change will be examined.

Figure 3 shows the ratio of left- to right-lever responding as a function of the ratio of left- to right-lever programmed reinforcers. The positively sloped curves show that as more reinforcers were programmed to derive from the left lever, most responses occurred on that lever, but some responding always occurred on both levers. The value of zero on the horizontal axis corresponds to an equal programmed reinforcement rate on each lever: this produces a ratio of 1 and the logarithm of 1 is 0. A ratio of 1 (whose logarithm is 0) on the ordinate corresponds to responding that is equally distributed between the two levers, and the point (0,0) indicates unbiased performance.

Figure 3 shows the relationship between left-lever responding and left-lever reinforcers (programmed) for three control monkeys and others exposed during gestation to methylmercury or lead. A dashed line showing a slope of 1 and no lever bias is presented for visual reference. The parameters describing slope and bias were estimated by least-squares regression and revealed that both the slope and the intercept were affected by these developmental neurotoxicants. Controls

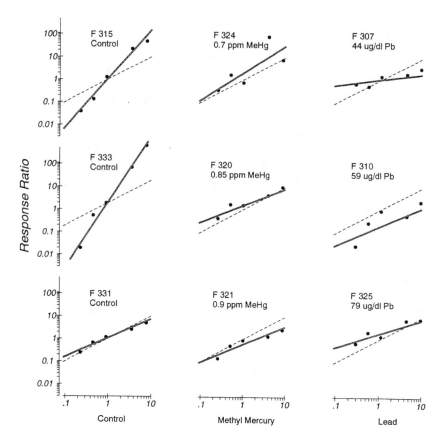

Programmed Reinforcement Ratio

Figure 3 Steady-state performance for control, lead-exposed and methylmercury-exposed monkeys. The ratio of left- to right-lever response rates is expressed as a function of the ratio of reinforcement rates programmed for responding on those levers (note log scales). Filled circles are the average of the four sessions just before a transition for an individual monkey at single schedule parameters. The solid lines indicate best fit linear regressions. The dashed line shows a line of identity between response and reinforcer ratios to provide a visual reference. Sensitivity to reinforcement ratios is indicated by the slope of the line, bias by its displacement from the origin. Exposed monkeys tended toward indifference between the two levers, as indicated by lines that tend toward the horizontal. (From Newland, et al., *Toxicol. Appl. Pharmacol.*, 125: 9 (1994). With permission.)

had a slope ≥ 1 — indicating that the programmed rates of reinforcement exerted considerable influence over the allocation of responses. The shallow slopes in the exposed monkeys indicate little sensitivity of behavior to the different rates of reinforcement. In addition, their behavior was heavily biased, some toward the left and others toward the right lever, indicating that their behavior was influenced by something other than the source of primary reinforcement until reinforcement density was greatly increased. The presence of relative insensitivity in steady-state behavior raises questions pertaining to what is required to change the behavior of those monkeys whose behavior did not change and to what acquisition looked like in animals whose steady-state performance resembled controls.

Not evident in Figure 3 is that during steady state, animals exposed to lower levels of lead looked the same as controls, but their performance during transitions was quite different. Figure 4 illustrates how this was determined. The proportion of responses on the newly rich lever was plotted through the course of a transition as a function of the cumulative reinforcers obtained on that lever. The choice of cumulative reinforcers to mark the transition, rather than session number was used because of the important role that consequences play in shaping behavior.

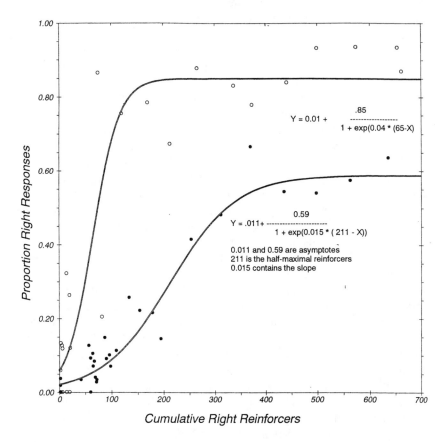

Figure 4   Transition 2 for monkeys F371 (unexposed, open circles) and F351 (exposed to 37 µg/dl maternal blood lead) showing the shift in response allocation from the left (formerly "rich") lever to the right (newly "rich") lever. Data are shown as a function of cumulative reinforcers derived from the rich lever. For example, the first nonzero point for F351 is over 41 and indicates that 41 reinforcers were derived from right-lever responding during the first session after a transition. The next point, over 56, shows that 5 more right-lever reinforcers occurred in the second transition session. Points over zero show the last three sessions before the transition. A logit function was fit to the data and the resulting equation is shown on the figure. From Newland, et al., *Toxicol. Appl. Pharmacol.*, 125: 11(1994). With permission.)

Figure 4 illustrates the approach to quantifying behavior change by showing data from a control (open symbols) and a lead-exposed (closed symbols) monkey. The magnitude of the transition, which is the difference between the two asymptotes, is about 58%. The transition for the exposed monkey was halfway complete after 211 reinforcers had been delivered, so this value represents half-maximal reinforcers. As this value changes the S-shaped curve slides laterally to the left or right. The steepness of the rising portion of the curve represents the pace of the transition once it begins. The lines show logistic functions fitted to these data using least-squares techniques. The beginning is represented in the first number, the steady-state asymptote is the numerator. The values associated with the exponential in the denominator represent the half-maximal reinforcers and the slope.

Both monkeys accomplished the transition, but the magnitude of the transition was greater for the control in the example (in general, however, the groups exposed to lower levels did not differ in steady-state performance). More reinforcers were required to support the transition in the lead-exposed monkey, as evidence in the value for the half-maximal reinforcers (65 vs. 211). Once the transition began, it moved at a lower rate, as evident in the different slopes. In general, the magnitude of the transition (representing steady-state performance) for the lead-exposed monkeys was indistinguishable from controls. It was the course of acquisition that separated the groups. Lead-exposed monkeys required about twice as many reinforcers to complete half of the transition and rate of transition was about ten times slower for that group. Concurrent schedule performance in transition identified effects in subjects whose behavior was indistinguishable form controls in steady state.

There is a wealth of literature describing behavior under concurrent schedules in human and nonhuman species.[44] An advantage of the concurrent schedule is the improved quantification of behavior it makes possible, as illustrated above. In recent years there has been an appreciation of the use of these schedules in behavioral phamracology.[46] One recent application pertinent here entailed administering the psychomotor stimulant *d*-amphetamine to monkeys performing under concurrent random interval random interval schedules.[47] In that experiment amphetamine, a psychomotor stimulant that can mimic some of the characteristics of ADHD, greatly disrupted the microstructure of behavior in a way that in other settings might appear as impaired vigilance. In a dose-related fashion, amphetamine increased the amount of time spent operating the rich lever, at the expense of the lever that produced reinforcers less frequently. This illustrates a subtle effect of this drug that could be of relevance to the effects resembling impaired vigilance sometimes associated with indoor air.

In the experiments described by Newland et al.,[45] acquisition was described in quantifiable and behaviorally meaningfully ways. The important components of behavior required to initiate and complete behavior change could be examined and used to increase an understanding of the behavior under study. Insofar as behavioral mechanisms are understood, the time required to complete transitions can be shortened without losing significant sensitivity. In those experiments, a transition required about a month of 30-min sessions to complete, in large part because of the retarded transition rates in the exposed monkeys, but also because of procedural considerations that could be changed. Currently we are developing a procedure in which a transition can be accomplished in a single 3-h session without losing the benefits provided by Newland et al.,[45] and it appears that even more rapid transitions will be possible (unpublished data). Such acceleration of the pace of experimentation will greatly enhance the utility of the preparation.

Concurrent schedules have been used with human subjects and in these cases the general findings conform to those reported with nonhuman species. The proportion of responses approximately matched the proportion of reinforcers obtained for a response. This was found by Baum[48] with people playing a stimulated video game in which subjects had to "shoot down" targets of different colors by operating a telegraph key. The frequency of reinforcement was adjusted by changing the way in which an observing response "detected" a target.

Human eye movements have been used as an operant in several studies of human performance under concurrent schedules, or other more conventional schedules.[43] Schroeder and Holland[49] reported an experiment in which human eye movements were placed under a concurrent schedule in a close approximation of a vigilance task. Subjects were instructed to monitor a display containing four dials and to report changes in the dials by pressing on a switch. Eye movements were recorded with a corneal reflection technique using a television-based digitizer and were found to reflect the relative rate at which the dial moved. Changes in a dial were placed under a concurrent schedule, i.e., if a change set up on a dial, then the next saccade toward that dial resulted in its change. The rate of eye movements toward each dial was sensitive to the relative rates of movement of a dial.

The conceptualization of FI schedule performance as concurrent performance (with the other schedule usually unspecified) is supportable by several studies of human FI schedules. A notable example is provided by Frazier and Bitetto[42] who placed human observing responses under three-ply concurrent schedules involving FR, FI, DRL, or VI components. The subjects were confronted with three response devices which could be operated to light a meter that they had to monitor. The subjects were instructed to detect 43 degree movements of the meter, and meter movements were placed under different schedules. For example, under the FI 1 min, a meter deflection followed the first response to occur after 1 min. Response patterns typical of nonhuman species performing under these schedules occurred in 18 of the 22 subjects studied within 10 sessions. This approach to vigilance does not directly reproduce the vigilance task, where changes in the object to be detected are not contingent upon the subject's response. Instead, it is a design in which the experimenter can explicitly control the relationship between looking at the meter and the occurrence of a meter movement, and in the sense that it provides greater experimental control it could be an improvement over conventional procedures.

Mechanisms of Action and the Link Between Human and Animal Research

A major difficulty in relating animal studies to humans has been the difficulty of linking these research domains along lines of mechanism or procedure. Such links could be greatly facilitated by designing human studies to be compatible with what we know of animal behavior. Naturally, the challenge could also apply in the other direction, but in many ways such an endeavor is unrealistic until the problem is approached functionally. A literal profile of mood states or even addition of numbers will not be a useful animal study, but the important components of these tasks might be. In view of the great replicability seen with these procedures in animal studies it seems that the time is ripe to think seriously about including in human test batteries procedures that have shown species generality.[50] There is every reason to believe that the mechanisms supporting schedule-controlled behavior are similar across species, including humans.[51]

One benefit of animal studies is that hazard assessment can be conducted more precisely because of the control possible when designing experiments. The generalization to humans and to risk assessment can be greatly assisted by the use of procedures that are functionally relevant to human settings. To support generalization it is necessary to identify a mechanism of action, and these mechanisms can be neural or behavioral.[52] A neural mechanism could involve relating neurochemical or structural changes in the brain to impairments in behavior. Mechanisms can also be *behavioral*. For example, the effects of a toxicant might be reduced to insensitivity to changes in reinforcement contingencies, aversive properties of stimuli, or impaired control by infrequent events. If a mechanism of toxicity is identified then generalization to other events with similar mechanisms, that produce similar signs and symptoms, is possible, and this can guide both clinical and scientific inference.

SENSITIVE POPULATIONS

Even in structures designated as sick buildings not everybody experiences discomfort or other effects. A complicating component of the problems associated with adverse effects of indoor air is the sensitivity of some individuals. Henry et al.[3] identified three ways in which such sensitivity appears, and they all present a challenge to the toxicologist. For classical toxicants, the distributions of effects in a population follows a normal or log-normal distribution. At low levels, some individuals will show an effect, and as exposure increases more show the effect. It is on such a relationship that such designations as ED 10 (effective dose in 10% of the population), ED 50, and ED 90, are based. Sensitive individuals will appear at the left end of this bell-shaped curve where exposures are low. Other forms of sensitivity can appear. Atopic or allergic individuals, whose sensitivity may be influenced by genetic predispositions or prior exposures might form a separate distribution centered on low concentrations. Others, with multiple or dynamic sensitivities may respond to such compounds as formaldehyde differently with repeated exposure (see note added in press), but in general this population, too, would fall on the low end of the dose-effect curves. Whatever the mechanism and the population, it is the low end of the dose-effect curve, and the possibility of separate distributions of sensitivity in subpopulations, that is of concern. Sensitive populations may simply comprise the low end of the curve or may represent a special population with differently shaped distributions.

Weiss et al.[53] and Weiss[54] provide striking examples of sensitive populations. Weiss et al.[53] studied 22 children suspected of being sensitive to artificial food colors. On close examination, 20 of the 22 children revealed no identifiable sensitivity but two showed robust sensitivity to a drink containing artificial food colors. This was clearly revealed in a double-blind, single-subject experimental design in which the child was presented, on separate occasions, with a drink containing a placebo or the food color.

Weiss[54] examined how sensitive individuals get lost in group averages. For example, Williams et al.[55] examined the hypothesis that food colors produce hyperactivity by giving hyperactive children a cookie containing a food color or a placebo. Sometimes the children were on medication and other times they were not. Group statistics revealed an effect, according to conventional statistical procedures, based on teacher ratings ($p = 0.025$) but not upon parent ratings. However, examination of parent ratings revealed a consistent effect of the cookie on a subgroup of children. One, and perhaps two clusters of children responded to the challenge when they were not on medication, an effect undetected by conventional group statistics (Figure 5). Identification of sensitive populations requires examination of individual subjects and statistical procedures that will detect such populations (see also Dews[56] and Glowa[57]).

The use of large numbers of subjects enhances the likelihood of detecting sensitive populations if the appropriate analytical strategies are used, but such effects have been detected in smaller studies. "Outliers" have been identified with a variety of toxicants, including ozone,[58] lead,[26,59] and toluene.[60] The report of Wood and Colatla[60] is particularly instructive. They examined the locomotor activity of mice in response to varying concentrations of toluene using a repeated measures design. This provided the ability to compare an individual mouse's activity against its own control level and greatly enhanced the sensitivity of the experiment. Figure 6 shows two concentration effect curves. One is formed by averaging activity counts under different conditions. In the other, information from individual animals is retained by expressing counts as a percentage of a subject's control value. The curve containing individual animals contains two interesting features. First, it is sharper and the error bars are smaller, which gives it greater precision in detecting a lowest observed adverse effect level. An effect at 560 ppm is visible only when data are compared against an animal's own control. The other notable characteristic is the large increase in the size of the error bars at the middle concentrations. This increased variability almost always reflects individual susceptibility. Wood and Colatla[60] examined the performance of individual animals and found a few whose response to the toluene differed greatly from the rest. The increase in locomotor activity was as many as 7 to 10 standard deviation units removed from the mean. There was no clue to this susceptibility in the group averages.

PARENTS RATINGS

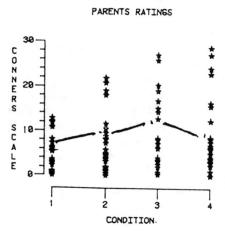

Figure 5 Parent ratings of individual children after being given a placebo or cookie containing a food-color challenge. During conditions 1 and 2 the children were given medication (methylphenidate or amphetamine) and during conditions 3 and 4 they were not. During conditions 2 and 4 they were given a food-color challenge and during conditions 1 and 3 they were given a placebo. (From Weiss, B., *J. Am. Acad. Child Psychiatr.*, 21: 144-152 (1982). With permission.)

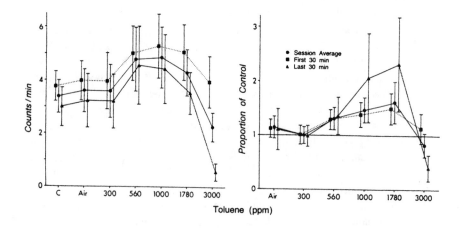

Figure 6 Effects produced by toluene on the locomotor activity of mice during the first and last halves of the session and for the session as a whole. The left panel shows data expressed as average activity counts taken from different conditions. On the right the data for each mouse are expressed as a percent of its control performance, and the percentages were averaged. Therefore, the right represents group averages and the left averages of individual effects. (From Wood, R.W. and Colatla, V.A., *Fund. Appl. Toxicol.*, 14: 6-14 (1990). With permission.)

SUMMARY AND CONCLUSIONS

The problems posed by reports of neurotoxicity associated with indoor air present a special challenge to the behavioral toxicologist. Many of the reports have included subjective complaints and reports of discomfort or difficulties in concentration. That these complaints have been reproduced in experimental exposures gives them weight. The design of animal studies to characterize this potential hazard in greater detail can be accomplished by drawing from protocols that are reasonably well understood, sensitive, involve behavior that may be impaired in human settings, and permit the identification of sensitive populations. When setting up protocols, consideration should be given to identifying behavioral actions that can be characterized in both human and nonhuman species.

NOTE ADDED IN PROOF

Wood and Coleman[61] described how individual sensitivity to the irritancy of formaldehyde can increase after a history of exposure to this irritant. A mouse could terminate exposure to formaldehyde and initiate a facial shower of clean, humidified air by placing its nose into a conical shaped opening. A full concentration-effect curve was determined twice in each of eight mice. Some mice received an ascending, and then a descending series of concentrations, while other mice received the opposite. The range of doses that a mouse terminated on half the occasions varied about three- to fivefold across six mice. Two mice were so sensitive that an ineffective concentration could not be determined.

Especially pertinent to the topic of sensitivity to indoor atmospheres is the observation by Wood and Coleman that the mice became more sensitive to formaldehyde during the second determination of the concentration-effect curve. Moreover, the elevated sensitivity was attributed directly to two behavioral mechanisms: respondent conditioning due to the repeated pairings of a novel stimulus with the irritation produced by formaldehyde and conditioned positive reinforcement associated with the reinforcing effects of the facial shower. The shift in the concentration effect curve was a consequential one. Prior experience to formaldehyde decreased the concentration at which one-half of the deliveries were terminated about threefold. Concatenating the individual variation with the conditioned sensitivity could result in a range of a full order of magnitude in sensitivities associated with a combination of individual sensitivities due to factors existing prior to the experiment and conditioning factors occurring during the experiment. Wood and Coleman's experiment replicated, with behavior, an observation made by Kane and Alarie[62] using respiratory endpoints. Tepper and Wood[63] also reported a similar observation with ozone. The behavioral mechanisms advanced by Wood and Coleman are plausible and, if they are truly operating, fundamental in understanding the dynamics of behavioral responses to atmospheres. Experiments aimed at reproducing these effects with the full range of control conditions required to confirm the operation of these behavioral mechanisms are called for.

REFERENCES

1. Wood, R.W., Stimulus properties of inhaled substances: an update. In *Nervous System Toxicology,* Mitchell, C.L. Ed. 199-212 (1982).
2. Berglund, B., Brunekreef, B., Knoppel, H., Lindvall, T., Maroni, M., Mølhave, L. and Skov, P., Effects of indoor air pollution on human health. *Indoor Air,* 22: 2-25 (1992).
3. Henry, C.J., Fishbein, L., Meggs, W.J., Ashford, N.A., Schulte, P.A., Anderson, H., Osborne, J.S. and Sepkovic, D.W., Approaches for assessing health risks from complex mixtures in indoor air: a panel overview. *Environ. Health Perspect.,* 95: 133-143 (1991).
4. Mølhave, L., The sick building — a subpopulation among the problem buildings. In *Indoor Air 1987: Proceedings of the IV International Conference on Indoor Air Quality and Climate,* Seifert, B., H. Esdorn, and M. Fischer, Eds., (Berlin: Inst. Water, Soil, Air Hygiene, 1987) pp. 469-473.
5. Bolla, K.I., Neuropsychological assessment for detecting adverse effects of volatile organic compounds on the central nervous system. *Environ. Health Perspect.,* 95: 93-98 (1991).
6. Marbury, M.C. and Woods, J.E., Building-related illnesses. In *Indoor Air Pollution: A Health Perspective,* Samet, J.M. and J.D. Spengler, Eds., (Baltimore, MD: Johns Hopkins University, 1991), p. 306-322.
7. Welch, L.S., Severity of health effects associated with building-related illness. *Environ. Health Perspect.,* 95: 67-69 (1991).
8. Baker, E.L. and Letz, R., Neurobehavioral testing in monitoring hazardous workplace exposures. *J. Occup. Med.,* 28: 987-990 (1986).
9. Laties, V.G. and Merigan, W.H., Behavioral effects of carbon monoxide in animals and man. In *Annual Review of Pharmacology and Toxicology,* Ed., (Annual Reviews, Inc., 1979) pp. 357-392.
10. Dick, R.B. and Johnson, B.L., Human experimental studies, In *Neurobehavioral Toxicology,* Annau, Z., Ed., (Baltimore, MD: Johns Hopkins, 1986) pp. 348-390.
11. O'Hanlon, J.F., Preliminary studies of the effects of carbon monoxide in man. In *Behavioral Toxicology,* Weiss, B. and V.G. Laties, Eds., (New York: Plenum, 1974) pp. 61-72.
12. Beard, R.R. and Grandstaff, N.W., Carbon monoxide and human function. In *Behavioral Toxicology,* Weiss, B. and V.G. Laties, Eds., (New York: Plenum, 1975) pp. 1-28.
13. Benignus, V.A., Dose-effect functions for carboxyhemoglobin and behavior. *Neurotoxicol. Teratol.* 12: 111-118 (1990).
14. Stein, N. and Landis, R., Differential reinforcement of low rate performance by impulsive and reflective children. *J. Exp. Psychol.,* 97: 28-33 (1975).
15. Skov, P. and Valbjorn, O., The sick-building syndrome in the office environment: The Danish town hall study. In *Indoor Air '87: Proceedings of the 4th International Conference on Indoor Air Quality and Climate,* Seifert, B., H. Esdorn, and M. Fischer, Eds., (Berlin: Inst. for Water, Soil and Air Hyg., 1987) pp. 439-443.
17. Hudnell, H.K., Otto, D.A., House, D.E. and Mølhave, L., Exposure of humans to a volatile organic mixture. II. Sensory. *Arch. Environ. Health,* 47: 31-38 (1992).
18. Otto, D.A., Hudnell, H.K., House, D.E., Mølhave, L. and Counts, W., Exposure of humans to a volatile organic mixture. I. Behavioral Assessment. *Arch. Environ. Health,* 47: 23-30 (1992).

19. Dick, R.B. and Johnson, B.L., Human experimental studies, In *Neurobehavioral Toxicology,* Annau, Z., Ed., (Baltimore, MD: Johns Hopkins, 1986) pp. 348-390.

20. Dick, R.B., Setzer, J.V., Taylor, B.J. and Shukla, R., Neurobehavioral effects of short duration exposures to acetone and methyl ethyl ketone. *Br. J. Ind. Health,* 46: 111-121 (1989).

21. Wood, R.W., Behavioral evaluation of sensory irritation evoked by ammonia. *Toxicol. Appl. Pharmacol.,* 50: 157-162 (1979).

22. Wood, R.W., Determinants of irritant termination behavior. *Toxicol. Appl. Pharmacol.,* 61: 260-268 (1981).

23. Schuster, C.R. and Balster, R.L., The discriminative stimulus properties of drugs. In *Advances in Behavioral Pharmacology,* Thompson, T. and P.B. Dews, Eds., (New York: Academic Press, 1977) pp. 86-139.

24. Rees, D.C., Knisley, J.S., Breen, T.J., and Balster, R.L., Toluene, halothane, 1,1,1-trichloroethane and oxazepam produce ethanol-like discriminative stimulus effects in mice. *J. Pharmacol. Exp. Ther.,* 243: 931-937 (1987).

25. Cory-Slectha, D.A., Widzowski, D.V. and Newland, M.C., Behavioral differentiation of the stimulus properties of a dopaminergic D1 agonist from a D2 agonist. *J. Pharmacol. Exp. Ther.,* 2250: 800-808 (1989).

26. Rice, D.C., Quantification of operant behavior. *Toxicol. Lett.,* 43: 361-379 (1988).

27. Cory-Slechta, D.A., Bridging experimental animal and human behavioral toxicity studies. In *Behavioral Measures of Neurotoxicity,* Russell, R.W., P.E. Flattau, and A.M. Pope, Eds., (Washington, D.C.: National Academy Press, 1990) pp. 137-158.

28. Glowa, J.R., *Behavioral Pharmacology: The Current Status,* Seiden, L.S. and R.L. Balster, Eds., (New York: Alan R. Liss, 1985).

29. Laties, V.G. and Weiss, B., Effects of alcohol on timing behavior. *JCCP,* 55: 85-91 (1962).

30. Sanger, D.J., Key, M. and Blackman, D.E., Differential effects of chlordiazepoxide and d-amphetamin on responding maintained by a DRL schedule of reinforcement. *Psychopharmacologia,* 38: 159-171 (1974).

31. Sanger, D.J. and Blackman, D.E., Operant behavior and the effects of centrally acting drugs. In *Neuromethods. Vol 13: Psychopharmacology,* Bouldon, A.A. and G.B. Baker, Eds., (Clifton, NJ: Humana, 1989) pp. 299-348.

32. Laties, V.G., Weiss, B., Clark, R.L. and Reynolds, M.D., Overt "mediating" behavior during temporally spaced responding. *J. Exp. Anal. Behav.,* 8: 107-116 (1965).

33. Rosenfarb, I.S., Newland, M.C., Brannon, S.E. and Howey, D.S., Effects of self-generated rules on the development of schedule-controlled behavior. *J. Exp. Anal. Behav.* 58: 107-121 (1992).

34. Stein, N. and Landis, R., Differential reinforcement of low rate performance by impulsive and reflective children. *J. Exp. Psychol.,* 97: 28-33 (1975).

35. Gordon, M., The assessment of impulsivity and mediating behaviors in hyperactive and nonhyperactive boys. *J. Abnorm. Child Psychol.,* 7: 317-326 (1979).

36. McClure, F.D. and Gordon, M., Performance of disturbed hyperactive children and nonhyperactive children on an objective measure of hyperactivity. *J. Abnorm. Child Psychol.,* 12: 561-572 (1984).

37. Barkley, R.A., The latest on DSM-IV and the disruptive behavior disorders. *ADHD Rep.,* 1: 3-5 (1993).

38. Rice, D.C. and Gilbert, S.G., Low lead exposure from birth produces behavioral toxicity (DRL) in monkeys. *Toxicol. Appl. Pharmacol.,* 80: 421-426 (1985).

39. Schrot, J. and Thomas, J.R., Multiple schedule performance changes during carbon monoxide exposure. *Neurobehav. Toxicol. Teratol.,* 8: 225-230 (1986).

40. Moser, V.C., Coggeshall, E.M. and Balster, R.L., Effects of xylene isomers on operant responding and motor performance in mice. *Toxicol. Appl. Pharmacol.,* 80: 293-298 (1985).

41. Colatla, V.A., Bautista, S., Lorenzana-Jimenez, M. and Rodriguez, R., Effects of solvents on schedule-controlled behavior. *Neurobehavioral Toxicology and Teratology (Suppl. 1): Test Methods for Definition of Effects of Toxic Substances on Behavior and Neuro,* 1: 113-118 (1979).

42. Frazier, T.W. and Bitetto, V.E., Control of human vigilance by concurrent schedules. *J. Exp. Anal. Behav.,* 12: 591-600 (1969).

43. Baron, A., Myerson, J. and Hale, S., An integrated analysis of the structure and function of behavior: aging and the cost of dividing attention. In *Human Operant Conditioning and Behavior Modification,* Davey, G. and C. Cullen, Eds., (New York: John Wiley, 1988) pp. 139-166.

44. Davison, M. and McCarthy, D., *The Matching Law* (Hillsdale, NJ: Lawrence Erlbaum, 1988).

45. Newland, M.C., Sheng, Y., Logdberg, B. and Berlin, M., Prolonged behavioral effects of in utero exposure to lead or methyl mercury reduced sensitivity to changes in reinforcement contingencies during behavioral transitions and in steady state. *Toxicol. Appl. Pharmacol.,* 125: (1994).

46. Heyman, G.M. and Monaghan, M.M., Contributions of the matching law to the analysis of the behavioral effects of drugs. In *Advances in Behavioral Pharmacology,* Barrett, J.E., T. Thompson, and P.B. Dews, Eds., (Hillsdale, NJ: Lawrence Erlbaum, 1990) pp. 39-78.

47. Ziriax, J.M., Snyder, J.R., Newland, M.C. and Weiss, B., Amphetamine modifies the microstructure of concurrent behavior. *Exp. Clin. Psychopharmacol.,* 1: 121-132 (1993).

48. Baum, W.M., Time allocation in human vigilance. *J. Exp. Anal. Behav.,* 23: 45-53 (1975).

49. Schroeder, S. and Holland, J.G., Reinforcement of eye movement with concurrent schedules, *J. Exp. Anal. Behav.,* 12: 897-903 (1969).

50. Paule, M.G., Forrester, T.M., Maher, M.A., Cranmer, J.M. and Allen, R.R., Monkey versus human performance in the NCTR operant test battery. *Neurotoxicol. Teratol.,* 12: 503-507 (1990).

51. Perone, M., Galizio, M. and Baron, A., The relevance of animal-based principles in the laboratory study of human operant conditioning. In *Human Operant Conditioning and Behavior Modification.* Davey, G. and C. Cullen, Eds., (New York: Wiley, 1988) pp. 59-85.

52. Marr, M.J., Behavioral pharmacology: issues of reductionism and causality. In *Advances in Behavioral Pharmacology,* Barrett, J.E., T. Thompson, and P.B. Dews, Eds., (Hillsdale, NJ: Lawrence Erlbaum, 1990) pp. 1-12.

53. Weiss, B., Williams, J.H., Margen, S., Abrams, B., Caan, B., Citron, L.J., Cox, C. and McKibben, J., et al., Behavioral responses to artificial food colors. *Science,* 207: 1487-1489 (1980).

54. Weiss, B., Food additives and environmental chemicals as sources of childhood behavior disorders. *J. Am. Acad. Child Psychiatr.,* 21: 144-152 (1982).

55. Williams, J.I., Cram, D.M., Tausig, F.T. and Webster, E., Relative effects of drugs and diet on hyperactive behaviors: An experimental study. *Pediatrics,* 61: 811-817 (1978).

56. Dews, P.B., On the assessment of risk. In *Advances in Behavioral Pharmacology: Vol 5, Developmental Behavioral Pharmacology,* Krasnegor, N.A., D.B. Gray, and T. Thompson, Eds., (Hillsdale, NJ: Lawrence Erlbaum, 1986) pp. 53-65.

57. Glowa, J.R., DeWeese, J., Natale, M.E., Holland, J.J. and Dews, P.B., Behavioral toxicology of volatile organic solvents. I. Methods: Acute effects. *J. Am. Coll. Toxicol.,* 2: 175-185 (1983).

58. Weiss, B., Ferin, J., Merigan, W., Stern, S. and Cox, C., Modification of rat operant behavior by ozone exposure. *Toxicol. Appl. Pharmacol.,* 58: 244-251 (1981).

59. Cory-Slechta, D.A., Weiss, B. and Cox, C., Delayed behavioral toxicity of lead with increasing exposure concentration, *Toxicol. Appl. Pharmacol.,* 71: 342-352 (1983).

60. Wood, R.W. and Colatla, V.A., Biphasic changes in mouse motor activity during exposure to toluene. *Fund. Appl. Toxicol.,* 14: 6-14 (1990).

61. Wood, R.W. and Coleman, J.B., Behavioral evaluation of the irritant properties of formaldehyde. *Toxicol. Appl. Toxicol.,* 130: 67–72 (1995).

62. Kane, L.E. and Alarie, Y., Sensory irritation to formaldehyde and acrolein during single and repeated exposures in mice. *Am. Ind. Hyg. Assoc. J.,* 38: 509–522 (1977).

63. Tepper, J.S. and Wood, R.W., Behavioral evaluation of the irritating properties of ozone. *Toxicol. Appl. Pharmacol.,* 78: 408–411 (1985).

CHAPTER 15

Effects of Irritants on Human Task Performance and Vigilance

Vernon A. Benignus

INTRODUCTION

It is frequently reported, anecdotally, that indoor air pollutants (IAP) reduce productivity and job performance in addition to a host of other complaints, e.g., irritation of nasal mucosa and eyes.[1] There is, however, a paucity of evidence from controlled experiments for the reduction in task performance (TP). Furthermore, there is only conjecture as to the mechanism by which TP may be affected by IAP.

The present chapter addresses the above issues, admittedly in a conjectural manner combined with a review of the few IAP studies, in an effort to ground the guesswork in theory by comparison to other related lines of inquiry. Before such an effort is begun, however, it will be convenient to establish a framework for discussion of TP.

FRAMEWORK FOR DISCUSSION OF TP

Task performance has two aspects which are frequently conflated — task specification and performance classification. The failure to distinguish which aspects is being discussed can readily lead to confusion. In this chapter, tasks will be classified according to popular categories, but, specified by the contingencies which exist for the control of behaviors (performance), i.e., tasks will be explicitly classified according to job specifications. The same category names are frequently applied to the behaviors which the contingencies control, but it may be advantageous to instead consider the kinds of behaviors which are required of the persons doing the tasks in a more topologically descriptive manner.

Researchers have spent much effort devising descriptive classifications for tasks frequently performed by workers. No attempt will be made in the present chapter to pursue a complete discussion of the various schemes. Rather, only three classifications which are frequently studied in the context of related research will be described. These (overlapping) classifications (according to behavior controlling contingencies) are (1) vigilance, (2) continuous performance, and (3) divided attention: Memory tasks, which are also frequently studied will not be covered in this chapter. Classical references are cited for each kind of task.

Vigilance

A vigilance task requires the subject to rapidly detect infrequently presented events or signals over long periods of time.[2,3] A vigilance task may require detection either with or without simultaneous presentation of unrelated (nontarget or nonsignal) stimuli. Even if nonsignal stimuli are not formally presented in the experiment, such stimuli are frequently posited as part of the external or internal environment of the subject. There may also be another related or unrelated task which the person must perform simultaneously.

A vigilance task is frequently said to require alertness or vigilance on the part of the subject,[4,5] but there are several problems with an uncritical use of the word vigilance for both the task specifications and for the cognitive, physiological, or behavioral processes needed to perform the task. Most obviously, while some behavioral or cognitive construct such as alertness or vigilance can be invoked as necessary, that is not the only behavior required. The whole-task performance may decline due to some intervention, e.g., indoor air pollution, but the decrement may be due to some other behavior than alertness being affected. Before a claim of alertness effects can be defended, appropriate measurements and experimental controls must be implemented to exclude other possible problems, e.g., decrements in discrimination, response speed, sensory or motor problems, etc.

Another problem with the use of the term "vigilance" or its synonyms as an attribute of the subject or his behavior is that such an attribute ought to be measurable independently of the task which requires it. Such attempts have involved electrical or biochemical measurements of central or peripheral nervous system or overt incidental behaviors. The single common denominator of efforts to independently measure alertness has been the lack of good predictability of vigilance performance. While some correlations have been found, they are not very high, are usually task specific, and very sensitive to experimental design.[2,3]

Unless one is primarily interested in the physiological or behavioral theories of how the body maintains vigilance task performance, it would seem more profitable to concentrate on the topology of the behaviors required to perform a vigilance task than to postulate "explanatory" entities. Certainly, if one is primarily interested in the effects of IAP on TP, such theory may well lead to a nonproductive morass. The behaviors required to perform vigilance tasks may be roughly classified as long-term maintenance of (1) high probability of a response to the stimulus, (2) discriminative behavior (deciding on the difference between signals and nonsignals), and (3) rapid response speed, because most signals are short in duration and must be detected before they disappear. In this terminology, vigilance is an attribute of the task description, not of the subject.

Continuous Performance

Continuous performance tasks require the person to respond to signals presented at a high rate for short periods of time.[6] As with vigilance tasks, there may be more than one event (signals and nonsignals) and more than one task simultaneously required of the person. The differences between vigilance and continuous performance tasks are on the continua of task duration and signal rate. The latter differences are based on the observation that performance which is well sustained for short periods of time may become decremented if the task is performed for longer periods with low signal rates. Because temporal continua define the difference between the two classes of tasks, there are obviously some ambiguities involved (deciding when to call the task vigilance and when to call it continuous performance). Because the different classification for the two kinds of tasks originated in the behavioral effects which they (usually) produce, it is not surprising that the behaviors required to perform the two kinds of tasks are the same, except that they need to be sustained over shorter periods of time for the continuous performance task. The difference in behavioral effects is, in fact, the only justification in maintaining a separate category and is frequently cited as justification for the explanatory constructs of cognitive, physiological, or behavioral alertness.

Divided Attention

Divided attention tasks require the subject to perform more than one task simultaneously.[7,8] In such situations, the two tasks may be highly similar, e.g., detection of visual events in the center of the visual field while simultaneously detecting events in the periphery. In contrast, the tasks may be dissimilar, e.g., tracking a moving target in the center of the visual field while simultaneously detecting events in the periphery. Any kind of task may be used, e.g., arithmetic calculations, motor tasks, color discriminations, etc. Obviously, a divided attention task may be used as a vigilance task if the rate of signal presentation is low and the duration of performance is long.

Again, the responses required of the person in a divided attention task are similar to those in vigilance or continuous performance tasks. The difference is in the emphasis or purpose of the study. Cognitive constructs of attentional processes have been proposed to classify the behaviors/processing required of the person performing such a task, but it may be hazardous to ignore the fact that many of the behaviors required in such performance are common with other tasks and have nothing to do with attention as a cognitive process.

BEHAVIORAL EFFECTS OF IRRITANTS

Mechanisms and Restrictions

Conceptually, an irritant may produce its effects by either of two broad classes of "mechanisms." On the one hand, effects may be produced organically by temporary or permanent impairment of the central or peripheral nervous system due to the substance having entered the body, e.g., via the blood stream. In the other case, effects may be produced purely due to the peripheral irritant effects of the substance. In the latter case, effects could still be due to alterations or reductions in the nervous system's capacity or activity, but the mechanism is functional, not organic. Functional effects may be either temporary or long term, depending on how they were produced, e.g., if they were learned or associated with other stimuli.

The following reviews and discussions will be restricted to the kinds or intensities of irritants which may be argued to have produced only functional effects. Irritants which may have produced part of their effects by organic impairment will be excluded from consideration because it is not usually possible to proportionally attribute effects to functional vs. organic mechanisms. In practical situations, both organic and functional effects may result from IAP, but for understanding of the nature of the overall effects, it would be profitable to conceive of and to study the two kinds of effects separately.

A number of mechanisms of action or hypotheses have been proposed to account for functional effects of some condition, e.g., irritants.[2,3] Each of the hypotheses fits into two broad classes. One class posits some general nervous system or cognitive process generally called arousal, activation, drive, or some synonym, which is either increased or decreased by the presence of the condition. An "optimum" level of arousal is presumed to exist for each person/task/time-of-day/etc., at which the task is best performed. Lower than optimal arousal is presumed to reduce the organism's sensitivity (alertness or vigilance), while higher levels of arousal are presumed to produce random responses which interfere or compete with appropriate behavior. The other broad class of hypotheses deals with such constructs as selective attention, expectancy, decision theory, or some synonym, which are supposed to be interfered with or disrupted by the presentation of some condition.

In the cases of all hypotheses of functional effects, except that of underarousal, the supposed mechanism is one of interference or disruption by whatever condition is being studied, in the present case, irritants. This hypothesis was parsimoniously explicated by Teichner[9] as the "distraction" hypothesis. It may be more testable to hypothesize that the disrupting condition produces irrelevant *behavioral* responses which compete with appropriate performance of the task at hand. The latter (behavioral) version of the distraction hypotheses will be called the "competing response" hypothesis.

For most of the performance effects of irritants, it is probable that underarousal (if an arousal hypothesis were to be espoused) would not be the problem. It may be argued (see above) that all other hypotheses about mechanisms of effects assume disruption of some constructed cognitive or nervous system process, e.g., attention, decision making, etc. It would appear that the competing response hypothesis is a behavioral competitor for all of them. Whether or not behavioral, cognitive, or nervous system interference is supposed, may be relevant only to experiments designed to "explain" or choose between competing hypotheses. The idea of interference or disruption is the common thread that runs throughout most accounts of the effects of conditions like irritants on the performance of tasks.

Kinds of Irritants

Of primary interest in the present chapter are the behavioral effects of chemical irritants. As yet there are, however, only a few experimentally based studies in the literature concerning behavioral effects of chemical irritants, and the body of evidence is inconclusive. Anecdotal reports are not experimentally controlled and therefore unique conclusions cannot be drawn.

If the effects of irritants is to disrupt task performance via distraction or by producing competing responses, then comparable effects should ensue from a number of other, more extensively studied environmental conditions. It may be instructive to review the literature of the performance-disrupting effects of such other conditions to learn about the nature and extent of the findings. It would not be unreasonable to expect similar findings in the cases of chemical irritation. To this end, the literature on the disrupting effects of cold or noisy environments has been reviewed in addition to the meager literature on chemical irritation.

Some Common Problems

A number of common problems occur in the research literature on the effects of IAP on TP. Some of these are avoidable, others are only partly avoidable. Attention to these problems during the experimental design phase of a study is well worth the effort.

Dose-Effects Studies vs. Point Estimates

Many experiments in the IAP literature employ only two experimental conditions — exposure and control. Furthermore, the experimental level is often selected to be representative of existing levels of IAP which have been reported. It is probable that effects at any existing non-laboratory environmental level will be small (if nonzero) because the exposure would not be tolerated if patent effect would result. Given the small expected effect of such an exposure, the experimenter is faced with either studying a very large number of subjects in order to detect the effect or finding unreliable or no effects.

The alternative to the two-condition study is to perform a dose-effects study in which more than one exposure level is well above typical environmentally occurring exposures. Thus the results will form a dose-related pattern if the substances being studied have an effect at all. The low-level effects in the range of the environmentally relevant levels can then be interpolated between clean air and the larger exposure values (assuming certain dose-effects curve characteristics). Alternatively, if a two-condition study must be done (e.g., due to economic or time constraints) the exposure level in the first study should be considerably larger than environmentally relevant so that a well-defined effect may be observed for later study in a dose-effects manner. If the large exposure level does not have an effect, then further study is probably not needed. It seems that the single, environmentally relevant exposure level is almost never justified because the results are almost never reliable unless large numbers of subjects are studied. A dose-effects study might require an equally large number of subjects, but the results would be much more informative.

The reasons for studying a single environmentally relevant exposure level are the need to economize and to get quick results. The results of such simple experiments do not logically lead to the desired results, however. In the interest of good science, researchers should attempt to resist pressure for quick, economical but probably inconclusive experiments.

Blinding

If the subject is not informed of the conditions of the exposure condition, the subject is said to be kept "blind." Similarly, the experimenter who interacts with the subject is also sometimes blinded. When both experimenter and subject are blinded, the experiment is said to be conducted in a double-blind manner. Blinding is to avoid the possibility that exposure effects are due to suggestion. It has been reported that in experiments which were conducted in a single- or non-blind manner, there were many more statistically significant effects reported than in double-blind studies of the effects of carbon monoxide exposure.[10]

The experimenter may usually be kept blinded in an IAP study by simply observing some precautions and having an independent person controlling exposure levels. The subject might be uninformed of the exposure level, but in a two-condition (control and exposure) experiment, the pollutant can usually be detected and thus the subject cannot be kept truly blind and free of suggestion. In a dose-effects study, the subject might not be able to discriminate with any degree of certainty between levels of exposure and thus may be kept at least partly blind. This is another argument for multiple exposure-level studies. An alternative is to use two nonzero exposures in which one level is below the level known to produce any effects but yet is detectable by the subject (if such a level is known or can be selected).

Exploratory vs. Hypothesis-Testing Strategy

It is possible to measure a wide variety of dependent variables in any given experiment. When an experimenter is uncertain of the effects of a pollutant, it may be good strategy to "cast a broad net" to be sure to include variables which may be affected. Such an exploratory strategy, however, has analytic and inferential consequences which must be carefully considered to avoid possible errors.[11]

To statistically test results of an experiment measuring more than one variable, the experimenter must consider the relationship between the dependent variables in the analysis. So-called "univariate" statistical tests (e.g., the Student t and analysis of variance) are built around the assumption that each time a test is conducted, the variable under test is independent of other variables being tested. If this is not true, either multivariate tests or corrections must be applied.[11] The latter requires more subjects in the study in proportion to the number of variables measured. If a series of uncorrected univariate tests is used to evaluate a group of independent variables measured in the same subjects, the probability of a falsely significant result is increased.[11] When an experiment is reported in which multiple variables were measured in the same subjects and the results were analyzed with uncorrected univariate tests, the results must be viewed with suspicion until replicated.

Small sized experiments with many dependent variables may be analyzed with uncorrected univariate tests if the experimenter only uses the results to design another experiment. Otherwise the risks of a falsely declared significant result are great, in proportion to the number of variables so tested. One strategy is to decide before the experiment which single or small group of variables to test (based on best information) and make decisions about those only. Exploratory analyses for further experimentation would not be reported until replication.

Behavioral Effects of Chemical Irritants

The first experimental report of the effects of IAP on TP was published by Mølhave et al.[12] in which subjects who reported that they were adversely affected by IAP (sensitive subjects) were exposed to a mixture of volatile organic compounds (VOCs) in a controlled situation. The subjects performed more poorly on the digit span test (in which they were required to repeat a series of digits spoken to them by an experimenter) during VOC exposure than during clean air. This experiment was clearly a survey or exploratory type design in which many other variables were measured and tested. No blinding was attempted and the exposure level was typically of what was found in various buildings.

The Mølhave group repeated the experiment described above with two groups of subjects: sensitive subjects and another group who did not report such effects (normal subjects).[13] Neither experimenters or subjects were informed about exposure conditions, but the exposure condition was clearly detectable by the subjects. While behavioral effects were found, they were different than in the first experiment and were difficult to interpret. Using only normal subjects, Otto et al.[14] approximately repeated the exposures of Mølhave and found no behavioral effects even though a wide variety of behaviors were explored. The behavioral tests were not, however, identical to those used in the earlier work. Low levels of exposure and a nonsubject-blind design were used.

In a double-blind, dose-effects study of formaldehyde exposure, the Mølhave group[15] reported multivariate, statistically significant behavioral effects. Dose-related decrements in speed and reaction time during an arithmetic (addition) task were demonstrated along with less systematic results for other behaviors. Subjects who had been occupationally exposed to formaldehyde had more extreme effects.

It appears that at some as yet undetermined level of exposure, chemical irritants produce dose-related TP decrements. Far too few data are available to speculate about the behaviors which are affected, however. It would also be difficult to relate the irritation level from the formaldehyde experiment to the irritation level in the multiple VOC experiments, but if this could be done, some inferences about typical IAP might be made. A scale of irritancy for various substances would permit such comparison. Magnitude estimation experiments[16] would help in constructing comparability scales.

Behavioral Effects of Noise

The effects of noise on behavior is well studied and, although not all researchers agree in detail, the principle effects are known. It may be possible to learn something about the TP effect of chemical irritants on behavior by reviewing the effects of noise on TP. Accordingly, the noise literature was reviewed and eight studies were selected for meta analysis[17] of results, using methods developed elsewhere.[18] The studies[19-27] were selected because the data were reported in such a manner as to make meta analysis possible.

Each of the studies used a slightly different task. The tasks were classified (by the present author) as either vigilance (V), continuous performance (CP), divided attention (DA), or two of the above where appropriate. The objective was to explore which kind of TP was most affected by noise.

To make the scale of the dependent variable in each of the seven studies comparable, all dependent variables were transformed to a new standard response metameter (SRM).[18] The scale properties of the SRM are that (1) descending scores represent a decrement, (2) baseline performance is represented by a value of 1.0, and (3) entirely decremented performance is represented by a value of 0.0. To avoid problems in curve fitting, only data from 70 dBA and up were plotted and analyzed. Because the effect of the maximum noise tested (114 dBA) was to produce an SRM value of slightly greater than 0.7, a linear scale was used to fit the data (a relatively linear portion of a probably ogival curve). Not enough data were available to have fitted a nonlinear function in that portion of the curve.

Figure 1 is a plot of the SRMs of the seven studies (dashed lines), and the linear equation fitted to the pooled group of points regardless of the study from which the data originated (heavy solid line). There is no discernible pattern in the data to indicate that any kind of task provides more sensitivity to disruption than another. From Figure 1, all that can be seen is a spread of data around a central trend line. For the pooled group-fitted line in Figure 1, r = 0.90.

It would appear that the points in Figure 1 all belong on the same line, regardless of the task being performed. This could imply that some behavior which was common to all of the tasks was being affected. Smith[23-25] argues persuasively that common behavior is attentional. His experiments and extensive review of the literature[27] lead him to believe that in a noisy environment, subjects' attention becomes more narrowed on the most important task and attention to other simultaneously performed tasks is decreased. This may be a cognitively stated variant of the distraction hypothesis.

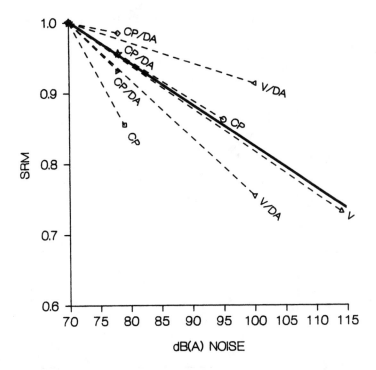

Figure 1   Plot of the results of a meta analysis of the literature on the TP effects of noise. The dashed lines represent the results of individual studies. Letters by each of the solid lines indicate the kind of task being tested (V = vigilance, CP = continuous performance, DA = divided attention). The heavy solid line is a least-squares fit to the data pooled across studies.

One of the tasks used by Smith[25] is of particular interest in this discussion. The subject was required to perform a simple serial reaction time task with three signal lights and corresponding buttons to press. One of the lights was presented with a probability of 0.5 while each of the other two lights had a presentation probability of 0.25. The effect of the noise was to increase the reaction time to the two improbable lights, but to *decrease* the reaction tie to the more probable (more important) light.

Behavioral Effects of Cold Exposure

Effects of cold exposure on TP were reviewed and meta analyzed in the same way as for noise exposure. Only exposures which were sufficiently short to not result in an appreciable core temperature drop were analyzed. Thus the effects of cold were of a functional nature. A total of five experiments (four reports) were found which could be meta analyzed.[28-31]

Figure 2 is a plot of the linear equation fitted to the SRMs of the five studies (dashed lines), and the linear equation fitted to the pooled group of points regardless of the study from which the data originated (heavy solid line). Again, there is no discernible pattern in the data to indicate that any kind of task provides more sensitivity to disruption than another. For the pooled group line in Figure 2, $r = 0.71$.

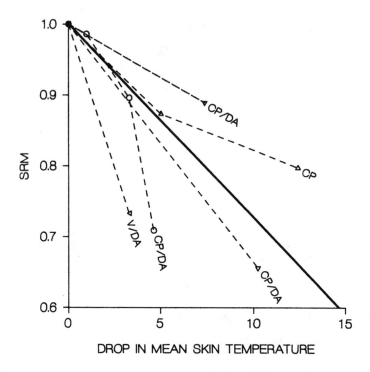

Figure 2   Plot of the results of a meta analysis of the literature on the TP effects of cold. All lines and letters have the same meaning as Figure 1.

It again appears that the points in Figure 2 all belong on the same line, regardless of the task being performed, although the scatter is wider than for noise. Perhaps the higher scatter is a consequence of the greater difficulty of exact control of the stimulus. From these results, it may again be argued that the kind of task is not as important as some common behavioral deficit which is produced by the cold and is affecting the subjects' ability to do all of the various tasks. Several of the authors[28,31] invoke the distraction (competing response) hypothesis which was developed by Teichner[9] for his cold-effects data. One author[29] found that his results did not agree with a distraction hypothesis, but it was not clear that those data could not have been interpreted as agreeing with the findings of Smith.[23-25,27]

CONCLUSIONS AND RECOMMENDATIONS

Not much is known about the effects of IAP on TP because (1) not many experiments have been done and (2) most of the ones which have been done used such low levels of exposure as to almost assure small and ambiguous effects (if any). In an effort to learn from other areas of research where some environmental condition was tested for effects on TP, the literature of noise and cold effects were reviewed. The effects of cold or noise were not peculiar to the kind of task which was being performed, implying that some behavior was being disrupted which was common in the performance of all of the tasks. A recurrent theme in the areas of cold and noise was that the effects were not due to organic effects, but rather functional effects, in particular, distraction or divided attention.

Chemical irritation is similar to cold or noise in that (1) no organic effects (effects due to increased blood levels of the IAP) are likely at the typical levels found in indoor air and (2) the stimulation produced is aversive. It may be argued that the irritative effects of IAP should also be expected to be due to distraction.

Recommendations

Future research in the effects of IAP on TP should be well planned and methodologically sound. The objective of the work should be to gain mechanistic understanding and general knowledge, not a quick "toxicity test" leading to a binary answer of effects/no effects. The latter usually leads to a series of studies with ambiguous results. The methods, theory, and data from related environmental areas should be considered in the planning of research. The following specific points should be considered.

1. Begin with either a wide range, dose-effects study or a high-level, single-exposure study. This strategy is more likely to yield unambiguous results and to typify the effects. Low-level results can be inferred from a continuum and, if necessary, tested in subsequent studies, once a continuum of knowledge has been established. Furthermore if high-level exposures have no effect on TP, further work is more unambiguously discouraged.

2. For any study, make explicit *a priori* hypotheses which can be statistically evaluated with ample test power, given the number of subjects which can be tested. *A priori* hypotheses should be based, insofar as possible, on theory or the findings of other related work. Exploratory measures may be profitably included in the design if it is recognized that any unexpected finding must be tested in a future experiment before reporting. If it can be demonstrated that statistical tests had high power, negative results should be made public.

3. Double-blind procedures should be meticulously followed insofar as possible, as should other principles of good experimental design.

4. Tasks should be selected based on the results of experiments in related environmental areas. Such tasks should always be well standardized before the experiment with IAP begins. This is to assure that the task is sufficiently challenging, but not too difficult, that the results are stable and that the task is sensitive to some other environmental condition (e.g., noise or cold).

5. Experiments should be designed to measure the effects of irritants by themselves. If odors or organic effects of IAP are also present, effects may be due to those. Once the effects of irritants are understood, an additive model could be constructed to account for all of the effects.

REFERENCES

1. World Health Organization. *Indoor Air Pollutants, Exposure and Health Effects Assessments.* Euro Reports and Studies No 78, Working Group Report. Copenhagen: Nordinger, (1982).

2. Davies, D.R. and G.S. Tune. *Human Vigilance Performance.* American Elsevier, New York, 1969.

3. Stroh, C.M. *Vigilance: The Problem of Sustained Attention.* Pergamon Press, New York, 1971.

4. Head, H. The conception of nervous and mental energy. II. Vigilance: a physiological state of the nervous system. *Br. J. Psychol.* 14:126-147 (1923).

5. Mackworth, N.H. Some factors affecting vigilance. *Adv. Sci.* 53:389-393 (1957).

6. Canestrari, R.E. The relationship of vigilance to paced and self-paced learning in young and elderly adults. *Diss. Abstr.* 24:2130-2131 (1963).

7. Broadbent, D.E. A mechanical model for human attention and immediate memory. *Psychol. Rev.* 64:205-215(1957).

8. Elliot, E. Perception and alertness. *Ergonomics.* 3:357-364 (1960).

9. Teichner, W.H. Reaction time in the cold. *J. Appl. Psychol.* 42:54-59 (1958).

10. Benignus, V.A. Importance of experimenter-blind procedure in neurotoxicology. *Neurotoxicol. Teratol.* 15:45-49 (1993).

11. Muller, K.E., C.N. Barton and V.A. Benignus. Recommendations for appropriate statistical practice in toxicologic experiments. *Neurotoxicology.* 5:113-126 (1984).

12. Mølhave, L., B. Bach and O.F. Pedersen. Human reactions to low concentrations of volatile organic compounds. *Environ. Int.* 12:169-175 (1986).

13. Kjaergaard, S.K., L. Mølhave and O.F. Pedersen. Human reactions to a mixture of indoor air volatile organic compounds. *Atmos. Environ.* 25A:1417-1426 (1991).

14. Otto, D.A., L. Mølhave, G. Rose, H.K. Hudnell and D. House. Neurobehavioral and sensory irritant effects of controlled exposure to a complex mixture of volatile organic compounds. *Neurotoxicol. Teratol.* 12:649-652 (1990).

15. Bach, B., O.F. Pedersen and L. Mølhave. Human performance during experimental formaldehyde exposure. *Environ. Int.* 16:105-113 (1990).

16. Stevens, J.C. and W.S. Cain. Aging and the perception of nasal irritation. *Physiol. Behav.* 37:323-328 (1985).

17. Glass, G.V. Primary, Secondary and meta-analysis of research. *Educ. Res.* 5:3-8 (1976).

18. Benignus, V.A. Behavioral effects of carbon monoxide: meta analyses and extrapolations. *J. Appl. Physiol.* 76: 1310-1316 (1994).

19. Hockey, G.R.J. Effect of loud noise on attentional activity. *Q. J. Exp. Psychol.* 22:28-36 (1970).

20. Hockey, G.R.J. Signal probability and spatial location as possible bases for increased selectivity in noise. *Q. J. Exp. Psychol.* 22:37-42 (1970).

21. Jerison, H.J. Effects of noise on human performance. *J. Appl. Psychol.* 43:96-101 (1959).

22. Mohindra, N. and J. Wilding. Noise effects on rehearsal rate in short term serial order memory. *Q. J. Exp. Psychol.* 35A:155-170 (1983).

23. Smith, A.P. The effects of noise and task priority on recall of order and location. *Acta Psychol.* 51:245-255 (1982).

24. Smith, A.P. Noise, biased probability and serial reaction. *Br. J. Psychol.* 76:89-95 (1985).

25. Smith, A.P. Noise and aspects of attention. *Br. J. Psychol.* 82:313-324 (1991).

26. Wright, J. von and L. Nurmi. Effects of white noise and irrelevant information on speeded classification: a developmental study. *Acta Physiol.* 43:157-166 (1979).
27. Smith, A. A review of the effects of noise on human performance. *Scand. J. Psychol.* 30:185-206 (1989).
28. Davis, F.M., A.D. Baddeley and T.R. Hancock. Diver performance: the effect of cold. *Undersea Biomed. Res.* 2:195-213 (1975).
29. Ellis, H.D. The effects of cold on the performance of serial choice reaction and various discrete tasks. *Hum. Factors.* 24:589-598 (1982).
30. Stang, P.R. and E.L. Wiener. Diver performance in cold water. *Hum. Factors.* 12:391-399 (1970).
31. Vaughn, W.S., Jr. Distraction effect of cold water on performance of higher-order tasks. *Undersea Biomed. Res.* 4:103-116 (1977).

CHAPTER 16

Conditioned Immune Alterations
in Animal Models

Donald T. Lysle

INTRODUCTION

Researchers are only beginning to determine the health consequences of exposure to indoor air pollutants. The common methodology for evaluating the health risks of exposure to pollutants involves identifying potential irritants or pathogens in the environment and making assessments of the effect of exposing individuals to those substances. This chapter proposes that a complete assessment of the impact of environmental pollutants on health should take into account the psychological consequences of exposure to aversive environmental stimuli. The reported studies indicate that exposure to aversive environmental stimuli, and the learning that results from the exposure, can adversely influence the immune system and affect physical health.

Most of the research using animals has focused on the health consequences of exposure to physically aversive stimulation, commonly referred to as unconditioned aversive stimuli. There are numerous animal studies indicating that exposure to unconditioned aversive stimuli can induce alterations of immune status. For example, exposure to loud noise can have profound effects on immune status in mice.[1] Furthermore, presentations of aversive stimulation in the form of electric footshock to rats has been reported to decrease the proliferative responsiveness of splenic and blood lymphocytes to mitogens,[2-4] natural-killer cell activity,[5] and the production of antibodies.[6] Taken together, these studies clearly establish that aversive environmental stimulation can have pronounced effects on a number of immunologic measures. These findings also suggest that the health of the individual may be affected by the immune alterations associated with the presentation of unconditioned aversive stimuli. In support of this suggestion, investigations have shown that presentations of electric shock to rats can increase

1-56670-144-9/96/$0.00+$.50

susceptibility to tumor challenge,[7-8] and alter the course of an experimentally induced autoimmune condition, experimental autoimmune encephalomyelitis.[9]

This research has been extended by findings showing that exposure to unconditioned aversive stimuli can lead to the development of conditioned or learned aversive stimuli that have acquired immunomodulatory properties. Conditioned aversive stimuli are innocuous stimuli that acquire aversive properties by predicting the occurrence of events that are inherently aversive. For example, the presentation of an environmental stimulus, such as an auditory, visual, or contextual cue, which has previously been paired with electric shock, has been shown to suppress the responsiveness of lymphocytes to T- and B-cell mitogens, decrease natural-killer cell activity, decrease the number of antibody forming cells, diminish production of γ-interferon and interleukin-2, and decrease the development of adjuvant-induced arthritis.[10-14] These studies clearly demonstrate that the immune alterations are the result of a learned state induced by the conditioned stimulus and are not due to prior electric shock experience itself, handling, or exposure to the type of stimulus used as the conditioned stimulus. Moreover, the suppression of lymphocyte responsiveness can be attenuated by two behavioral manipulations known to reduce the aversive conditioning, extinction, and pre-exposure, confirming that these immune alterations represent an effect of learning.[11]

Although there is convincing evidence that both conditioned and unconditioned aversive environmental stimuli can modulate the immune system, there are only a few studies directed at investigating the mechanisms underlying the immunomodulatory effects. There is no shortage of potential neural and endocrine mechanisms for these effects. The central nervous system can signal the immune system via the hypothalamic-pituitary-adrenal axis and direct sympathetic nerve fiber connections with cells of the immune system in primary and secondary lymphoid compartments.[15] Furthermore, both adrenocortical hormones and catecholamines have been shown to modulate lymphocyte function, providing evidence for the influence of both systems on immunoregulation.[16-21] Thus, it is likely that multiple physiological systems play a role in the immunomodulatory effect of aversive stimulation.

The research in my laboratory has provided evidence that both the opioid system and the sympathetic nervous system are involved in the immunomodulatory changes elicited by a conditioned aversive stimulus. We have also collected evidence showing that conditioned immune alterations involve the modulation of nitric oxide production by macrophages.[22]

EVIDENCE FOR THE INVOLVEMENT OF OPIOID ACTIVITY IN CONDITIONED IMMUNOMODULATION

Substantial attention has been given to the possibility that endogenous opioids can modulate immune status by interacting directly with cells of the immune system. Opioid receptors have been identified on cells of the immune system.[23-25] These receptors provide a structural basis for the effects of opioids on immune function. Furthermore, studies have indicated that endogenous opioids can modulate *in vitro* immune responses.[26-32] There is a paucity of studies, however, directly linking the immunomodulatory effects of aversive stimulation with endogenous opioid activity. Laudenslager and colleagues showed that presentations of inescapable electric shock to rats not only produce opioid-mediated stress-induced analgesia, but also reduce the proliferative responsiveness of splenic lymphocytes to mitogen.[33] Furthermore, studies utilizing a similar electric shock paradigm showed that electric shocks which induce elevations in opioid activity can also enhance tumor development.[34] However, given that electric shock presentations induce many different neuroendocrine alterations, the correlational design of these studies makes it difficult to conclude that opioid activity is directly responsible for the immune alterations observed following shock. Shavit and colleagues provided more direct evidence for opioid involvement in unconditioned immunomodulatory effects.[5] They showed that the opioid-receptor antagonist, naltrexone, blocked the suppression of natural-killer cell activity induced by inescapable electric shock.

Research in my laboratory has focused on examining the role of endogenous opioid activity in the immunomodulatory effects of exposure to a conditioned aversive stimulus developed through pairings with electric footshock.[13] The effectiveness of the opioid-receptor antagonist, naltrexone, in blocking the immunomodulatory effect of the conditioning procedure was examined to provide evidence for the involvement of opioid receptors in the conditioned effect. As discussed previously, the opioid-receptors involved in the effect might be located directly on the surface of cells of the immune system,[23-25] however, opioid receptors located in the central nervous system involved in the regulation of neuroendocrine factors could be responsible for the immune alterations. To begin to assess these alternatives, we investigated the effect of administration of naltrexone and *N*-methylnaltrexone, a quaternary form of naltrexone that does not readily cross the blood-brain barrier, on the conditioned immunomodulatory effect in rats. The comparison of the effectiveness of naltrexone and *N*-methylnaltrexone provides information about the location of the opioid receptors involved in the conditioned immunomodulatory effects.

The basic design of the experiment entailed the development of a conditioned aversive stimulus by pairing a contextual stimulus (a standard rodent chamber) with presentations of electric footshock. On a subsequent test day, rats received a subcutaneous injection of either saline or naltrexone (1.0, 5.0, 10.0 mg/kg) or *N*-methylnaltrexone (1.0, 10.0, and 100.0 mg/kg) prior to exposure to the conditioned aversive stimulus or home cage treatment. Immunologic assessments were performed following exposure to the conditioned stimulus or home cage treatment. The mitogenic responsiveness of lymphocytes from the spleen was measured using the T-cell mitogen concanavalin-A (Con-A) and the B-cell mitogen lipopolysaccharide (LPS). The cytotoxic activity of natural-killer cells derived from the spleen was also assessed.

The results showed that the administration of naltrexone prior to presentation of the conditioned aversive stimulus, dose-dependently attenuated the conditioned stimulus and induced suppression of the proliferative response of splenic lymphocytes to Con-A and LPS. Naltrexone also attenuated the conditioned reduction in natural-killer cell activity. In contrast, the administration of the quaternary form of naltrexone. N-methylnaltrexone, prior to presentation of the conditioned aversive stimulus did not significantly attenuate the conditioned immunomodulatory effects, even at doses 1 log unit higher than those used for naltrexone. These results indicate the exposure to conditioned aversive stimuli activates opioid activity in the central nervous system which induces physiological changes that can influence the immune system.

EVIDENCE FOR THE INVOLVEMENT OF β-ADRENERGIC ACTIVITY IN CONDITIONED IMMUNOMODULATION

Although our research has provided evidence for the involvement of central opioid receptors in conditioned alterations of immune status, the peripheral mechanism of these immunomodulatory effects is less clear. The sympathetic nervous system is a likely candidate to serve as a peripheral mediator of the conditioned effects. There is considerable experimental evidence linking the opioid system and the sympathetic nervous system. For example, central administration of opioids to rats has been shown to increase central sympathetic outflow to the adrenal medulla and from peripheral sympathetic nerve terminals,[34] and primary and secondary lymphoid organs are densely innervated by sympathetic nerve terminals.[15] Furthermore, radioligand binding studies indicate that α- and β-adrenergic binding sites are present on immunocyte membranes,[36-38] and *in vitro* studies suggest that these binding sites are functional adrenergic receptors.[20,39]

Taken together, the above findings, along with data from my laboratory, suggest that exposure to a conditioned aversive stimulus may release centrally acting endogenous opioids which influence the release of catecholamines in the periphery, thereby inducing alterations of immune status. My laboratory has investigated this possibility by evaluating the role of peripheral catecholamines in the conditioned immunomodulatory effects elicited by exposure to a conditioned aversive stimulus. Luecken and Lysle evaluated the effect of β_1- and β_2-selective adrenergic antagonists on the conditioned immunomodulatory effects.[10] The β_2-selective antagonist ICI-118,551 was chosen to determine whether the immunomodulation is primarily mediated by action at β_2-adrenergic receptors.[40] The β_1-selective antagonist atenolol was similarly investigated for its ability to block conditioned immune alterations.[41-42] The basic design of these experiments is similar to those described for the opioid antagonists. Briefly, rats received conditioning sessions during which electric shock was paired with a contextual stimulus, establishing the context as a conditioned aversive stimulus. Then, prior to exposure to the conditioned stimulus on the test day, the rats received subcutaneous injections of either saline or ICI-118,551 (0.125, 0.5, 2.0, 8.0 mg/kg) or atenolol (0.125, 0.5, 2.0, 8.0 mg/kg). The immunologic measures included an assessment of splenic lymphocyte proliferation to the T-cell mitogens phytohemagglutinin (PHA), Con-A, and the B-cell mitogen LPS, as well as a combination of ionomycin and phorbol myristate acetate. Blood lymphocyte proliferation to Con-A and PHA was assessed, along with natural-killer cell activity, γ-interferon, and interleukin-2 production. The results showed that both the β_2-adrenergic antagonist ICI 118,551 and the β_1-adrenergic antagonist atenolol were highly effective at blocking the conditioned suppression of the mitogenic response of splenic T cells to Con-A and PHA. Furthermore, both antagonists dose-dependently blocked the conditioned reduction in γ-interferon production. However, neither drug was effective at blocking suppression of the mitogenic response of splenic B cells to LPS, splenic natural-killer cell activity, interleukin-2 production by splenic lymphocytes, or the proliferative response of blood lymphocytes to Con-A and PHA. The fact that β_1- and β_2-selective adrenergic antagonists block a subset of the conditioned immune alterations suggests that both subtypes of β-adrenergic receptors are involved in conditioned immune alterations. However, these results indicate that additional neural and endocrine systems are involved in the conditioned alterations of immune status.

Thus, significant progress has been made in recent years in describing the impact of stressful or conditioned stimuli on immune endpoints, and some progress has been made toward relating neural and endocrine events to those immunomodulatory effects. However, few studies have focused on how stressful or conditioned aversive stimuli alter specific interactions between the cells of the immune system to induce changes in immune endpoints. In this regard, our very recent work indicates that conditioned alterations of the proliferative response of lymphocytes is the result of modulation of nitric oxide production by macrophages.[22]

EVIDENCE FOR THE INVOLVEMENT OF NITRIC OXIDE IN CONDITIONED IMMUNOMODULATION

Nitric oxide is a reactive nitrogen intermediate that serves an important role in numerous biological processes. The spectrum of nitric oxide-mediated activities ranges from vasodilation[43] and synaptic plasticity within the brain,[44] to a role in the cytostatic activity of macrophages on *Trypanosoma brucei gambiense* and *T. brucei brucei*,[45] tumor growth,[46] and defense responses against a number of intracellular microbial infections including those caused by *Leishmania* and *Schistosoma mansoni*.[47-49] It has also been suggested that nitric oxide contributes to the toxic effect of the human immunodeficiency virus type 1 protein, gp120, in primary cortical cultures.[50]

One of the main functions proposed for the inducible form of nitric oxide found in macrophages is the regulation of lymphocyte proliferation. For example, the presence of nitric oxide in cultures of Con-A stimulated rat splenocytes limits the degree of lymphocyte proliferation.[51-55] There is also evidence that nitric oxide mediates the inhibitory effect of "suppressor" macrophages on mitogen-induced proliferation of splenic lymphocytes in rodents.[56-60] Excessive nitric oxide production has been shown to be involved in the inability of splenic lymphocytes from certain rat strains to proliferate in response to alloantigen.[61] Furthermore, macrophage-derived nitric oxide appears to mediate the depressed mitogenic responsiveness of splenic lymphocytes from both spontaneously hypertensive rats[62] and low responder, inbred Brown Norway rats.[54]

Taken together, these findings led us to hypothesize that nitric oxide production by macrophages may be involved in the suppression of the proliferative response of lymphocytes to mitogen observed following exposure to a conditioned aversive stimulus developed through pairings with electric shock.[22] The results of that study show that the conditioned suppression of the mitogenic responsiveness of splenocytes to Con-A is accompanied by a marked increase in nitrite accumulation. Nitrite is a stable form of nitrogen intermediate that is formed nonenzymatically when nitric oxide is exposed to oxygen. Nitrite is easily measured using the Greiss reagent assay. Removal of adherent cells from these cultures results in elimination of the conditioned stimulus-induced suppression of lymphocyte proliferation, suggesting that nitric oxide production by splenic macrophages is responsible for the suppression.

To provide additional support for the involvement of nitric oxide in conditioned immunomodulation, we assessed the effect of the addition of a competitive inhibitor of oxidative L-arginine metabolism, N^G-mono-methyl-L-arginine (L-NMMA), to cultures of unfractionated spleen cell suspensions from animals exposed to the conditioned stimulus and home cage control animals. Figure 1 shows the results of this manipulation. The addition of L-NMMA dose-dependently attenuates the suppression of lymphocyte proliferation to mitogen induced by the aversive conditioned stimulus, indicating the involvement of the L-arginine-dependent nitric oxide synthesizing pathway. To determine the pharmacological specificity of the effect of L-NMMA, we also assessed the effect of the inactive enantiomer, D-NMMA. As shown in Figure 2, D-NMMA does not have a significant effect on the suppression of lymphocyte proliferation induced by the conditioned stimulus indicating that the effect of L-NMMA is stereospecific. Furthermore, Figure 3 shows that the attenuation of the conditioned suppression afforded by addition of L-NMMA to culture is countered by addition of excess L-arginine, the preferred substrate for nitric oxide synthase, providing additional confirmation of the involvement of the L-arginine-dependent nitric oxide synthesizing pathway. Collectively, these results provide some of the first evidence that the neuroendocrine response to a conditioned stimulus induces alterations in immune status through changes in the level of nitric oxide production by macrophages.

CONCLUSION AND FUTURE DIRECTIONS

The present research provides evidence that exposure to either conditioned or unconditioned aversive stimuli can, through a complex cascade of neural and endocrine changes, evoke alterations in the immune system. The immunomodulatory effects involve the activation of opioid receptors in the central nervous system and the sympathetic nervous system. Moreover, our recent work indicates that the modulation of immune status by a conditioned aversive stimulus involves

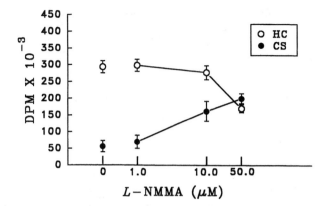

Figure 1 Proliferative response of cultured splenocytes to the optimal concentration of Con-A (5.0 μg/ml) and increasing concentrations of L-NMMA. Thymidine (^{3}H) incorporation is expressed as the mean (±SE) of the averaged triplicate disintegrations per minute (DPM). CS represents animals re-exposed to the conditioned stimulus on the test day, and HC (home cage) represents cultures from animals that remained in their home cages on the test day.

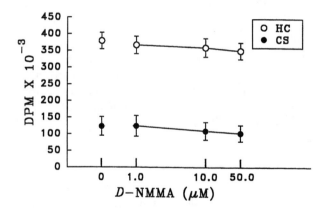

Figure 2 Proliferative response of cultured splenocytes to the optimal concentration of Con-A (5.0 μg/ml) and increasing concentrations of D-NMMA. Thymidine (^{3}H) incorporation is expressed as the mean (±SE) of the averaged triplicate disintegrations per minute (DPM). CS represents animals re-exposed to the conditioned stimulus on the test day, and HC (home cage) represents cultures from animals that remained in their home cages on the test day.

the alteration of the production of nitric oxide by macrophages. These studies are just beginning to identify the immune alterations and define the mechanisms of the effects.

Although the present research does not directly address the health consequences of exposure to indoor pollutants, the findings suggest that exposure to aversive indoor pollutants may result in the activation of neuroendocrine responses that can adversely influence health. In addition, the results indicate that

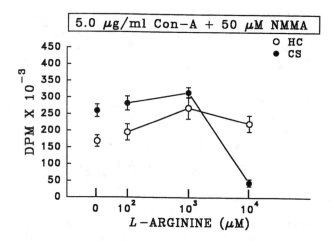

Figure 3 Proliferative response of cultured splenocytes to the optimal concentration of Con-A (5.0 µg/ml) together with 50 µM L-NMMA and increasing concentrations of L-arginine. Thymidine (^{3}H) incorporation is expressed as the mean (±SE) of the averaged triplicate disintegrations per minute (DPM). CS represents animals re-exposed to the conditioned stimulus on the test day, and HC (home cage) represents cultures from animals that remained in their home cages on the test day.

the initial exposure to an aversive indoor air pollutant may produce lasting health consequences through learning processes. Therefore, the present findings suggest that a complete evaluation of the health consequences of exposure to indoor air pollution take into consideration that indoor pollutants can be aversive stimuli which act upon the neural and endocrine systems and produce learned responses that have immunomodulatory effects.

ACKNOWLEDGMENT

The writing of this chapter and the reported studies were supported by a grant from the National Institute of Mental Health (MH46284). Address correspondence to Donald T. Lysle, Department of Psychology, Davie Hall, CB# 3270, University of North Carolina, at Chapel Hill, Chapel Hill, NC 27599-3270.

REFERENCES

1. Monjan, A.A., and M.J. Collector. Stress induced modulation of the immune response. *Science,* 96: 307-308 (1977).
2. Keller, S.E., J.M. Weiss, S.J. Schleifer, N.E. Miller, and M. Stein. Suppression of immunity by stress: Effect of a graded series of stressors on lymphocyte stimulation in the rat. *Science,* 213: 1397-1400 (1981).
3. Keller, S.E., J.M. Weiss, S.J. Schleifer, N.E. Miller, and M. Stein. Stress-induced suppression of immunity in adrenalectomized rats. *Science,* 221: 1301-1304 (1983).

4. Lysle, D.T., M. Lyte, H. Fowler, and B.S. Rabin. Shock-induced modulation of lymphocyte reactivity: Suppression, habituation, and recovery. *Life Sci.*, 41: 1805-1814 (1987).

5. Shavit, Y., J.W. Lewis, G.W. Terman, R.P. Gale, and J.C. Liebeskind. Opioid peptides mediate the suppressive effect of stress on natural killer cell cytotoxicity. *Science*, 223: 188-190 (1984).

6. Laudenslager, M.L., M. Fleshner, P. Hofstadter, P.E. Held, L. Simons, and S.F. Maier. Suppression of specific antibody production by inescapable shock: Stability under varying conditions. *Brain Behav. Immun.*, 2: 92-101 (1988).

7. Lewis, J.W., Y. Shavit, G.W. Terman, R.P. Gale, and J.C. Liebeskind. Stress and morphine affect survival of rats challenged with a mammary ascites tumor (MAT 13762B). *Nat. Immun. Cell Growth Regul.*, 3: 43-50 (1983/84).

8. Lewis, J.W., Y. Shavit, G.W. Terman, L.R. Nelson, R.P. Gale, and J.C. Liebeskind. Apparent involvement of opioid peptides in stress-induced enhancement of tumor growth. *Peptides*, 4: 635-638 (1983).

9. Bukilica, M., S. Djordjevic, I. Maric, M. Dimitrijevic, B.M. Markovic, and B.D. Jankovic. Stress-induced suppression of experimental allergic encephalomyelitis in the rat. *Int. J. Neurosci.* 59: 167-175 (1991).

10. Luecken, L.J., and D.T. Lysle. Evidence for the involvement of β-adrenergic receptor in conditioned immunomodulation. *J. Neuroimmunol.*, 38: 209-220 (1992).

11. Lysle, D.T., J.E. Cunnick, H. Fowler, and B.S. Rabin. Pavlovian conditioning of shock-induced suppression of lymphocyte reactivity: Acquisition, extinction, and preexposure effects. *Life Sci.*, 42: 2185-2194 (1988).

12. Lysle, D.T., J.E. Cunnick, B.J. Kucinski, H. Fowler, and B.S. Rabin. Characterization of immune alterations induced by a conditioned aversive stimulus. *Psychobiology*, 18: 220-226 (1990).

13. Lysle, D.T., L.J. Luecken, and K.A. Maslonek. Modulation of immune status by a conditioned aversive stimulus: Evidence for the involvement of endogenous opioids. *Brain Behav. Immun.*, 6: 179-188 (1992).

14. Zalcman, S., M. Richter, and H. Anisman. Alterations of immune functioning following exposure to stressor-related cues. *Brain Behav. Immun.*, 3: 99-109 (1989).

15. Felten, D.L., S.Y. Felten, K.D. Ackerman, D.L. Bellinger, K.S. Madden, S.L. Carlson, and S. Livnat. Peripheral innervation of lymphoid tissue. In: S. Freier (Ed.), *The Neuroendocrine-Immune Network*, CRC Press, Boca Raton, FL, pp. 9-18 (1990).

16. Crabtree, G.R., A. Munck, and K.A. Smith. Glucocorticoids and lymphocytes. II. Cell-cycle dependent changes in glucocorticoid receptor content. *J. Immunol.*, 125: 13-17 (1980).

17. Onsrud, M., and E. Thorsby. Influence of *in vivo* hydrocortisone on some human blood lymphocyte populations. I. Effect on natural killer cell activity. *Scand. J. Immun.*, 13: 573-579 (1981).

18. Robbis, D., and M.E. Gershwin. Identification and characterization of lymphocyte subpopulations. *Semin. Arthritis Rheum.*, 7: 245-277 (1978).

19. Sanders, V.M., and A.E. Munson. Kinetics of the enhancing effect produced by norepinephrine and terbutaline on the murine primary antibody response *in vitro*. *J. Pharmacol. Exp. Ther.*, 231: 527-531 (1984).

20. Sanders, V.M., and A.E. Munson. Norepinephrine and the antibody response. *Pharmacol. Rev.*, 37: 229-247 (1985).

21. Felten, D.L., S.Y. Felten, D.L. Bellinger, S.L. Carlson, K.D. Ackerman, K.S. Madden, J.A. Olschowki, and S. Livnat. Noradrenergic sympathetic neural inter-actions with the immune system: Structure and function, *Immunol. Rev.*, 100: 225-260 (1987).

22. Coussons-Read, M.E., K.A. Maslonek, K. Fecho, L. Perez, and D.T. Lysle. Evi-dence for the involvement of macrophage-derived nitric oxide in the modulation of immune status by a conditioned aversive stimulus. *J. Neuroimmunol.*, 50: 51-58 (1994).

23. Carr, D.J.J., B.R. DeCosta, C.H. Kim, A.E. Jacoben, K.C. Rice, and J.E. Blalock. Opioid receptors on cells of the immune system: Evidence for delta and kappa classes. *J. Endocrinol.*, 122: 161-168 (1989).

24. Madden, J.J., R.M. Donahoe, J. Zwemer-Collins, D.A. Shafer, and A. Falek. Binding of naloxone to human T-lymphocytes. *Biochem. Pharmacol.*, 36: 4103-4109 (1987).

25. Wybran, J., T. Appleboom, J.P. Famaey, and A. Gavaerts. Suggestive evidence for receptors for morphine and methionine enkephalin on normal human peripheral blood T-lymphocytes. *J. Immunol.*, 123: 1068-1070 (1979).

26. Froelich, C.J., and A.D. Bankhurst. The effect of β-endorphin on natural cytotox-icity and antibody dependent cellular cytotoxicity. *Life Sci.*, 35: 261-265 (1984).

27. Gilmore, W., and L.P. Weiner. Beta-endorphin enhances interleukin-2 production in murine lymphocytes. *J. Neuroimmunol.*, 18: 125-130 (1988).

28. Kayu, N., J. Allen, and U.J.E. Morley. Endorphins stimulate normal human periph-eral blood lymphocyte natural killer activity. *Life Sci.*, 35: 53-59 (1984).

29. Mandler, R.N., W.E. Biddison, R. Mandler, and S. Serate. β-endorphin augments the cytolytic activity and interferon production of natural killer cells. *J. Immunol.*, 136, 934-936 (1986).

30. Mathews, P.M., C.J. Froelich, W.L. Sibbitt, and A.D. Bankhurst. Enhancement of natural cytotoxicity by β-endorphin. *J. Immunol.*, 130 :1658-1662 (1983).

31. McCain, H.W., I.B. Lamster, J.M. Bozzone, and J.T. Grbic. β-endorphin modulates human immune activity via non-opiate receptor mechanisms. *Life Sci.*, 31: 1619-1624 (1982).

32. Van Epps, D.E., and L. Saland. Beta-endorphin and met-enkephalin stimulate human peripheral-blood mononuclear cell chemotaxis. *J. Immunol.*, 132, 3046-3053 (1984).

33. Laudenslager, M.L., S.M. Ryan, R.C. Drugan, R.L. Hyson, and S.F. Maier. Coping and immunosuppression: Inescapable but not escapable shock suppresses lympho-cyte proliferation. *Science*, 221: 568-570 (1983).

34. Visintainer, M.A., J.R. Volpicelli, and M.E.P. Seligman. Tumor rejection in rats after inescapable or escapable shock. *Science*, 216: 437-439 (1982).

35. Appel, N.M., J.A. Kiritsy-Roy, and G.R. Van Loon. Mu receptors at discrete hypothalamic and brainstem sites mediate opioid peptide-induced increases in central sympathetic outflow. *Brain Res.*, 378: 8-20 (1986).

36. Fuchs, B.A., K.S. Campbell, and A.E. Munson. Norepinephrine and serotonin content of the murine spleen: Its relationship to lymphocyte β-adrenergic receptor density and the humoral immune response *in vivo* and *in vitro*. *Clin. Immunol.*, 117: 339-351 (1988).

37. Williams, L.T., R. Snyderman, and R.J. Lefkowitz. Identification of β-adrenergic receptors in human lymphocytes by (—)[³H]alprenolol binding. *J. Clin. Invest.*, 57: 149-155 (1976).

38. McPherson, G.A., and R.J. Summers. Characterization and localization of [³H]-clonidine binding in membranes prepared from guinea-pig spleen. *Clin. Exp. Pharmacol. Physiol.,* 9: 77-87 (1982).

39. Hadden, J.W., E.M. Hadden, and E. Middleton. Lymphocyte blast transformation. I. Demonstration of adrenergic receptors in human peripheral lymphocytes. *Cell. Immunol.,* 1: 583-595 (1970).

40. Bilski, A.J., S.E. Halliday, J.D. Fitzgerald, and J.L. Wale. The Pharmacology of a β_2 selective adrenoceptor antagonist (ICI 118,551). *J. Cardiovasc. Pharmacol.,* 5: 430-437 (1983).

41. Ablad, B., E. Carlson, and L. Ek. Pharmacological studies of two new cardioselective adrenergic β-receptor antagonists. *Life Sci.,* 12: 107-119 (1973).

42. Barrett, A.M. The pharmacology of atenolol. *Postgrad. Med. J.,* 53: Suppl 3, 58-64 (1973).

43. Palmer, R.M.J., A.G. Ferrige, and S. Moncada. Nitric oxide release accounts for the biological activity of endothelium-derived relaxing factor. *Nature,* 327, 524-526 (1987).

44. Shibuki, K., and D. Okada. Endogenous nitric oxide release required for long-term synaptic depression in the cerebellum. *Nature,* 349: 326-328 (1991).

45. Vincendeau, P., S. Daulouede, B. Veyret, M.L. Darde, B. Bouteille, and J.L. Lemesre. Nitric oxide-mediated cytostatic activity on *Trypanosoma brucei gambiense* and *Trypanosoma brucei brucei. Exp. Parasitol.,* 75: 353-360 (1992).

46. Mills, C.D., J. Shearer, R. Evans, and M.D. Caldwell. Macrophage arginine metabolism and the inhibition or stimulation of cancer. *J. Immunol.,* 149: 2709-2714 (1992).

47. Green, S.J., C.A. Nacy, and M.S. Meltzer. Cytokine-induced synthesis of nitrogen oxides in macrophages: A protective host response to *Leishmania* and other intracellular pathogens. *J. Leukoc. Biol.,* 50: 93-103 (1991).

48. Green, S.J., M.S. Meltzer, J.B. Hibbs, and C.A. Nacy. Activated macrophages destroy intracellular *Leishmania major* amastigotes by an L-arginine-dependent killing mechanism. *J. Immunol.,* 144: 278-283 (1990).

49. James, S.L., and J. Glaven. Macrophage cytotoxicity against schistosomula of *Schistosoma mansoni* involves arginine-dependent production of reactive nitrogen intermediates. *J. Immunol.,* 143: 4208-4212 (1989).

50. Dawson, V.L., T.M. Dawson, G.R. Uhl, and S.H. Snyder. Human immunodeficiency virus type 1 coat protein neurotoxicity mediated by nitric oxide in primary cortical cultures. *Proc. Natl. Acad. Sci. U.S.A.,* 90: 3256-3259 (1993).

51. Albina, J.E., J.A. Abate, and W.L. Henry. Nitric oxide production is required for murine resident peritoneal macrophages to suppress mitogen-stimulated T cell proliferation. *J. Immunol.,* 147: 144-148 (1991).

52. Albina, J.E., C.D. Mills, W.L. Henry, and M.D. Caldwell. Regulation of macrophage physiology by L-arginine: role of the oxidative L-arginine deiminase pathway. *J. Immunol.,* 143: 3641-3646 (1989).

53. Albina, J.E., and W.L. Henry. Suppression of lymphocyte proliferation through the nitric oxide synthesizing pathway. *J. Surg. Res.,* 50: 403-409 (1991).

54. Fu, Y., and E.P. Blankenhorn. Nitric oxide-induced anti-mitogenic effects in high and low responder strains. *J. Immunol.,* 148: 2217-2222 (1992).

55. Stuehr, D.J., and C.F. Nathan. Nitric oxide: A macrophage product responsible for cytostasis and respiratory inhibition in tumor target cells. *J. Exp. Med.,* 169: 1543-1555 (1989).

56. Allison, A.C. Mechanisms by which activated macrophages inhibit lymphocyte responses. *Immunol. Rev.,* 40: 3-27 (1978).

57. Folch, H., and B.H. Waksmann. Regulation of lymphocyte responses *in vitro.* Suppressor activity of adherent and non-adherent rat lymphoid cells. *Cell. Immun.,* 9: 12-20 (1973).

58. Johnston, R.B. Current concepts: Immunology. Monocytes and macrophages. *N. Engl. J. Med.,* 318: 747-752 (1988).

59. Mills, C. Molecular basis of "suppressor" macrophages: Arginine metabolism via the nitric oxide synthetase pathway. *J. Immunol.,* 146: 2719-2723 (1991).

60. Unanue, E.R., and P.M. Allen. The basis for the immunoregulatory role of macrophages and other cells. *Science,* 236: 551-557 (1987).

61. Hoffman, R.A., J.M. Langrehr, T.R. Billiar, R.D. Curran, and R.L. Simmons. Alloantigen-induced activation of rat splenocytes is regulated by the oxidative metabolism of L-arginine. *J. Immunol.,* 145: 2220-2226 (1990).

62. Pascual, D.W., V.H. Pascual, K.L. Bost, J.R. McGhee, and S. Oparil. Nitric oxide mediates immune dysfunction in the spontaneously hypertensive rat. *Hypertension,* 21: 185-194 (1992).

CHAPTER 17

Mechanisms of Stressor-Induced Immune Alteration in Humans

Bruce S. Rabin

INTRODUCTION

Homeostasis is the desire to remain in an undisturbed state but there are many ways in which homeostasis can be disturbed. If there is a disturbance in one's environment, and the environmental disturbance produces discomfort, one can restore homeostasis by moving to a new environment that is not uncomfortable or alternatively, the cause of the discomfort can be removed. Thus, when the brain senses the presence of a situation that will disturb homeostasis, an action is taken that will restore homeostasis. Often, as part of this process, there is an activation of the sympathetic nervous system and the hypothalamic-pituitary-endocrine system. Immune system activation is associated with activation of the sympathetic nervous system and the hypothalamic-pituitary-endocrine system. It is therefore suggested that the immune system is part of the homeostatic process.

HOMEOSTASIS AND THE IMMUNE SYSTEM

The presence of a foreign microorganism in the body produces a disturbance of homeostasis. If there is no effective action to remove the microorganism (whether it be viral, fungal, bacterial, or parasitic) it is possible that death will result. To prevent this outcome, the defense mechanisms of the body will become activated once the microorganism is detected. An acute inflammatory response rapidly occurs in an attempt to remove the microorganism. This happens prior to the immune system becoming activated. If the acute inflammatory response does not eliminate the microorganism, the immune system will become activated. A

1-56670-144-9/96/$0.00+$.50

successful reaction to the foreign microorganism will result in removal of the microorganism and the restoration of homeostasis.

The normal functioning of the immune system involves bi-directional communication with the central nervous system. Thus, just as environmental disturbances of homeostasis can activate the central nervous system, so too does the immune system in its attempt to restore homeostasis.

IMMUNITY AND THE BRAIN

Why is the brain involved in the functioning of the immune system? One possibility is that as the immune system becomes increasingly active, the potential of unregulated proliferation of lymphocytes may present an increased risk for the development of lymphomas or leukemias. The activated immune system emits chemical signals (possibly cytokines or neuropeptides) that are sensed by the brain which then activates hormonal systems to suppress lymphocyte proliferation. However, although this hypothesis appears to have some logic to it, there is no reason to believe that there is any fact.

It is even harder to understand why a psychological or physical stressor alters the functioning of the immune system with a resultant negative implication for health. Although there is much anecdotal information suggesting that stress is associated with a predisposition for the development of viral infections, autoimmune diseases, or malignancy, there is only data to support the possibility that stress predisposes to viral infections in humans.[1] However, on careful examination, the association is not as clear as first appears, because causality has not been shown. It is possible that at times of stress individuals alter their behavior with a resultant increased exposure to viral infections or a decrease in immune system function (for example, inadequate amounts of sleep or a decreased food intake). It is even harder to suggest a biological advantage which could be associated with a stressor-induced alteration of immune system function predisposing to disease development.

This chapter will review information regarding the mechanism of stressor-induced immune alteration emphasizing studies in humans and nonhuman primates. In addition, studies in rodents will be referred to when information that has been obtained from the study of rodent organs is appropriate.

THE ROLE OF THE SYMPATHETIC NERVOUS SYSTEM IN ACUTE STRESS

Utilizing a variety of stressors which can be delivered in a laboratory setting, significant information regarding the role of the sympathetic nervous system (SNS) in stressor-induced immune alteration has been learned. The tasks given to humans are structured to be frustrating. In one such study, healthy male subjects aged 18 to 30 were presented with a frustrating mental task and several physiologic and immunologic measures determined at baseline and after 20 min of the task. Some of the subjects had an elevated heart rate, systolic blood pressure, and catecholamine measures in response to the stressor. The other subjects undergoing the mental task had only slight alterations of these parameters. This allowed the experimental subjects to be grouped into those who were reactive to the stressor or those who were not reactive to the stressor. A third group of subjects comprised the control group which underwent the same experience of being prepared for the test and sitting in the room where the test was administered, but they did not experience the psychological stressor.[2]

Those individuals showing high sympathetic reactivity to the stressor had a significant increase in the numbers of circulating CD8 lymphocytes whereas those individuals who did not react to the stressor did not have the elevation of CD8 cells. In response to stimulation with the nonspecific mitogen phytohemagglutinin (PHA), lymphocyte proliferation was significantly decreased in those subjects who were reactors to the stressor while those subjects who failed to react to the stressor showed no change in mitogenic responsiveness to PHA. The association of an elevation of the CD8 lymphocyte population which has the functional ability to suppress immune responses, and the decreased mitogenic responsiveness to PHA, suggests that the elevation of CD8 cells is related to the decreased ability to respond to PHA. Cortisol measurements did not show an alteration during the stressor task. Therefore, it is unlikely that glucocorticoids participated in the immune alterations.

In another study,[3] female subjects aged 21 to 41 were administered a psychological stress task. Twelve minutes after initiation of the task there was an increase in the number of CD8 lymphocytes and natural killer (NK) cells and NK cell activity was increased. Interestingly, in older female subjects, aged 65 to 85 years, although both CD8 and NK lymphocyte numbers increased in response to the stress task, NK cell function did not increase. The female subjects experienced increases in SNS activity as indicated by significant increases in heart rate, blood pressure, and plasma levels of catecholamines. Thus, these studies suggest that an acute psychological stressor in humans is associated with activation of the SNS and alteration of both quantitative and qualitative aspects of immune function as measured in blood.

A further indication that cortisol does not participate in the acute immunologic changes associated with stress has been obtained from studies in which we examined peripheral blood of individuals exposed to a psychological stressor 5 min after initiation of the stressor.[4] At 5 min after stressor initiation, in both males and females, CD8 and NK lymphocytes were quantitatively increased in blood while the proliferative response of lymphocytes to nonspecific mitogen was decreased. The alterations that were present at 5 min did not show further alteration 16 min later. Thus, the reaction is rapid and fully manifest within the first 5 min of experiencing the stressor. As the hypothalamic-pituitary-adrenal axis does not respond quickly enough to lead to an elevation of plasma cortisol within this time frame, cortisol cannot participate in these acute changes.

Another approach in determining the importance of the SNS in stress-induced immune alterations is to pretreat normal subjects with an adrenergic antagonist prior to exposing them to a psychological stressor. We performed such a study,

pretreating subjects with an alpha- and beta-adrenergic antagonist prior to exposing them to mental stress.[5] Controls consisted of subjects receiving the antagonist
but not receiving stress and subjects injected with saline and then exposed to the
stressor.

Quantitative changes of CD4, CD8, and NK lymphocytes indicated significant
increases in the peripheral blood of the CD8 and NK lymphocytes in subjects
receiving saline and stress. Subjects who received the adrenergic antagonist who
were not stressed, or the adrenergic antagonist with stress or saline without stress,
did not differ from each other in regard to alterations of cell numbers. The
adrenergic antagonist prevented the alterations in cell numbers. This further
confirms the important role of the sympathetic nervous system in stressor-induced
immune alteration of lymphocyte numbers in the peripheral blood in humans.

In regard to lymphocyte proliferation to nonspecific mitogens, only the subjects receiving saline and stress had a significant suppression of mitogenic function. The three other groups of subjects did not differ from each other. Thus, both
quantitative and functional alterations of immunity induced by stress are related
to catecholamines.

Although activation of the SNS is reported to be associated with an alteration
of immune function and chronic stress in humans, there is one report of an
association of neuropeptide Y (NPY) elevation, rather than catecholamine elevation with alteration of the function of NK cells.[6] The presence of NPY receptors
on lymphocytes has not been clearly documented, however. Therefore, the association of an elevation of NPY with altered lymphocyte function needs to be
confirmed and clarification of the presence of NPY receptors, as well as the
functional alteration of lymphocytes, in vitro, when incubated with NPY,
determined.

GLUCOCORTICOIDS

Short term studies of stress on immune function are terminated before cortisol
becomes elevated in humans as cortisol becomes elevated approximately 30 min
after stress onset. Cortisol changes in chronic stress have not adequately been
related to the lymphocytic changes which are present in chronically stressed
individuals and, as has been indicated, the changes in acute and chronic stress
differ.

There are studies which do suggest that glucocorticoids can have an effect
on lymphocytes. Cortisol receptors are present in the cytoplasm of cells and when
receptors bind to glucocorticoids they move to the nucleus where they bind to
DNA. The receptors are of two types and are termed type I and type II. In rats,
ligands for the type I and type II receptor significantly decrease the white blood
cell count and the number of lymphocytes and monocytes which are in the
peripheral blood. Binding of a ligand to the type I glucocorticoid receptor
decreases the number of NK cells in the circulation while a ligand binding to the
type II receptor decreases the numbers of all lymphocyte subsets present. As there
are different concentrations of type I and type II receptor in different lymphocyte

subpopulations, the effect of glucocorticoid elevation which occurs in chronic stress may have a varied effect upon the immune system, depending on the lymphocyte subpopulation which is being studied.[7]

Although there is only limited data available in humans regarding the role of glucocorticoids in altering lymphocyte function, there is more substantial data available in rodents. It has been reported that type I glucocorticoid receptor binding in the spleen is substantially reduced in rats that are exposed to restraint stress in the light phase of the light/dark cycle, whereas binding in rats which are exposed to restraint in the dark is not effected. As circulating levels of corticosterone are lower during the light phase, the reduced binding in rats stressed in the light is likely due to occupation of the glucocorticoid receptors by the stressor elevated corticosterone levels. In addition, the lack of a stress effect on binding in the dark is likely due to already occupied type I glucocorticoid receptors. Thus, even though corticosterone levels are elevated by restraint imposed in the dark, the ability to bind type I glucocorticoid receptors on lymphoid cells in the spleen is prevented by prestress high levels of circulating corticosterone.[8] The physiologic implications of this have not been adequately determined yet.

CHRONIC STRESS AND IMMUNE FUNCTION

The above studies utilizing acute stressors may not necessarily be applicable to immune alterations which occur in association with chronic stress. A clear example is obtained by evaluating numbers of CD8 lymphocytes which are present in the blood of individuals experiencing high vs. low levels of life event stressors. In such subjects, there is a negative correlation of CD8 lymphocytes with stress.[9] This observation has significant implications regarding the interpretation of the biologic significance of immune alterations associated with acute vs. chronic stress. For example, there is extensive epidemiological data indicating that autoimmune diseases are more likely to have their onset at a time of emotional stress or a patient in remission with an autoimmune disease is more likely to have a relapse at a time of stress. Although the mechanism for the etiology or pathogenesis of autoimmune disease is not fully understood, there are certain testable hypotheses which can be generated from the current theories for the onset of autoimmune disease and stressor-induced immune alteration.

For example, if an autoimmune disease occurs because of a decrease of the suppressor lymphocyte population with a resultant increased activity of autoreactive B lymphocytes, the decrease of CD8 cells occurring with chronic stress would be consistent with this pathogenesis. Alternatively, if an autoimmune disease occurs secondarily to a viral infection altering tissue antigens, a decrease of the CD8 cytotoxic lymphocyte population would be compatible with increased viral susceptibility. Indeed, the suppressor lymphocyte population and the cytotoxic lymphocyte population each have the CD8 surface marker. Further, CD8 lymphocytes are active in antiviral immunity and increased upper respiratory viral infections may then relate to the decrease of the CD8 cell population in association with chronic stress.

Chronic stress conditions are difficult to produce experimentally in humans, but there are life event situations which are associated with chronic stress. These life situations, such as marital difficulties or caring for a relative with Alzheimer's disease, have been studied for their association with alteration of immune function. The data clearly indicate that chronic stress associated with aversive life events is associated with activation of the SNS and an increased susceptibility to upper respiratory infection.[1] It is also of interest that the subjects who have good social interactions and social support are less likely to develop alterations of immune function and upper respiratory viral infections. Thus, it appears as if there are ways in which the effect of stress on the immune system can be buffered and that social support is a prominent factor.

THE BUFFERING OF STRESS

Studies consistent with the buffering effects of social support on stressor-induced immune alterations have been performed in medical students at Ohio State University. At times of examination stress, many of the students have a decrease of *in vitro* measures of cellular immunity with an increase of antibody to the Epstein-Barr (EB) virus. It is believed that the decrease of cellular immunity allows virus activation with a subsequent increase of antibody production. In addition to immune measures, the students were assayed for a number of psychological measures. Included among these were loneliness. It was those students who were lonely who had the greatest decreases of cellular immune function and increases of antibody production to the EB virus. Thus, this provides evidence that social interaction may participate in buffering the effects of stress on immune system function.[10]

In our studies, we have used social reorganization of nonhuman primates to study the effect of this stressor on cellular immune function. In these studies, we found no relationship between social rank and measures of cellular immune function.[11] We do, however, find several associations between agonistic behavioral patterns and immune function. Higher lymphocyte counts were present in the monkeys that were most aggressive and the most aggressive monkeys had lower baseline function of NK cells. Monkeys that had high levels of fear had a decreased responsiveness of lymphocytes to nonspecific mitogenic stimulation, which is consistent with the hypothesis that social stressors and subsequent emotional states can impair cellular immune responses.

As has been suggested, life events can alter the function of the immune system in normal individuals. There is also information to indicate that alterations of immune function can occur in an individual depending upon events which occurred during the developmental stage of a fetus. This may involve alterations of the number of receptors for various ligands which are available on the external surface of a membrane and which become quantitatively altered subsequent to an elevation of the ligand during gestation. Changes in the number of receptors on cells may be associated with changes in behavior or biochemical processes. Studies have demonstrated that exposing a pregnant animal to stress will cause changes in the ways the offspring respond to stressful conditions.[12] In addition, alterations in the function of the hypothalamic-pituitary-adrenl axis of the off-spring of stressed animals have been reported.[13] Even the postnatal handling of rodents can alter the development of the glucocorticoid receptor system in the hippocampus and frontal cortex.[14]

In nonhuman primates, the response of lymphocytes to mitogenic stimulation differs between nursery and maternally reared infants,[15] and these differences persist for at least two years. Other immune functional differences are induced by the time of weaning and by multiple separations of the offspring from the mother. These studies indicate that early rearing conditions can have a long-lasting effect on immune function.

CONSIDERATIONS FOR STRESS-IMMUNE INTERACTION STUDIES IN HUMANS

Most studies evaluating the effect of stressors on immune function in humans are done in adults. Recently, we studied 17 nondepressed adolescents, each of whom had a conduct disorder. A variety of immunologic assays were performed in the subject and in 20 normal control adolescents. In the study population, NK cell activity showed a significant negative correlation with both past year and lifetime adverse life events. There were no significant correlations between adverse life events and lymphocyte proliferation to nonspecific mitogens or to quantitative measures of lymphocytes in the peripheral blood. Thus, an increase in adverse life events in adolescents is associated with low NK cell activity.[16]

The above raises the question of what should be considered normal. If a subject is being included in a control group of individuals and passes the criteria of having no immunologically related diseases, taking no medications that could effect the immune system, and being in good health, is that individual indeed "normal"? As these data suggest, events occurring in an individual's life can affect parameters of immune function and therefore, should be controlled for when control populations are established. This exemplifies the difficulty in performing psychoneuroimmunologic studies in humans.

Another difficulty in humans is evaluating the effect of a stressor on immune function in tissues which are inexcessible for study. It is thus justifiable to question whether the limited insight which is gained by evaluating peripheral blood lym-phocytes in humans will contribute to our understanding of health-related changes which stress may cause. This question can best be answered by considering some basic aspects of immune function.

When an infectious agent contacts the body, it is the skin and mucosal surfaces which form the first line of defense. Secretory IgA is present on all of the mucosal surfaces of the body and acts to prevent infectious agents from adhering to the mucosa and subsequently, gaining entrance into the body. Studies of the amount of total IgA and specific IgA antibody have indicated that there is a decrease in their quantity on the mucosal surfaces following a stressor.[17] If this reduction of IgA predisposes to the development of infections, such as viral infections, it is likely that a resultant increase in upper respiratory viral infections will occur. Indeed, it has recently been reported that subjects who are under stress are more susceptible to upper respiratory viral infection.[1] In addition, an increase in susceptibility to the development of autoimmune diseases has been reported in individuals who have IgA deficiency. Whether a transient decrease in the amount of IgA antibody present on mucosal surfaces may be associated with a predisposition to develop an autoimmune disease has not been established. However, if a transiently reduced concentration of IgA allows viruses to enter the body, an association between the viral infection and autoimmune disease is possible.

The principle means by which protection against bacterial, fungal, viral, and parasitic infections occurs is through an inflammatory response. An acute inflammatory response occurs prior to activation of the immune system and involves the attraction of cells such as polymorphonuclear leukocytes and monocytes to the site of an infectious agent. If these cells are unable to remove the infectious agent, the chronic inflammatory response begins with activation of the immune system. If the neuroendocrine hormonal response to a stressor interferes with the activation of the complement system, the ability of acute inflammatory cells to respond to the chemotactic stimulus is impaired and the inflammatory response will be unable to eliminate many infectious agents. Indeed, once infectious agents have been phagocytized by polymorphonuclear leukocytes, their ability to kill bacteria may also be impaired. Indeed, we have recently found that the ability of polymorphonuclear leukocytes in rats to kill bacteria is decreased when the rats have been exposed to an aversive event.[18]

Lymphocytes which are localized at a site of an inflammatory reaction migrate to that site from the blood. Part of the process of the lymphocytes arriving at the site where they are needed involves infectious agents which may be ingested by macrophages, producing cytokines which act upon endothelial cells to increase the concentration of adhesion molecules allowing lymphocytes to more readily adhere near where an infectious agent is localized. Alteration of this interaction may interfere with the ability of lymphocytes to localize where an infectious agent is and thus, the ability of cytotoxic lymphocytes or lymphocytes involved in cellular-mediated immune reactions may be impaired. However, all of these lymphocytes are derived from the blood and studying functional alterations of the lymphocyte populations in the blood is appropriate.

The study of peripheral blood lymphocytes will indeed, reveal much information on the ability of an individual to resist infection subsequent to a stressor. Even though functional changes of lymphocytes may be occurring in the spleen and/or lymph nodes, the lymphocytes that will eventually reach the location where an infectious agent is, are found within the peripheral circulation. Obviously, if stress is able to alter the ability of lymphocytes and lymphocytic tissue to respond to an antigen which has localized in the lymphatic tissue, the frequency of antigen reactive cells in peripheral blood will be decreased. Thus, it is important to know the immune and stress-related events that occur in lymphatic tissue in order to obtain a complete understanding of the various mechanisms of stressor-induced immune alteration. However, even with this restriction, significant information will still be obtained by studying the changes in the various functional components of the immune system present in peripheral blood.

The type of immune response which is being studied, following a stressor, is important to consider when determining which components of the immune system have become functionally altered. For example, if an antigen is injected into muscle, the antigen will migrate through lymphatics to the draining lymph node. At this point the antigen will be processed either by macrophages or other antigen-presenting cells which will present the antigen to helper T lymphocytes. The helper T lymphocytes will then interact with the antigen-processing cell to stimulate the proliferation and antibody production by B lymphocytes. The production of cytokines by different classes of T lymphocytes is important in determining which class of antibody will be produced by the B lymphocyte. If a stressor is capable of eliciting the production of neuropeptides or other hormones which interfere with either antigen uptake, antigen processing, antigen presentation, T-lymphocyte recognition, cytokine production, or B-lymphocyte proliferation, there will be an impairment of antibody production. Studying the variety of events which occur within organized lymphatic tissue is obviously not going to be an easy task in the human system. It is important that a basic understanding of immune system function be employed when designing studies to evaluate the effect of stressors on the immune system.

In summary, functioning of the immune system, in humans, is likely altered by events which occur as early as during gestation. Life events which the individual experiences can further influence the functioning of the immune system, and such events may occur during the earliest stages of life. Study of lymphocytes within the peripheral blood of humans likely reflects events related to resistance to disease. Stressor-induced alteration of immune function appears related to the activation of the SNS. Methodologies to optimize the functioning of the immune system to resist diseases such as infections, autoimmunity, and malignancy, may have benefit in optimizing the health of healthy individuals.

REFERENCES

1. Cohen S., Tyrrell D.A.J., Smith A.P. Psychological stress and susceptibility to the common cold. *N. Engl. J. Med.,* 325: 606-612, 1991.
2. Manuck S.B., Cohen S., Rabin B.S., Muldoon M.F., Bachen E.A. Individual differences in cellular immune responses to stress. *Psychol. Sci.,* 2: 1-5, 1991.
3. Naliboff B.D., Benton D., Solomon G.F., Morley J.E., Fahey J.L., Bloom E.T., Makinodan T., Gilmore S.L. Immunological changes in young and old adults during brief laboratory stress. *Psychosom. Med.,* 53: 121-132, 1991.
4. Herbert T., Cohen S., Marsland A.L., Bachen E.A., Rabin B.S., Muldoon M.F., Manuck S.B. Cardiovascular reactivity and the course of immune response to an acute psychological stressor. *Psychosom. Med.,* 56: 337-344, 1994.
5. Bachen E.A., Manuck S.B., Cohen S., Muldoon M.F., Raible R., Herbert T.B., Rabin B.S. Adrenergic blockade ameliorates cellular immune responses to mental stress in humans. *Psychosom. Med.,* 54: 366-372, 1995.

6. Irwin M., Brown M., Patterson T., Hauger R., Mascovich A., Grant I. Neuropeptide Y and natural killer cell activity: findings in depression and Alzheimer caregiver stress. *Fed. Soc. Exp. Biol. Med. J.,* 5: 3100-3107, 1991.

7. Miller A. Presented at the Fourth Research Perspectives in Immunology Meeting, Boulder, CO, (1993).

8. Spencer R.L., Miller A.H., Moday H., Stein M., McEwen B.S. Diurnal differences in basal and acute stress levels of Type I and Type II adrenal steroid receptor activation in neural and immune tissues. *Endocrinology,* 133: 1941-1950, 1993.

9. Herbert T., Cohen S. Stress and immunity in humans: A meta-analytic review. *Psychosom. Med.,* 55: 364-379, 1993.

10. Kiecolt-Glaser J.K., Glaser R. Interpersonal relationships and immune function. In: *Mechanisms of Psychological Influence on Health.* Carstensen L, and Neale J. Eds., Plenum Press: New York, pp. 43-60, 1989.

11. Cohen S., Kaplan J.R., Cunnick J.E., Manuck S.B., Rabin B.S. Chronic social stress, affiliation, and cellular immune response in non-human primates. *Psychol. Sci.,* 3: 301-304, 1992.

12. Fride E., Han Y., Feldon J., Halevy G., Weinstock M. Effects of prenatal stress on vulnerability to stress in prepubertal and adult rats. *Psychol. Behav.,* 37: 681-690, 1986.

13. Weinstock M., Matlina E., Maor E.I., Rosen H., McEwen B.S. Prenatal stress selectively alters the reactivity of the hypothalamic-pituitary adrenal system in the female rat. *Brain Res.,* 595: 195-203, 1992.

14. Wakshlak A., Weinstock M. Neonatal handling reverses behavioral abnormalities induced in rats by prenatal stress. *Physiol. Behav.,* 48: 289-297, 1990.

15. Coe C.L., Lubach G.R., Ershler W.B., Klopp R.G. Influence of early rearing on lymphocyte proliferation responses in juvenile Rhesus monkeys, *Brain Behav. Immun.,* 3: 47-55, 1989.

16. Birmaher B., Rabin B.S., Garcia M.R., Jain U., Whiteside T.L., Williamson D.E., Al-Shabbout M., Nelson B.C., Dahl R.E., Ryan N.E. Cellular immunity in depressed, conduct disorder, and normal adolescents — Role of adverse life events. *J. Am. Acad. Child Adolesc. Psychiatr.,* 33: 671-678, 1994.

17. Stone A.A., Valdimarsdottir H., Jandorf L., Dox D.S., Neale J.M. Evidence the secretory IgA antibody is associated with daily mood. *J. Pers. Soc. Psychol.,* 52: 988-993, 1987.

18. Shurin M.R., Kusnecov A., Hamill E., Kaplan S., Rabin B.S. Stress induced alteration of polymorphonuclear function in rats, *Brain Behav. Immun.,* 8: 163-169, 1994.

Part IV
Cancer

Overview: Lung Cancer

Jonathan M. Samet

Indoor air may be contaminated by diverse carcinogens, some causing lung cancer and others causing malignancies of other sites, e.g., benzene and leukemia. The chapters in this section focus on three agents linked to lung cancer: asbestos, environmental tobacco smoke (ETS), and radon. The agents are known human carcinogens with shared scientific and nonscientific dimensions. For each, carcinogenicity was amply documented by exposures at high doses; for asbestos, through the experience of miners, millers, and workers involved in manufacturing and using asbestos products; for ETS, through the documentation of a causal link between lung cancer and mainstream smoke in active smokers; and for radon, through the excess lung cancer in underground miners of uranium and other ores. Each of these agents has been controversial with questioning of the scientific evidence and each has received widespread media attention. For asbestos and radon, the benefits and costs of federal control programs have been questioned. For ETS, control of exposures of nonsmokers has raised complex questions concerning individual rights, regulatory approaches, liability, and public health.

Chapter 18 by Dr. Rodricks presents risk assessment for low-level exposures. Dr. Rodricks reviewed the origins of risk assessment as a tool for assessing food safety in the early 1970s and its formalism as a four-step process in the 1983 report of the National Research Council. These steps include hazard identification, exposure assessment, dose-response assessment, and risk characterization. Risk assessment provides a useful framework for evaluating uncertainties and Dr. Rodricks predicted further development of approaches for incorporating uncertainties into risk assessments.

Concern that ETS may cause lung cancer in never-smokers began with the publication of findings of two epidemiologic studies in 1981: a cohort study of women in Japan and a case-control study in Greece. Dr. Dockery reviewed these studies and the subsequent evolution of the epidemiologic evidence, using meta-analysis to provide a summary estimate of risk. As a basis for considering ETS as a cause of lung cancer in never-smokers, he cited the evidence from active smokers, the composition of ETS, the documentation of exposure with biomarkers such as cotinine, and the epidemiologic studies of nonsmokers married to smokers. Synthesis of the epidemiologic findings was accomplished by a meta-analysis,

1-56670-144-9/96/$0.00+$.50

which showed a statistically significant increment in risk for never-smoking women married to smokers of approximately 20%. It was the view of Dr. Dockery that the excess could not be attributed to confounding.

Radon and lung cancer was one of the selected topics in the 1984 symposium sponsored by the Oak Ridge National Laboratory Life Science. Dr. Samet focused on the evolution of risk assessment approaches for indoor radon since the previous symposium. The development of risk assessment models has been facilitated by new statistical methods for longitudinal data analysis and by the completion of a number of studies of lung cancer in underground miners. Beginning in 1984, risk assessment models were published by the National Council on Radiation Protection and Measurements (NCRP), the International Commission for Radiological Protection (ICRP), and the Biological Effects of Ionizing Radiation (BEIR) IV Committee of the National Research Council. The latter model was based on analysis of data from four epidemiological studies. In 1994, the National Cancer Institute reported an analysis of data from 11 studies involving over 68,000 underground miners. The risk models developed from these data showed that the effect of exposure waned with increasing time since exposure and increasing age, as in the BEIR IV model, and that the effect increased as exposure rate declined, the "inverse dose-rate effect." Epidemiologic studies of indoor radon directly assess the risk of indoor radon. The power of these investigations is likely to be limited and pooling of the case-control studies has been advocated.

During the 1980s, there was widespread removal of asbestos from schools and mounting concern that asbestos would also need to be removed from public and commercial buildings. Dr. Shaikh reviewed work accomplished by the Health Effects Institute–Asbestos Research, an institution mandated by Congress to document exposures in public and commercial buildings and assess management approaches. Available data assembled by the institute showed extremely low exposures in public and commercial buildings to fibers of 5 μm or greater in length. Case studies presented by Dr. Shaikh indicated that custodial and maintenance workers may have significant exposures and that an operations and maintenance program may reduce concentrations.

Dr. Samet reviewed the presentations on the three agents — ETS, radon, and asbestos — in a risk assessment framework. Each agent had completed the hazard identification step with classification as a human carcinogen. Variable data were available on exposure. For radon, population surveys had been conducted and smoking patterns are well-documented, thorough surveys. Biomarkers, e.g., cotinine, provide another index of exposure to ETS. By contrast, little information is available on indoor exposures to asbestos and bias toward lower levels is likely to affect available data. Exposure-response relationships have been variably characterized for the agents. The pooled analysis of data from studies of miners has made possible relatively sophisticated analyses directed at factors determining risk. For ETS emphasis has been placed on the point estimate of risk associated with marriage to a smoker. Risk assessments for asbestos have been based primarily on conventional epidemiologic analyses of data from worker groups. The future is likely to bring risk assessment approaches based on both epidemiologic data and understanding of mechanisms of carcinogenesis.

CHAPTER 18

ASSESSING CARCINOGENIC RISKS ASSOCIATED WITH INDOOR AIR POLLUTANTS

Joseph V. Rodricks

INTRODUCTION

A method of scientific analysis called risk assessment has assumed a central role in both industrial and governmental decision making regarding permissible human exposure to chemicals present in the environment, at least in the U.S. Over the past 15 to 20 years, regulatory officials in the principal governmental agencies responsible for placing restrictions on chemical uses and exposures — the U.S. Food and Drug Administration (FDA; responsible for food chemicals and contaminants, human and veterinary drugs, and cosmetics), the Environmental Protection Agency (EPA; responsible for pesticides, air and water pollutants, hazardous wastes, and industrial chemicals), the Occupational Safety and Health Administration (OSHA; responsible for workplace exposures), and the Consumer Product Safety Commission (CPSC, responsible for household products not dealt with by FDA or EPA) — have gradually enlarged the uses of risk assessment, so that now virtually no decision on limiting chemical uses and exposures is taken without explicitly considering the question of human health risks. Industrial decision makers are now increasingly applying the risk assessment approach to their own products and processes, without waiting for governmental actions.

These various practical applications have given rise to major research initiatives. Risk assessment remains an uncertain scientific enterprise, and improvements are needed. Because of its practical importance, as well as its role in issues of public health concern, governmental and industrial support for research and study to improve risk assessment is at an all-time high (NRC, 1994).

1-56670-144-9/96/$0.00+$.50

The purpose here is to present a brief sketch of the risk assessment process, its relationship to regulatory and public health decision making, and to provide a glimpse of its possible utility in dealing with indoor air pollutants. This presentation is by no means an exhaustive treatment of the subject; it is rather an attempt to illuminate certain general principles and to suggest what needs to be done to further understanding.

NATURE OF RISK ASSESSMENT

People are exposed daily to thousands of synthetic and naturally occurring chemicals. Some of these exposures are intended (e.g., dietary ingredients and intentional additives, pharmaceutical agents, cosmetics), some are incidental (e.g., pesticide residues in food, exposures from the use of household products, workplace exposures), and some are unintended (contaminants of food, air, water). It is a fundamental principle of toxicology that every one of these chemical exposures could, under some conditions, damage the health of those exposed. No matter how seemingly benign some chemicals appear to be, all become hazardous at some level of exposure.

It is similarly the case that, for every one of these chemicals there is an exposure level at and below which harm will not occur, or will occur with only a very remote probability — what might be called *risk free* or, preferably, *negligible risk* exposures. (Scientists can never determine that any exposure to any substance is absolutely without risk to health. This would require proving the absolute absence of an adverse effect, which is scientifically impossible.)

The goal of *risk assessment* is to identify the relationships between exposure and the risk of toxic harm, ideally in quantitative terms. Once these relationships are understood, it becomes possible to place limits on human exposure to avoid creating a significant risk. This second step — deciding where to draw the line between tolerable and intolerable exposures — is called *risk management*. Managing risks generally requires several types of both technical and policy judgments, many of which are dictated by the various laws dealing with chemicals in our environment. The risk assessment does not reveal whether and to what degree limits should be placed on exposures to avoid excessive health risks; the risk assessor provides information on the relationship between risk and chemical exposure, and leaves to the policy maker, whether in government or in industry, the risk management decision.

CONTENT OF RISK ASSESSMENT

Risk assessment is not an experimental science.* It is not concerned with gathering original data. Rather, it is concerned with the evaluation of available data in a specific context. Risk assessment proceeds in steps, and asks the following questions (NRC, 1983):

Step 1   What types of *hazards* are known to be associated with the chemical of interest? Chemical hazards include flammability, explosivity, radioactivity, and any of a large number of forms of toxicity. Any of these could be the subject of a risk assessment. Here, we are concerned with *toxicity* (more specifically, carcinogenicity).

Step 2   What is the relationship between the risk of toxicity and exposure — the *dose-response* relation? Typical measures of toxic response include the frequency of occurrence (incidence) and severity of effect. By exposure is meant the amount of chemical entering the body of the exposed person, per unit of time. This is technically known as the *dose*.

Step 3   What *exposure*, or dose of chemical, is or could be experienced by the population of interest?

Step 4   What *risk of toxicity* exists for the population of interest? This question is answered by combining the results of the step 1 and 2 evaluation (risk of toxicity as a function of dose) with the results from step 3 (the dose experienced by the population of concern).

Information for the evaluation of hazards and dose-response for a specific chemical derives from the scientific literature or from other sources (e.g., unpublished original toxicity studies). Information for the exposure assessment derives from an evaluation of how the chemical is used, enters the environment, and reaches the population of concern. Exposure assessment might involve, for example, identifying how much of a pesticide reaches people because of its use on food crops, or the amount of intake of a food additive resulting from its use in various foods. It could also involve estimation of the intake of an air pollutant by individuals living near or working in a manufacturing facility, or that received by people living near a hazardous waste site. In the latter case, several chemicals may reach people through air, their drinking water, and even through contaminated soils. Exposure assessment may take place prior to the introduction of a chemical into commerce, or only after exposure has already occurred. Before a risk assessment is embarked upon, the assessor needs to have a clear understanding of the population group or subgroup of interest.

Toxic hazard data derive primarily from two sources: epidemiology studies in groups of exposed people and animal toxicity experiments. The former provides data of clearest relevance to humans, but is difficult to acquire and oftentimes even more difficult to interpret. Epidemiology studies are, by their nature, not controlled experiments, so establishing whether a chemical exposure can *cause* an adverse health effect is often very difficult. It is almost always the case that several studies, involving different populations exposed to the same chemical, studied using different epidemiologic methods, and yielding similar results, are needed to establish causation with a sufficient degree of reliability. The International Agency for Research on Cancer (IARC), a unit of the World Health

---

* It is not incorrect to use the phrase risk assessment to describe investigations intended directly to measure risks in human populations. It is most common, however, to use the phrase to describe those activities devoted to estimating risks that are not directly measurable using currently available scientific procedures.

Organization (WHO), assembles groups of experts to judge the quality and meaning of epidemiology data on chemicals (or, in many cases, chemical mixtures). IARC publishes the results of these deliberations; the epidemiological evidence regarding the carcinogenicity of chemicals and chemical mixtures is ranked as "sufficient" only when a consensus exists that the weight of all the available evidence points toward a causal relationship. IARC has thus far listed only 39 chemicals or chemical mixtures (e.g., benzene, arsenic, tobacco smoke, etc.) in the "sufficient" category; many others are listed as having "limited" or "inadequate" evidence of human carcinogenicity (IARC, 1987).

There are many more animal carcinogens listed by IARC. This is in part because there are hundreds more animal toxicity studies available than there are epidemiology studies. Moreover, properly performed animal studies allow causal relationships to be easily recognized (they are truly controlled studies). Animal toxicity data can also be obtained prior to the introduction of a chemical into commerce, whereas epidemiology data are obtainable only after exposure has occurred. In the case of cancer-causing chemicals, meaningful epidemiology data are typically obtainable only after exposure has occurred for several decades, because cancer takes many years to develop to a detectable disease.

Animal studies provide one further advantage over epidemiology studies: the toxicologists can examine thoroughly the health status of the test animals, so that the full range of toxic effects produced by a chemical can be identified. This is generally not possible with epidemiology studies, because of the absence of full medical information on the study populations (EPA, 1986).

All of these advantages of animal data are to be placed against the obvious disadvantage that animals are not people. Much could be said about the difficult problem of interspecies extrapolation — applying results from animal studies to people — but this would move us too far from our main purpose. Suffice it to say that, in the absence of compelling evidence that a particular result from an animal toxicity experiment is *not* applicable to people, toxicologists generally assume that it is. The scientific basis for such an assumption is generally accepted, but only with important qualifications; the assumption is made in part because of prudence; it could be dangerous to the public health to assume without good reason that a particular animal toxicity finding is not potentially relevant to people. There are, of course, important exceptions to this general principle. In the regulatory setting, such assumptions represent a mix of scientific and policy choices; the NRC report of 1983 noted the need for such "science policy" choices to compensate for gaps in scientific knowledge.

Although much emphasis has been given to chemicals that cause cancer, it needs to be kept in mind that chemicals can cause other adverse effects. Some can damage the nervous system, the reproductive system, or the immune system. Some can harm the lungs, the liver, or the kidneys. Some can cause birth defects (teratogens) or damage to the genes. The purpose of the hazard evaluation step is to acquire as full a picture as possible of the toxic effects a chemical can cause.

The comparative advantages and disadvantages of epidemiology and animal studies are summarized in Table 1.

Table 1   Comparison of Epidemiology and Animal Studies for Identifying Toxic Properties

	Epidemiology studies	Animal studies
Opportunity to conduct study	Often not possible	Always possible
Opportunity to obtain information prior to human exposure	No	Yes
Time requirements	Years to decades after exposure begins	Weeks to years after exposure initiated
Species of interest	Yes[a]	No
Cause-effect determination	Difficult	Usually easy
Opportunity to obtain quantitative dose-response data	Not frequently	Always

[a] Note that epidemiology studies may not provide data on both sexes or on all relevant subgroups of the human population.

THE LOW-DOSE PROBLEM

Dose-response evaluation (step 2) is concerned with understanding what the available scientific literature reveals about the probability of a toxic effect occurring (the risk) as a function of the size and duration of the dose. At the present time, scientists are capable of measuring dose-response relations over a limited range of doses and risks in both epidemiology and animal studies. Only relatively large risks, usually occurring at high doses, can be readily measured. No means are available to measure risks at low doses, such as those that may result from human exposures to carcinogenic air pollutants (NRC, 1994). This limitation is primarily the result of the fact that all studies necessarily involve limited numbers of subjects. This is a statistical limitation. Unfortunately, the principal concerns of risk assessment relate to exposures at relatively low doses, occurring in large populations of people. Epidemiology can rarely measure such low probability events (most epidemiology data are collected in relatively small populations, usually workers, experiencing relatively intense exposures).

To say anything at all about low-dose risks, the assessment must adopt certain untested hypotheses about the nature of the dose-response relationship in the low-dose/low-risk region. Various mathematical models are applied to the measured dose-response data, obtained either from human or animal studies, and parameters are derived that can be used to create a mathematical relationship that might hold in the low-dose region. Several such models are available. Each is based on certain biological or statistical principles that are thought to apply to dose-response phenomena. None is based on a thorough empirical test and all must therefore be considered to be unproven hypotheses about low-dose risks (NRC, 1983).

In the regulatory context, two classes of models are typically used. For all toxic effects except carcinogenicity, the dose-response model assumes that a *threshold dose* must be exceeded before any toxicity occurs. Several procedures exist for deriving an estimate of the threshold dose for large human populations when the available dose-response relations reveal the threshold dose for only a relatively small human population or for a test animal population (Rodricks,

1992). There is widespread use of these procedures for specifying "acceptable daily intakes" (or toxicity reference doses, RfDs, as they are called by the EPA) for chemicals causing noncarcinogenic forms of toxicity.

Carcinogens, at least in the U.S., are treated differently. Based on various biological theories regarding the carcinogenic process, it is assumed that carcinogenic effects do not require a threshold dose to be exceeded before they become manifest. Any dose greater than zero is assumed to increase the probability (risk) of a carcinogenic process; as the dose increases, so does the risk. Borrowing from radiation biology, chemical risk assessors further assume that, at low doses, the risk increases in direct proportion to dose — this is the so-called *linear, no-threshold model* (EPA, 1986).

Alternative models of carcinogen dose-response models exist, including threshold models. Most show equal or lower risks at a given dose than does the linear model. The linear, no-threshold model thus yields what EPA terms an *upper bound* on risk; actual risks almost certainly do not exceed the upper bound, are probably lower, and could for some carcinogens be zero. In the face of uncertainty regarding the true dose-response relationship, and because of the concern that risk not be underestimated, regulatory officials usually choose the upper bound estimate of risk as a basis for decision making. Much controversy surrounds this regulatory choice, and other such choices in the risk assessment process that tend to yield "worst-case" pictures of risk. In the absence of scientific certainty, this approach — choosing from a range of possible assumptions and models that yield the highest estimate of risk — will likely remain the practice of regulators. There are specific cases, however, in which data become available to provide a basis for departure from the usual, worst-case approach. A recent report from the National Research Council (NRC) urges the EPA to incorporate more clearly biologically based models of risk assessment (NRC, 1994). Indeed, alternative models are much in discussion among risk assessors, and considerable research is under way to identify appropriate alternatives (OTA, 1993; see below).

CHARACTERIZING RISKS

For carcinogens, risks are presented as an upper bound on the excess probability of developing cancer over a lifetime. These probabilities are obtained from combining the dose-response evaluation (which yields values for upper bound on lifetime risk per unit of dose) with the dose estimates for the exposed population. The probability is a unitless fraction; most environmental risks carry probabilities below one in 100, and many seem to fall in the one in 1000 to one in 1,000,000 range. Recall these are upper bounds on risk, at least as long as they are based on the linear, no-threshold model and other assumptions that reflect the adoption of cautious assumptions in the face of scientific uncertainty (Rodricks, 1992).

For noncarcinogenic forms of toxicity, the ratio of the estimated dose experienced by the exposed population to the estimated threshold dose (RfD) is calculated. This so-called hazard quotient, while not a direct measure of risk of the type used for carcinogens, nevertheless provides decision-makers a guide to the potential public health impact of an exposure. As the hazard quotient rises above 1.0, it is expected that human risk (i.e., the fraction of the population experiencing doses greater than the threshold dose) also rises, while values less than 1.0 suggest the absence of a significant risk.

Risk assessments typically conclude with a description of all relevant data and their limitations, a restatement of the critical assumptions used, and the quantitative outcomes. This is the information sought by risk managers to help them decide whether, in specific contexts, actions should be taken to reduce risks and, if so, the degree of risk reduction needed to ensure public health protection.

It will be useful now to discuss the origins of this form of risk assessment and the various uses to which it is put.

EVOLUTION OF CARCINOGENIC RISK ASSESSMENT

Until the early 1970s chemical carcinogens were regulated either by imposition of a complete ban on their use or, for those substances thought to be unavoidable, by specifying that exposures should not exceed the detection limits of whatever analytical methods happened to be available to measure these agents in the environment. These approaches to regulation were based on the notion that no safe level of exposure to a chemical carcinogen could be identified. Adequate protection from a carcinogen could be assured only if exposures were eliminated. A complete ban could, of course, guarantee such a result. Use of what might be called the "below analytical detection limit" criterion could not, however, ensure zero exposure, and thus could not ensure safety in the same sense that a ban could. The failure to find a carcinogenic contaminant in an environmental medium with a given method of analysis means only that the contaminant is not present above the detection limit of whatever analytical method is used. It may well be present at any level up to the detection limit. Detection limits vary greatly among chemicals and among the media in which the chemicals are present; and, more importantly, analytical detection limits bear no relationship to a carcinogen's biological potency, and hence to the risks it may pose. For both these reasons, the risks of cancer associated with different carcinogens could vary greatly under the "below analytical detection limit" approach to standard setting.

Risk assessment was introduced in 1973 by the FDA in an effort to remedy the difficulties associated with the below analytical detection limit criterion. The FDA proposed that, for a certain limited class of indirect food additives not necessarily subject to the "ban" provisions of the U.S. food laws (carcinogenic drugs used in food-producing animals that could yield residues in meat, milk, or eggs), additive-specific risk assessments should be carried out using a particular methodology. The food residue concentration corresponding to an excess lifetime cancer risk of one in one million would be calculated from the results of this risk assessment. In order to gain FDA approval, those who proposed to use such a drug would then have to show that, under the conditions of its use, "no residue" of the drug could be found in edible tissue at or above the concentration calculated to pose an excess lifetime cancer risk of one in one million. In effect the FDA was specifying, on an additive-specific basis, the detection limits that would have to be reached. If no method having adequate detection limits was available, those who sought FDA approval would have to engage in method improvement as the first step in showing compliance with the no residue requirements (Rodricks, 1988).

Although this approach retained some of the characteristics of the older below analytical detection limit approach, it had, at least in theory, distinct advantages. Foremost among these was that it ensured a maximum level of risk (one in one million over a lifetime) that would be uniformly applied to all members of the class of regulated chemicals. It ensured that more potent carcinogens were held to lower levels than less potent ones (the more potent the carcinogen, the greater the risk it poses per unit dose). It also ensured that the "detection limit" requirement was tied first of all to the carcinogenicity data and only secondarily to the capabilities of the analytical chemist.

The most significant aspect of the 1973 FDA initiative was, however, the introduction of the notion that a certain level of human exposure to carcinogens could be tolerated without creating a significant public health problem, even if the no-threshold hypothesis were accepted. The one in one million lifetime risk level was stated by FDA risk managers to have no significant impact on public health. Moreover, the agency added that the risk assessment methodology used to derive the dose corresponding to one in one million risk level involved the use of a number of so-called "conservative" assumptions, the cumulative effect of which was the production of a risk estimate highly likely to overstate the true risk. If this were true, then the degree of health protection provided by a one in one million risk level was actually greater than that implied by the quantitative figure.

The FDA's idea caught on. Beginning in the mid-1970s the EPA applied similar approaches to the regulation of certain pesticides. The FDA expanded the use of risk assessment to other classes of regulated agents. After first rejecting quantitative risk assessment as a component of decision making for occupational carcinogens, OSHA was encouraged to reconsider by a U.S. Supreme Court decision on benzene, and they did so. Risk assessment has now become a feature of decision making in all the EPA programs, and its use has even penetrated its sister agencies in the states. It should be noted that not all regulatory uses have adopted the FDA's one in one million risk management goal.

The use of risk assessment in regulation got a boost in 1983 by a mostly encouraging report from the National Academy of Sciences (NRC, 1983). The report also served to create some uniformity in the definitions used by risk assessors, clarified the distinctions between risk assessment and risk management, and urged continued improvements in the conduct of both activities. (These definitions and distinctions are the ones introduced earlier.) In 1985 the White House's Office of Science and Technology Policy, following up on an earlier interagency effort, published a comprehensive review and evaluation of carcinogen risk assessment methodology, and further fueled its use in regulation (OSTP, 1985). EPA guidelines on the conduct of carcinogen risk assessment were issued in 1986 and now form the basis for much of the agency's scientific activity.

Working under a mandate contained in the Clean Air Act Amendments of 1991, the NRC has recently issued a report on the EPA's risk assessment approaches to air pollutants. The NRC report is generally supportive of the EPA's approaches, but also contains more than 70 recommendations for improvement (NRC, 1994). The EPA is currently responding to these recommendations.

CARCINOGENIC RISKS OF INDOOR AIR POLLUTANTS

Against this general background let us turn to the specific issue of indoor air pollutants. For the risk assessor, the problem of assessing risks from chemical contaminants of indoor air is no different in kind from any other problem in risk assessment. The risk assessor needs to understand the chemical characteristics of the contaminants, their toxic properties (hazard assessment), and how those manifestations of toxicity change with dose (dose-response evaluation). The next sets of questions pertain to the issue of dose: what indoor air environments contain the chemical, what is the range of concentrations found therein, how much contaminated air is inhaled and by whom, and what intake of the contaminant results, over what period of time? The dose (intake) data are then combined with hazard and dose-response information to derive a characterization of the risk. Let us illustrate the process by applying current EPA methods of carcinogenic risk assessment to some selected examples of indoor air pollutants that have been identified by the agency as human or animal carcinogens.

One source of reliable information on a class of indoor and outdoor air pollutants — the volatile organic chemicals (VOCs) — is the EPA's national VOC database (Shah and Singh, 1988). The EPA's systematic assembly of data on concentrations of VOCs found in outdoor and indoor air* began in 1980, and has resulted in a reasonably comprehensive profile of the distributions of this class of air pollutants. The database includes information on a total of 320 VOCs, with 261 of them measured in outdoor air and 66 in indoor air. The limitations on the quality and representativeness of this national database are discussed by Shah and Singh (1988), and their report should be consulted before any definitive conclusions are drawn from the assessment to be offered here.

Most of the VOCs identified in indoor air are hydrocarbons of relatively low toxicity, and only a few have been identified as carcinogens. Still, several recognized carcinogens are regularly found, and a few of these are listed in Table 2.

Table 2 also shows that the EPA has evaluated the available evidence of carcinogenicity for the four selected chemicals and has classified the evidence according to its so-called "strength." The agency's "A" classification, as shown for benzene, is reserved for substances that have been demonstrated to be carcinogenic in humans; the cumulative epidemiological data on such substances are said to meet the IARC criteria for "sufficient evidence" of a causal relationship, as discussed earlier in this paper. There is no epidemiological evidence that

---

* Indoor air data derive from commercial and residential buildings, but not from industrial settings.

Table 2   Some Carcinogens Identified Among Volatile Organic Chemicals Commonly Found in Indoor Air[a]

Chemical	Approx. median indoor concentrations ($\mu g/m^3$)	EPA classification and inhalation unit risk ($\mu g/m^3)^{-1}$	
Formaldehyde	50	B1	1.3×10^{-5}
Carbon tetrachloride	0.6 (average value) ·	B2	1.5×10^{-5}
Chloroform	0.5	B2	2.3×10^{-5}
Benzene	10	A	8.3×10^{-6}

[a] National VOC database, as reported by Shah, J.J., Singh, H.B., *Env. Sci. Technol. 22*, 1381-1388 (1988). EPA information derived from the agency's Integrated Risk Information System (IRIS). See text for definitions of EPA terms.

benzene causes leukemia at or near the levels typically found in indoor air (ca. 10 $\mu g/m^3$); rather, the epidemiological evidence derives from studies conducted in certain occupational settings where exposures were ordinarily thousands of times more intense (Rinsky et al., 1987).

The EPA also assumes that the risk of benzene-induced leukemia is directly proportional to dose, and that there is no threshold in its dose-response relationship. Using this assumption, the agency has derived a so-called unit risk factor for benzene, with the low dose extrapolation beginning with the observed dose-response data from three different epidemiology studies. The result of this extrapolation is the unit risk factor shown in Table 2. The unit risk is upper bound on lifetime risk for a continuous lifetime exposure of 1 $\mu g/m^3$.

None of the other carcinogens selected for inclusion in Table 2 are "known" human carcinogens. The B_1 classification, as applied to formaldehyde, reflects the EPA's judgment that the evidence of animal carcinogenicity is convincing and that there is limited evidence of carcinogenicity from epidemiology studies. The evidence supporting the B_2 classification derives entirely from animal data (EPA, 1986). The unit risks shown in Table 2 for formaldehyde, carbon tetrachloride, and chloroform were derived by EPA in the same general way they were derived for benzene, although animal dose-response data were used as the starting points. The four unit risk estimates, while all intended to convey the same quantitative meaning (all are the upper bound risk for a continuous lifetime exposure of 1 $\mu g/m^3$), are in fact uncertain in substantially different ways and so are not truly commensurable values. They are, nevertheless, often treated as strictly commensurable, and this crude oversimplification is one of the current problems in risk assessment that the 1994 NRC study sets out to remedy (NRC, 1994).

Note also that there is no correspondence between the strength of the evidence of carcinogenicity and the "strength" (or potency) of the carcinogen, as reflected by the unit risk value (Table 2). There is no inherent reason why there should be any such correspondence, but confusing these two factors is common.

If the EPA estimates of unit risk are adopted and the agency's data on indoor and outdoor air levels of these chemicals are accepted as reasonably representative of population exposures in the U.S., then a first approximation to the assessment of cancer risk can be produced. It is first necessary to derive an estimate of the continuous lifetime exposure level incurred, on average, by members of the exposed population, presumed in this case to be the vast majority of residents of the U.S. If benzene is taken as the example, then, while indoors, individuals may be expected, on average, to inhale air containing about 10 $\mu g/m^3$ (Table 2). The same database suggests that outdoor air levels of benzene average about 5 $\mu g/m^3$. If it is assumed that, on average, people spend about 80% of their time indoors and 20% of their time outdoors (EPA, 1990), then their time-weighted, continuous lifetime exposure to ambient levels of benzene will be about 9 $\mu g/m^3$. If the unit risk for benzene-related leukemia is 8.3×10^{-6} for a continuous lifetime exposure of 1 $\mu g/m^3$, then the upper bound on lifetime risk from ambient (indoor and outdoor) levels of benzene is given by:

$$8.3 \times 10^{-6} \left(\mu g/m^3 \right)^{-1} \times \left(9\, \mu g/m^3 \right) \cong 70 \times 10^{-6} \tag{1}$$

The upper bound on extra lifetime cancer risk is thus 70 extra cases per million people, to be incurred over a lifetime (about one per million per year). Most of these are a result of exposures incurred indoors. Actual risks are unlikely to exceed the upper bound, could be substantially lower and, in fact, could be zero. Note that the lifetime risk for leukemias from all causes is about 7 per 1000; clearly, if indoor air pollution increased lifetime risk by about 7 per 100,000 it would not be possible to detect such an increase against a background that is 100 times larger.

The assessment presented here is nothing more than a crude "screening" analysis, and should be used only as a starting point for further analysis. A starting point for further analysis might be an examination of available experimental data bearing on the question of the biological mechanisms underlying benzene's role in leukemia induction. Such data may shed some light on the question of the plausibility of EPA's linear, no-threshold assumption, and may suggest alternative dose-response models. Alternative approaches to dose-response modeling for benzene have also been published (see e.g., Paxton et al., 1994, and references therein), based on differing views of the nature of the observed dose-response data and of how they should best be examined statistically.

A far more refined exposure assessment, in which attempts are made to identify the *distribution* of benzene exposures in the population, is also suggested. Crude "point estimates" of exposures, as derived above, are *not* very informative and fail to convey to decision makers an accurate picture of the risk profile for an entire population.

A screening analysis of the type set forth above is useful for priority setting and for deciding which issues require further attention. They may also reveal the areas in which additional research and data collection are most likely to bring improvement in understanding. Definitive public health and regulatory decisions, if based on such crude analyses, are not to be recommended.

RELIABILITY OF CURRENT RISK ASSESSMENT METHODS
AND SOME POSSIBLE IMPROVEMENTS

The fundamental problem in chemical risk assessment arises from the fact that information about risk can, with few exceptions, be collected only under exposure conditions far removed from the conditions of interest. There are no means available to reach conclusions about the size of the risk under the conditions of interest without engaging in several different types of extrapolation (the two most well known of which are extrapolation from experimental animals to humans and from high risk/high dose to low risk/low dose). There is no scientific consensus on the accuracy of available extrapolation models, especially concerning their quantitative features, and little useful scientific knowledge upon which a consensus might be built. Also, of course, use of different forms of extrapolation yield varying (sometimes widely varying) estimates of human risk. It should be obvious, then, that we cannot know which of several plausible risk estimates, if any, accurately portrays the true human risk. Moreover, there are in most cases no means currently available to test the accuracy of these estimates. The results of risk assessments are thus best characterized as scientific hypotheses that cannot as yet be subjected to empirical testing. This applies not only to carcinogenic risks, but also to hazard quotients for threshold agents.

A major premise of much regulation is that if these hypothesized risks are judged to be "too high," action should be taken to reduce them, in advance of the development of the types of scientific data that would be needed to test their accuracy. To put this idea into somewhat cruder terms, regulators will not await a reliable "human body count" to take action. This notion stems from a body of thought holding that the benefit of scientific doubt should, on balance, favor public health protection.

Although most laws and the regulatory policies that have flowed from their implementation, as well as ethical considerations, would generally support this proposition, it is fair to ask whether the risk assessment methodologies (i.e., the specific set of extrapolation models and assumptions) that have been adopted by regulatory agencies, from among all those available, yield results that so greatly overestimate the risk experienced by most people that the degree to which public health protection is favored by the uncertainty in risk assessment is excessive, perhaps greatly so. Most, but not all, expert commentators on the regulatory approach to risk assessment seem to agree that it yields what are termed "upper bounds" on risk. Upper bounds on risk may apply to some individuals, but exceed, perhaps greatly, the average risk. The postulated degrees of risk overstatement vary among experts and among different regulated substances, and no one can be sure about this matter (NRC, 1994).

Risk assessors within and without government are presently exploring new approaches to risk assessment in efforts to reduce uncertainty. Many of these explorations involve the use of data related to the molecular mechanisms by which chemicals produce their adverse effects. It is widely agreed that such knowledge can greatly improve extrapolation across species and doses. Even though knowledge of mechanisms is reasonably complete for only a few substances, there are definite signs that a basis will exist in the near future for greater reliance on these types of data. Although a strong case can be made that regulatory risk assessors should now be incorporating into their assessments as much of this new type of information as possible, giving due attention to the uncertainties in it, several years will probably be required before we begin to see substantial movement away from the generic approach to regulatory risk assessment that yields relatively crude upper bounds on risk.

Although these comments may suggest that the practice of risk assessment is inherently unsatisfactory, it is difficult to imagine better alternatives. Certainly a case can be made for an approach to risk assessment that encompasses a broader range of plausible assumptions and models than is typically allowed by regulatory guidelines, and that thus reveals more fully to decision-makers the uncertainties in the results of risk assessments, but it is hard, in the opinion of the author, to make a case that risk assessment involving a substantial degree of quantification not be undertaken. Some effort must be made to reveal the relative public health importance of various chemical exposures. This is what risk assessment attempts to achieve. Without such an effort there is no basis at all for distinguishing among different exposures and for separating important from less important or trivial public health problems. Public health policies that do not involve use of the discriminating power provided by the tools of risk assessment (for example, by regulating purely on the basis of certain toxicity characteristics), even though these tools are imperfect, are inconsistent with current scientific understanding and are almost certain to mislead us in our efforts to distinguish important from unimportant public health problems.

In respect to carcinogenic agents two types of information in particular are now beginning to influence the conduct of risk assessment: data on the so-called *pharmacokinetic* behavior of carcinogens and data on rates of *cell proliferation* caused by the carcinogen.

Exploration of the use of so-called pharmacokinetic data is especially vigorous at the present time. Here the risk assessor is seeking to understand the quantitative relationships between the size of the administered dose and the dose that actually reaches the site in the body, even in certain cells of the body, where tumor formation is initiated. Because the so-called "target site" dose is the ultimate determinant of risk, any nonlinearity in the relationship between administered dose and target site dose, or any quantitative differences in the ratio of the two between humans and test animals, could greatly influence the outcome of a risk assessment (which now relies only on the relationship between administered dose and toxic risk). The problem of obtaining adequate pharmacokinetic data in humans is being attacked by the construction of so-called physiologically based pharmacokinetic (PB-PK) models, whose forms depend upon the relative physiologies of humans and test animals, solubilities of chemicals in various tissues, and relative rates of metabolism. Several relatively successful attempts at predicting target site doses in humans and other species have been made based on PB-PK modeling, and greater use of this tool is being encouraged by the regulatory community (NRC, 1994).

A second major trend in risk assessment stems from the work of several academic investigators and regulatory scientists, who have pointed out that certain chemicals that increase tumor incidence do so only indirectly, by first causing biological changes of a kind that put cells at increased risk of becoming malignant. Until a dose of such a carcinogen sufficient to cause the necessary biological changes is reached, no significant risk of cancer exists. Such carcinogens, or their metabolites, show little or no propensity to damage genes (they are non-geno-toxic), but at sufficiently high doses cause biological changes that lead to increased rates of proliferation of the affected population of cells. It is relatively well documented that such a proliferative response creates an increased risk of progression to the malignant state (Cohen and Ellween, 1990).

A major implication of the "cellular proliferation" hypothesis is that there is a threshold in the dose-response curve for the carcinogenic effects of agents that act through such indirect mechanisms, so that the linear, no-threshold model now generically applied to all carcinogens is not suitable to estimate their low dose risks. Instead, the traditional approach to establishing insignificant risk doses — the application of safety factors to derive RfDs — is thought by many experts to provide adequate protection against this class of carcinogens. While this approach is not yet a significant component of regulatory risk assessments, several examples of its application have been published in the literature and discussed by regulatory scientists.

Several types of toxicity or physiological changes produced by certain non-genotoxic chemicals can lead to increased cellular proliferation. The antioxidant food ingredient butylated hydroxyanisole (BHA) is a prime example of such an agent. This substance produced excess benign and malignant tumors in the fore-stomach of rats fed 2% of the chemical in their diets. No excess tumors were produced when the feeding level was 0.5%. Substantial evidence exists to show that BHA produces cellular proliferation in the rat forestomach at the 2% dietary level, but does not do so at the lower (tumor-free) level. This type of information is necessary, but not sufficient, to establish that tumors would not occur in the absence of toxicity-induced cellular proliferation, but a number of other lines of evidence point to the same general conclusions (Clayson, 1989).

Tumors in the urinary bladder of rodents can be induced by chemicals that cause calculi (stones) to be formed and deposited there. The presence of stones in bladder tissue also causes a proliferative response, by some form of physical action. Again, no risk of bladder tumorigenesis is expected until a dose of the chemical is reached sufficient to induce stone formation. The FDA has accepted this mechanism as accounting for the rodent bladder tumors induced by the indirect food additive melamine, and the agency accepts the absence of a human risk from this chemical at the ordinary levels of human exposure, where doses are much too low to cause stone formation (Clayson, 1989).

Some non-genotoxic chemicals have hormonal activity, or act to increase the activity of certain normal body hormones. Once excessive hormonal activity is stimulated, cellular proliferation in tissues responsive to the affected hormone is enhanced, again putting the tissue's cells at increased risk of tumorigenesis. It is hypothesized by some that until a dose of such a chemical is reached sufficient to sustain an increase in hormonal activity, no significant risk of tumorigenesis is expected.

Perhaps the most well-studied examples of this "hormonal effect" are non-genotoxic chemicals that induce thyroid tumors and those that induce neoplasms in various tissues that are responsive to estrogens. Exposure to high doses of agents that bring about a sustained increase in the body's natural levels of thyroid-stimulating hormone (TSH, a hormone secreted by the pituitary gland to increase the thyroid's output of other hormones) can create an excess risk of thyroid tumorigenesis. Many of these thyroid tumorigens act by interfering with the body's production or use of the hormones secreted by the thyroid, thereby causing the pituitary to continue secreting an abnormally high output of TSH, which acts to maintain an abnormal proliferative response in the thyroid. Some scientists suggest that TSH output will not be affected unless a threshold dose of the causative agent is exceeded; with no excess of TSH the type of proliferation needed to increase cancer risk will not occur.*

As more data become available indicating that these and perhaps other indirect mechanisms of action play an important role in the carcinogenicity of many substances, interest has increased in generalizing the linearized multistage model of the dose-response relationship so that it will accurately reflect these additional elements of the carcinogenic process. In particular, the two-stage growth/death model proposed by Moolgavkar and his colleagues in the late 1970s has generated excitement because it has the capacity explicitly to account for both direct geno-toxic and other toxic effects in target tissues (Moolgavkar, 1986).

Critical to effective use of this new model is accurate determination of the dose-response relationships for cell death, differentiation, and division in target tissues. These processes may exhibit threshold-like dose responses in contrast to the presumed low-dose linear response of the conventional multistage model. Thus, the low-dose risks from exposure to carcinogens that act through indirect mechanisms of the type discussed above may be significantly lower than would be predicted by the multistage model favored by the EPA. Cellular dynamics processes may also be explicitly dependent upon age of the exposed individual as well as the level and duration of exposure. Furthermore, variation in the susceptibility of individuals to exposure-related effects on these processes needs to be adequately characterized. Thus, successful use of this model in the risk assessment process will require a greater variety and amount of information and mechanistic understanding than is typically available for most chemicals. As a consequence, initially at least, it would appear that this data-intensive alternative approach will likely be developed only for those carcinogenic substances that also have great economic value. In the long run, however, as noted earlier with regard to pharmacokinetic models, effective incorporation of relevant mechanistic

---

* The use of threshold models for these various classes of non-genotoxic carcinogens seems to have been more readily accepted in some countries than others. The most notable recent example concerns the non-genotoxic animal carcinogen 2,3,7,8-tetrachlorodibenzo-*p*-dioxin (2,3,7,8-TCDD), for which a NOEL-safety factor approach to regulation has been taken in Ontario, Canada, Germany, England, Switzerland, and The Netherlands, while the no-threshold model has been applied in the U.S.

data is essential to further improvements in the scientific basis for quantitative risk assessment.

RISK ASSESSMENT AS A POLICY GUIDE

Risk assessment can, in theory, be applied to any potentially hazardous activity or exposure. The topic of this chapter is chemical toxicity, primarily of the chronic type, unquestionably a hazard of concern. Still chemicals pose other types of hazards as well: some are so exceedingly toxic that single exposures resulting from an accidental release may be lethal, some are highly flammable, some are explosive, and some are radioactive. We need also to keep in mind that chemicals of natural origin display most of the same types of toxicity, over equally wide dose ranges, as those of industrial origin. Other hazardous agents, particularly pathogenic microorganisms, are prevalent in the environment. Substances having nutritional properties can, under some conditions, create hazards of many types, either because of their absence in adequate amounts or because people chose to ingest them excessively. People partake in many types of potentially hazardous activities, whether at work, at leisure, or as part of the daily business of living.

Our environments and our lives are, it seems too obvious to say, filled with potentially hazardous activities and substances. Sometimes our exposures to them are sufficient to create measurable risks — we can count, with varying degrees of accuracy, deaths and injuries due to transportation accidents, workplace and home accidents, certain infectious diseases, and even use of tobacco products. We can obtain rough estimates of risk from other activities and exposures, such as those associated with chemical accidents and spills that create immediately apparent injuries. Finally, we can estimate risks that occur with frequencies too low to be directly measured, the types of risks discussed in this chapter. These risks, which are not known with equal accuracy (indeed, many are not known at all), vary greatly over time and from place to place. Although most societies face the same array of risks, their relative importance in different countries is highly variable.

In theory, it would seem that public health resources should be devoted to reducing risks in at least rough proportion to the toll they take on human health; and risk assessment would seem to be the tool used to assign risks their rightful order. Seen from this perspective risk assessment should be a principal component of public health and regulatory programs everywhere. Risk management approaches will differ, perhaps greatly, depending upon local laws and customs, the availability of technical skills and the resources to deploy them, and political prerogatives. Establishing the relative needs for risk reduction programs, by continuing pursuit of comprehensive assessments of risks, should everywhere be an objective, and we as scientists should be willing to share our data, knowledge, and experiences to assist each other in achieving it.

We also need to work to convince governments and the governed that public health and regulatory objectives should be consistent with the best available scientific and medical knowledge regarding the relative threats to our health and well-being of the many risks we face, and that they should not be established primarily upon the politically attractive trends of the moment.

Although this principle may seem the correct one to those of us concerned with the problem of identifying and characterizing health risks, certain social and cultural issues that may run contrary to it need to be considered.

First, it is apparent from many studies that peoples' perceptions of risk do not match those of the experts (Slovic, 1987). In fact, when it comes to describing risks most people do not give the probability that an adverse outcome will occur — the principal concern of the technical expert — as much weight as many other features of the risk, most of which are not (and should not) be considered by the risk assessor. Thus, it is clear that most people feel greater anxiety about low probability events with catastrophic outcomes (such as an airplane crash) than they do about much riskier activities that take one or a few lives at a time (such as automobile accidents). People want to accept no risk, no matter how tiny, unless they feel that the risky activity or exposure provides some clear personal benefit; this appears to be true even if the benefit is only perceived and not objectively verifiable. Risks imposed by others are less well tolerated than those voluntarily assumed. Risks that scientists do not understand well, and over which they may publicly squabble, are more dreaded than those about which scientific consensus is strong. If the risk is of natural origin it is somehow less threatening than if it is one created by human beings. We could mention many more attributes of a risk that influence people's perceptions of it and the intensity of the dread or even outrage they feel about it. Observations by social psychologists help us to understand why people and their governments seem much more anxious about, and willing to act against, relatively small risks that may be associated with many industrial chemicals, while taking a more relaxed attitude about risks that scientists recognize as more important from a public health perspective. This is not to say that health risks from industrial products and byproducts are to be ignored — indeed many are of significant concern and require regulatory controls — but only that the risks created by many such products are often perceived as much greater than can be objectively demonstrated.

Another factor complicating efforts to devote resources to various risks to the public health in proportion to their importance pertains to the fact that our options for controlling different sources of risk vary greatly. Many large risks, such as those due to smoking, alcohol abuse, and poor nutrition are not readily susceptible to traditional and readily definable regulatory approaches, but require intensive and long-term efforts to influence the public's attitudes and to reallocate resources. Specifying a limit on benzene emissions from a petroleum refinery or PCB levels in fish is, technically speaking, a much easier undertaking (notwithstanding the fact that the economic consequences for manufacturers or fishermen might be quite severe). So in many societies risk reduction priorities are often based on the relative ease with which risk reduction objectives may be achieved, and this sometimes has little correspondence to the public health importance of the risks being attacked.

Other factors serve to place more emphasis on small, less well-established risks than on large, well-documented ones. Not least of these are the various food, drug, consumer product, environmental, and occupational laws that our legisla-tures have passed. These laws presumably reflect broad social concerns, such as public perceptions of the type discussed above. Most food safety laws insist, for example, that we need to be highly risk averse when we purposefully add a substance to food, while they remain relatively silent on the (almost certainly) much larger risks associated with natural components of the diet. Regulatory agencies everywhere have been denied opportunities to do much about tobacco use. These are but two examples of policies that do not reflect current scientific understanding, but which are based on other social values.

For these and several other reasons, including the fact that experts do not always agree on the scientific issues and do a wholly inadequate job in public education on these matters, we shall no doubt continue to see public health and regulatory priorities based on factors other than relative risk. It would, neverthe-less, seem essential that those of us involved in risk assessment continue to seek the development of the data necessary to improve understanding of risks to health and safety and to press government officials and the public they serve to work toward risk-based priorities.

REFERENCES

Clayson, D. Can a mechanistic rationale be provided for non-genotoxic carcinogens identified in rodent bioassays? *Mutat. Res.* 1989, 221:53-57.

Cohen, S.A., and L.B. Ellween. Cell proliferation in carcinogenesis. *Science*, 1990, 249:1007-1011.

EPA (Environmental Protection Agency). Guidelines for carcinogen risk assessment. 1986, Fed. Reg. 51:33992-34003.

EPA (Environmental Protection Agency). Exposure Factors Handbook. Exposure Assess-ment Group, Office of Health and Environmental Assessment, U.S. Environmental Protection Agency, Washington, D.C. EPA/600/8-89/043. March 1990.

IARC (International Agency for Research on Cancer). IARC Monographs on the Evalu-ation of Carcinogenic Risks to Humans. Overall Evaluations of Carcinogenicity: An Updating of IARC Monographs Volumes 1 to 42. Supplement 7. WHO, Lyon, France, 1987.

Moolgavkar, S. Carcinogenesis modeling: from molecular biology to epidemiology. Ann. Rev. Public Health 7, 1986, 151-169.

NRC (National Research Council). Risk Assessment in the Federal Government: Managing the Process. Committee on the Institutional Means for Assessment of Risks to Public Health, National Academy of Sciences Commission on Life Sciences. National Acad-emy Press, Washington, D.C. 1983.

NRC (National Research Council). Science and Judgment in Risk Assessment. National Academy Press, Washington, D.C. 1994.

OSTP (Office of Science and Technology Policy). Chemical carcinogens: review of the science and its associated principles. *Fed. Reg.*, 1985, 50:10372-10442.

OTA (Office of Technology Assessment). Researching Health Risks. OTA-BBS-570, Washington, D.C. U.S. Government Printing Office, November 1993.

Paxton, M.B. et al. Leukemia risk associated with benzene exposure in the pliofilm cohort. *Risk Analysis,* 1994, 14:147-154.

Rinsky, R.A. et al. Benzene and leukemia: an epidemiologic risk assessment. *N. Engl. J. Med.,* 1987, 316:1044-1050.

Rodricks, J.V. Origins of Risk Assessment in Food Safety Decision Making. *J. Am. College Toxicol.* 1988, 7:539-542.

Rodricks, J.V. Calculated Risks: The Toxicity and Human Health Risks of Chemicals in our Environment. Cambridge University Press. Cambridge 1992, chapters 10 and 11, (paperback, 1994).

Shah, J.J., and H.B. Singh. Distribution of volatile organic chemicals in outdoor and indoor air. *Environ. Sci. Technol.* 1987, 22(12), 1381-1388.

Slovic, P. Perception of risk. *Science,* 236, 1987, 250-285.

CHAPTER 19

ENVIRONMENTAL TOBACCO SMOKE AND LUNG CANCER: ENVIRONMENTAL SMOKE SCREEN?

Douglas W. Dockery

INTRODUCTION

On January 7, 1993, in one of his last acts as Administrator of the Environmental Protection Agency, William Reilly approved and released the EPA report "Respiratory Health Effects of Passive Smoking: Lung Cancer and Other Disorders."[1] Among the findings of this report was the conclusion that environmental tobacco smoke was a human carcinogen. While the EPA has no authority to regulate tobacco smoke or any other indoor pollutant, the finding that environmental tobacco smoke is a human carcinogen had immediate and substantial impact on public perception and on public policy. Numerous local governments have acted to ban or restrict smoking in restaurants, offices, and other public places. On March 25, 1994, the Occupational Safety and Health Administration (OSHA) announced a proposal to regulate environmental tobacco smoke in the workplace.[2] As would be expected, the tobacco industry has challenged the studies and methods which were used in the EPA report[1] in testimony at public hearings before the Clean Air Science Advisory Committee. These arguments have subsequently appeared in a series of published articles[3-5] and in paid advertisements. These articles have suggested that the EPA designation of environmental tobacco smoke as a human carcinogen is based on a flawed review of poor epidemiology.

Is there good scientific evidence showing that environmental exposures to tobacco smoke are associated with increased risk of lung cancer, or is control of environmental tobacco smoke only a smoke screen for an effort to restrict tobacco smoking? This chapter will review the specific arguments that have been presented for and against this designation, and discuss their relevance to the EPA decision.

1-56670-144-9/96/$0.00+$.50
© 1996 by CRC Press, Inc.

CRITERIA FOR CLASSIFYING ENVIRONMENTAL
TOBACCO SMOKE AS A CARCINOGEN

What is the basis for defining environmental tobacco smoke, or any other substance, as a carcinogen? The EPA finding is based on a weight of the evidence argument. Briefly, the argument is:

- Tobacco smoke is a known cause of lung cancer.
- Environmental tobacco smoke contains the same carcinogenic chemical found in mainstream tobacco smoke.
- Nonsmokers have been shown to have measurable exposures to tobacco smoke.
- Epidemiologic studies demonstrate that there is increased risk of lung cancer among nonsmokers associated with environmental exposures to tobacco smoke.

The epidemiologic evidence is crucial in this argument. The EPA has established guidelines for the categorization of an environmental pollutant as a carcinogen based on an evaluation of the weight of the evidence from human and animal studies.[6] Designation as a "Group A Human Carcinogen" is reserved for those substances for which there is sufficient evidence from epidemiologic studies to support a causal association between exposure to the agents and cancer. Before a causal association can be inferred, epidemiologic evidence (1) must be shown to be free of bias which could explain the association, (2) confounding must be considered and ruled out as an explanation of the association, and (3) the possibility that the observed association was due to random chance must be considered. Weight is also given to reports of associations in several independent studies (consistency), the strength of the association, evidence for a dose-response association, and indications that reduction in exposure leads to a reduction in cancer incidence.

Let us then briefly examine the first three elements of the EPA argument, and then consider in detail the critical epidemiologic evidence.

CARCINOGENICITY OF TOBACCO SMOKE

Tobacco smoke is a known cause of lung cancer.[7] The carcinogenicity of tobacco smoke has been demonstrated by all methods used to assess risk, that is, in animal bioassay studies, genotoxicity studies, and epidemiologic studies. Moreover, tobacco smoke is a strong carcinogen. Epidemiologic studies have shown that active smokers develop lung cancer at a rate at least ten times that of never-smokers. Moreover, epidemiologic studies have shown that the risk of lung cancer associated with tobacco smoke increases with exposure, measured either by number of cigarettes smoked per day, or years of cigarette smoking. There is no evidence from studies of smokers that even the smallest exposures to tobacco smoke are free of risk. Thus it is plausible that environmental exposures to low concentrations of tobacco smoke should be associated with increased risk of lung cancer.

COMPOSITION OF ENVIRONMENTAL EXPOSURES

For the nonsmoker, environmental exposures to tobacco smoke come from the exhalations of active smokers plus the side-stream smoke of the cigarette between puffs. Critics have argued that the composition of environmental tobacco smoke differs from mainstream tobacco smoke, and therefore the risks of tobacco smoke exposure cannot be extrapolated to nonsmokers environmentally exposed to tobacco smoke.[3-5] Yet the carcinogenic compounds found in mainstream tobacco smoke are also found in environmental tobacco smoke.[8] Samples of air collected in rooms containing tobacco smoke have been shown to be mutagenic.[9] Thus environmental exposures to tobacco smoke include the same carcinogenic compounds inhaled by the active smoker.

ENVIRONMENTAL EXPOSURES

There is substantial evidence showing that nonsmokers are passively exposed to tobacco smoke at nontrivial levels (e.g., Reference 10). A considerable mass of data has been developed estimating environmental exposures to tobacco smoke, direct measurements of ambient indoor concentrations of tobacco smoke, and direct measures of dose based on biologic samples. The estimates of dose of the tobacco smoke received by the nonsmoker vary considerably, but certainly are several orders of magnitude less than that of a typical smoker. Nevertheless, nonsmokers living or working with smokers are found to have elevated markers of tobacco smoke in biologic samples.[11] Particularly noteworthy is the observation that nonsmokers environmentally exposed to tobacco smoke have elevated urinary mutagenicity.[9,11]

Given that tobacco smoke is associated with increased incidence of lung cancer even to the lowest exposures among active smokers, and that there is widespread environmental exposure to tobacco smoke among nonsmokers, increased incidence of lung cancer should be expected among never-smokers chronically exposed to environmental tobacco smoke.

EPIDEMIOLOGIC EVIDENCE

Tobacco smoke consistently has been shown to be causally associated with lung cancer in epidemiologic studies.[7] These studies have been shown to be free from bias, confounding, and random chance. The association is strong and shows a clear dose-response association. Risk of lung cancer declines among former smokers with years since quitting. Thus tobacco smoke meets the EPA guidelines for designation as a Class A Human Carcinogen.

It has been argued that experience with high-dose exposures to tobacco smoke through active smoking cannot be extrapolated to environmental exposures. On the other hand, all other designations of environmental agents as Class A Human

Carcinogens are based on extrapolations of experience from high exposure situations. For example, the designation of asbestos as a Class A Human Carcinogen is based on the "observation of increased mortality and incidence of lung cancer, mesotheliomas, and gastrointestinal cancer in exposed workers."[13] That is, the designation is not based on evidence from environmental exposures. There is no differentiation between high occupational exposures to asbestos (or benzene or radon, etc.) and environmental exposures. The label "environmental tobacco smoke" has inappropriately suggested that these are exposures to new potential carcinogens. We do not differentiate environmental asbestos (or benzene or radon, etc.) from other (high dose) exposures, and there is no reason to differentiate environmental exposures to tobacco smoke from the high dose exposures of active smoking.

Nevertheless, tobacco smoke is unique in that epidemiologic studies have shown increased risk associated with environmental exposures. Given that environmental exposures to tobacco smoke are much lower than those in active smokers, much weaker association should be expected. The implication of this is that statistical power will be low, such that is large samples are required to exclude random chance as a possible explanation of observed associations, and secondly, that confounding and bias have to be carefully considered as alternative explanations for observed observations. Nevertheless, epidemiologic observations of increased lung cancer risk associated with environmental exposures to tobacco smoke are only confirmatory of the demonstrated causal association from high dose exposures.

In 1981, three epidemiologic studies appeared suggesting that environmental exposures to tobacco smoke were associated with increased risk of lung cancer in nonsmoking women. In January, Hirayama[14] reported that in a prospective follow-up of nonsmoking married women in Japan, husband's reported smoking was associated with a relative risk for lung cancer death of 1.61 for former or light per smokers (1 to 19 cigarettes per day) and 2.08 for heavy smokers (20+ cigarettes per day). This was followed by a preliminary report of a case control study from Greece[15] of nonsmoking women with lung cancer. The reported relative risk of lung cancer in nonsmoking women married to light smokers (1 to 20 cigarettes per day) was 2.4, and 3.4 for heavy smokers (21+ cigarettes per day). Both of these studies were questioned because the estimated risk ratios were much higher than would have been expected given the relatively low tobacco smoke exposures. A third study[16] reported lung cancer incidence among nonsmoking women in the American Cancer Society prospective follow-up study. Estimated relative risks were elevated among the wives of smoking husbands, but were much lower than in the previous two studies — 1.27 for husbands smoking 1 to 19 cigarettes per day, and 1.10 for husbands smoking 20+ cigarettes per day.

Following the publication of these seminal epidemiologic studies in 1981, a large number of epidemiologic studies of lung cancer in nonsmokers environmentally exposed to tobacco smoke have been reported. The EPA review included results from thirty studies, and several more have been published since.

CONSISTENCY OF EVIDENCE AND META-ANALYSIS

As enunciated by Hill[17] in his classic lecture on causality, an important consideration in the evaluation of epidemiologic data is the consistency of the association, that is the replication of results by independent investigators in different populations. The key consideration is similarity of effect estimates, and not, as some have suggested, findings of statistical significance. In order to assess the consistency of the effect estimates, it is first necessary to define the problem in terms of a single exposure and a single health endpoint common to all studies. To evaluate consistency the EPA considered all lung cancers in nonsmoking women associated with smoking by their husbands. These are the same exposure and endpoints chosen in previous reviews by the National Research Council[8] and the U.S. Surgeon General.[18] This is not to suggest that these are necessarily the best measures of environmental exposure to tobacco smoke, nor the best measures of lung cancer incidence. Nevertheless, they provide reasonable measures which are reported in almost all studies.

Figure 1 presents a summary of the epidemiologic studies reported to date showing the association of lung cancer in nonsmoking women with the cigarette smoking of their husbands. Results are presented from the thirty studies considered by the EPA plus two new studies[19,20] and one update[21] which appeared subsequent to the writing of the EPA report. For each study, the estimated relative risk of lung cancer associated with husband's cigarette smoking is presented along with the 95% confidence interval (CI) for that estimate. The studies are ordered by the size of the effect estimate. The center line corresponding to a relative risk of one implies no effect. An estimated relative risk greater than one (that is, on the right-hand side of the plot) indicates that lung cancer in these nonsmoking women is positively associated with their husbands' smoking. Risks to the left indicate a protective effect.

Our prior hypothesis must be that husband's smoking would be associated with an increased risk of lung cancer. The weight of the evidence in this figure clearly shows that almost all of these studies have found positive associations between husband's smoking and lung cancer. The consistency of these results across so many independent studies showing a positive association between lung cancer and husband's smoking is a very strong statement of the robustness of those findings.

A quantitative estimate of the effect of husband's smoking on lung cancer can be calculated as a weighted average of the logarithm of the study-specific relative risk estimates, where the weight is the inverse variance of the estimate for each study. The combination of results across individual studies is called a meta-analysis. The combined estimate of the effect of husband's smoking on lung cancer in their nonsmoking spouses is a relative risk of 1.27 with 95% CI of 1.18 to 1.38. In this weighted analysis, the greatest weights are given to those studies with the narrowest CI, that is the smallest variance of the estimate. In general these studies are those with the largest sample sizes, although studies with better designs and methods will tend to have smaller variances of effect estimates about a true association. Among the studies considered, the smallest variances, and therefore the greatest weights, are found in the more recent studies.[19-21] It is

Studies (Reference)

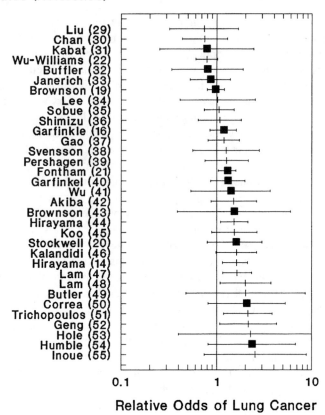

Relative Odds of Lung Cancer

Figure 1 Summary of study-specific relative odds (95% Confidence Interval) of lung cancer among nonsmoking women associated with husband's smoking. Studies (and reference numbers) are ordered by effect size. Studies from United States highlighted with box.

interesting to plot the weight given to an individual study against the effect estimate (Figure 2). The plotted data typically have a cone (or funnel) shape centered on the combined estimated effect.

Critics have suggested that the use of meta-analytic techniques, as in the EPA report, are inappropriate for epidemiologic data. However, meta-analysis is a well-developed technique for contrasting and combining results from different studies. Its concepts and methods are actually a simple extension of statistical analysis already used in epidemiology. These methods are commonly used in public health research, particularly in the evaluation of clinical trials.[23] Meta-analysis has been applied in epidemiologic research for many decades[24] and has recently been the subject of two scholarly reviews.[25,26] The use of these methods is not only appropriate but is a significant advance over traditional methods of evaluating epidemiologic studies individually.

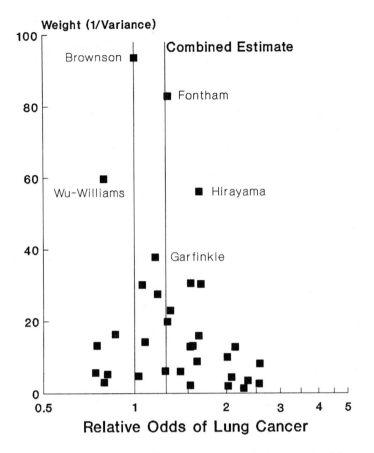

Figure 2   Study-specific weight (1/variance) vs. estimated relative odds of lung cancer associated with husband's smoking. Vertical lines show no effect level (1.0) and weighted mean (1.27) over all studies.

Meta-analysis is significant because it provides an objective method for combining data and for evaluating effects estimates from different studies. Moreover, it provides a method for contrasting various studies in the literature and for evaluating the effects of bias, confounding, and other issues of importance in epidemiological studies. For example, it has been argued that studies from outside the U.S. may not reflect living patterns in this country, or might be flawed by cultural differences. Meta-analysis allows a direct evaluation of this suggestion. The weighted mean of the 13 studies from the U.S. gives a combined estimate of 1.16 (95% CI 1.03 to 1.30). While this is a smaller risk than found for all the studies combined, the data from the U.S. studies is elevated and consistent with the hypothesis that environmental exposure to tobacco smoke is causally associated with lung cancer.

Reviews of the literature are often criticized, particularly for environmental exposures with weak effects, because of a potential bias toward only positive results being published. Such publication bias would result from negative studies not being accepted or not even being submitted for publication. The literature on lung cancer among nonsmoking women associated with husband's smoking is remarkable in that the majority of published studies (25 of the 31 studies in Figure 1) report associations which would not be considered statistically significant. Thus there is little indication of publication bias in these studies.

Another way to consider the issue of selection bias is through sequential meta-analyses. Typically, when a new association is postulated, the initial studies will show strong (and statistically significant) associations. Subsequent studies designed to test the original hypothesis find weaker associations compared to the initial findings. Thus, a sequential meta-analysis looking at the accumulated evidence by year will often show large combined effect estimates drifting back toward the null (relative risk of 1.0). For the lung cancer vs. husband's smoking literature, the two initial papers published in 1981 gave a combined effect estimate of 1.74 (95% CI 1.31 to 2.30). Figure 3 shows the cumulative combined relative risk estimate by publication year. While there was some attenuation of the effect estimate between the initial reports in 1981 and 1982, there has been very little change in the estimated effect size over the past decade, while the CIs have narrowed. In fact, the estimated relative risk from the 1981 American Cancer Society prospective study[16] of 1.17 is remarkably similar to the combined estimate for all U.S. studies of 1.16.

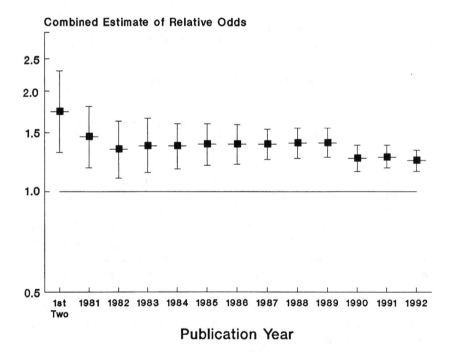

Figure 3   Cumulative combined effect estimate (weighted mean) of relative odds of lung cancer associated with husband's smoking by publication year.

Critics of the EPA analysis[3-5] have suggested that the results would have been different if two studies[19,20] published after the EPA review was completed had been included. The Brownson study is a large study, having 432 lifetime non-smokers, and is given a high weight in the meta-analysis. The weakness of this study is the relatively poor exposure data. There was no association between husband's smoking and lung cancer in this study (relative risk 1.0, 95% CI 0.8 to 1.2). However, a positive association was found between spouse smoking and lung cancer which increased with years of exposure. The authors conclude: "Ours and other recent studies suggest a small but consistent increased risk of lung cancer from passive smoking." The Stockwell study[20] reports a relatively large (relative risk 1.60) but less stable association (95% CI 0.80 to 3.00). The sequential meta-analysis (Figure 3) shows that the inclusion of these studies published in 1992 has very little effect on the combined effect estimate or its CI. The estimated effect of spouse smoke on lung cancer in nonsmoking women after including these studies was 1.24 with 95% confidence interval of 1.15 to 1.33, which was only slightly different from the estimate based on the studies included in the EPA review of 1.27 with 95% confidence interval of 1.18 to 1.38.

BIAS AND CONFOUNDING

Epidemiologic studies are observational, that is the investigator has little or no control over the various determinants of disease in the sample population being studied. Thus epidemiologic findings must be evaluated to determine if observed associations could be the result of bias or confounding in the data, that is, that the observed associations are due to some characteristic other than spouse smoking which is related to *both* lung cancer and spouse smoking. Inadequate exposure measurement, e.g., the fact that spouse smoking is only a crude measure of environmental tobacco smoke exposure in the home, at work, and in other settings, will produce underestimates of the true association, but is unlikely to produce an overestimate of the true association.

Critics of the EPA review[3-5] have suggested that the observed associations may be the result of uncontrolled confounding, that is failure to adequately consider a third variable associated with both environmental tobacco smoke and lung cancer which might be an alternative causative explanation. Huber and colleagues[4] highlight diet as one such risk factor. It is known that a diet deficient in anticarcinogenic nutrients will increase the risk of lung cancer in smokers.[27] However, there is no evidence that lack of these dietary nutrients will produce lung cancers in people who have no exposure to a causative agent. Thus while diet may reduce the risk of lung cancer in smokers, suggesting diet is the causative agent in nonsmokers is not a credible epidemiologic argument.

Fontham et al.[21] have recently reported the final results of a multi-center case-control study of the association of lung cancer in lifetime never-smoking women with environmental exposure to tobacco smoke. Preliminary results of this study[28] were included in the EPA review, and the final results are consistent with the earlier findings. Cases were 653 female lifetime never-smokers from five centers in the U.S. with histologically confirmed lung cancer. Random digit dialing supplemented by random sampling of Health Care Financing Administration records for women over 65 years of age identified 1253 population controls. This study is particularly important because it addresses directly the weaknesses suggested for previous epidemiologic studies. Lifetime smoking status of lung cancer cases was determined from medical records, from interviews of their personal physicians, and from interviews of study subjects (or next of kin). These data were verified by urinary cotinine measurements from consenting study subjects. This multi-tiered evaluation of personal smoking reduces the chance for bias due to misclassification of active smokers as nonsmokers. If bias by misclassification of smokers as nonsmokers was producing the observed associations, then restriction of the sample to those with the lowest urinary cotinine levels would be expected to reduce the estimated association of lung cancer with environmental exposures to tobacco smoke. In contrast, this restriction produced a slightly increased effect estimate.

The crude estimate of the relative risk for all lung cancers associated with any tobacco smoking by the husband in this multi-center study was 1.26 (95% CI 1.04 to 1.54). The effect of confounding was evaluated by comparison of the effect estimate, with and without adjustment for possible confounding variables. In this study, adjustment for age, race, study area, education, diet (fruits, vegetables, vitamin supplements, and cholesterol), family history of lung cancer, and occupational exposures, increased the effect estimate to 1.29 (95% CI 1.04 to 1.60). The comparability of the crude and adjusted effect estimates shows that these factors did not confound the association between lung cancer and husband's smoking. A dose response association was observed across categories of environmental exposure to tobacco smoke as measured by amount smoked or by years of exposure (Figure 4). Similar effects were observed for environmental exposures to tobacco smoke in the home, environmental exposures at work, and environmental exposures in social settings.

An alternative method for controlling bias and confounding in epidemiologic studies is to restrict the study sample to subjects with no variation in the postulated confounders or sources of bias. Such a sample would have no potential for active smoking and little or no variation in diet, occupational exposures, etc. While such severe restriction of subjects is impractical in human studies, a recently published case-control study of the association of lung cancer in pet dogs provides such control of confounding and bias.[56] Cases were 51 pet dogs diagnosed with lung cancer, and controls were 83 pet dogs with other cancers drawn from oncology records at two university veterinary teaching hospitals. Exposure was estimated from a questionnaire mailed to the household of each subject (dog). The estimated relative odds for lung cancer associated with cancer in the home was 1.6 (95% CI 0.7 to 3.7), remarkably similar to the estimated effect from human studies. The results were suggestive of an exposure response association with adjusted relative odds of 1.2 (95% CI 0.5 to 3.2) for less than 2 packs per day smoked in the home, increasing to 3.4 (95% CI 0.7 to 16.5) for 2 or more packs per day

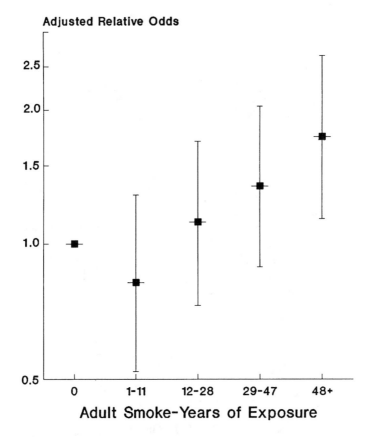

Figure 4   Adjusted relative odds for lung cancer associated with years of environmental exposure to tobacco smoke in the home, after data from Fontham et al.[21]

smoked in the home, both compared to nonsmoking homes. While there are clear and important differences between pet dogs and humans, this study has several important implications for the epidemiologic studies of environmental exposures to tobacco smoke and lung cancer. First of all, misclassification of active smoking is not possible for these pet dogs, that is, the observed associations cannot be explained by unreported active smoking by the dogs themselves, as has been suggested for the human studies. Secondly, it is unlikely that these results are confounded by the factors which have been suggested to affect human studies. For example, for diet to be a confounder in this study, smokers would have to feed their pets a diet which is different from that of the pets of nonsmokers, and this diet would have to be independently associated with lung cancer. Thus, while this study did not report statistically significant results, because of small study size, this study does make a very important contribution, in showing that two other potential sources of error, that is misclassification of active smoking and confounding are not important factors in the epidemiologic studies of environmental exposure to tobacco smoke and lung cancer.

RANDOM CHANCE

Critics of the EPA review[3-5] have questioned the use of 90% CIs in the meta-analysis. The CI reflects only one of several elements that must be considered in the evaluation of epidemiologic studies: the CI is a quantitative measure of the statistical or random error in the data and reflects, therefore, only the influence of chance in these data. It is inappropriate to define an adverse health effect by a statistical significance or a CI. Whether that definition is based on a 90% or a 95% CI is irrelevant.

Use of a 90% (rather than 95%) CI in this case corresponds to use of a one-tailed (rather than two-tailed) test of statistical significance at the $p = 0.05$ level. In the case of a known carcinogenic agent, such as tobacco smoke, it is entirely appropriate to use a one-tailed or one-sided test of significance for an adverse effect of tobacco smoke. We would not expect tobacco smoke to be a protective agent against cancer and therefore the use of a 5% one-tailed test, or equivalently a 90% CI, is justified. All of this discussion is moot, however, since equivalent results are obtained with the use of 95% CI about the combined effect estimates.

As Bradford Hill[17] stated almost thirty years ago regarding causal inference in environmental studies:

> No formal tests of significance can answer those questions. Such tests can, and should, remind us of the effects that the play of chance can create, and they will instruct us in the likely magnitude of those effects. Beyond that they contribute nothing to the "proof" of our hypothesis.

CONCLUSION

In summary, do environmental exposures to tobacco smoke meet the EPA criteria for designation as a Class A Human Carcinogen? The vast data on the effects of tobacco smoke at high doses (active smoking) meet all of the criteria for designating tobacco smoke as an environmental carcinogen. The epidemiologic studies showing lung cancer to be associated with environmental exposures to tobacco smoke only confirm this conclusion.

Nevertheless, these epidemiologic studies of environmental exposures also meet all the criteria for defining a human carcinogen. The recently reported multi-center case-control study[21,28] alone is sufficient to show that lung cancer in nonsmoking women is causally related to environmental exposures to tobacco smoke. This study most clearly demonstrates that the associations are not due to bias, confounding, or random chance. Quantitatively similar associations have been found by many investigators in many different settings (consistency). There is also consistent evidence for a dose-response association in these epidemiologic studies. The evidence from high dose exposures shows that tobacco smoke is very strongly associated with lung cancer, and that removal of exposure (i.e., smoking cessation) leads to a reduction in risk.

Thus the body of evidence for the designation of environmental exposures to tobacco smoke as a human carcinogen is large, consistent, and compelling. This is not an environmental smoke screen as has been suggested by the tobacco industry and other critics. Rather these critics are raising smoke to obscure the issues and suggest controversy. These criticisms have little or no scientific merit or relevance.

REFERENCES

1. Environmental Protection Agency *Respiratory Health Effects of Passive Smoking: Lung Cancer and Other Disorders.* EPA/600/6-90/006B, Washington, D.C. 1993.
2. Occupational Safety and Health Administration. Indoor Air Quality. Federal Register 1994; 59(65):15968-16039.
3. Luik JC. Pandora's box: the dangers of politically corrupted science for democratic public policy. Bostonia, Winter 93-94:50-60.
4. Huber GL, Brockie RE, Mahajan VK. Smoke and Mirrors: The EPA's Flawed Study of Environmental Tobacco Smoke and Lung Cancer. Regulation: The Cato Review of Business and Government. 1993; 3:44-54.
5. Sullum J. Passive reporting on passive smoke. Forbes Media Critic. 1994:41-47.
6. Environmental Protection Agency. Guidelines for carcinogen risk assessment. Federal Register 1986; 51:3392-34003.
7. International Agency for Research on Cancer. IARC monographs on the evaluation of the carcinogenic risk of chemicals to man. Vol. 38. Tobacco smoking. Lyon, France: World Health Organization, 1986.
8. National Research Council. *Environmental Tobacco Smoke: Measuring Exposures and Assessing Health Effects.* Washington, D.C.: National Academy Press, 1986.
9. Husgafvel-Pursiainen K, Sorsa M, Engstroem K, Einistoe P. Passive smoking at work: biochemical and biological measures of exposure to environmental tobacco smoke. *Int. Arch. Occup. Environ. Health.* 1987; 59:337-345.
10. Cummings KM, Markello SJ, et al. Measurement of current exposure to environmental tobacco smoke. *Arch. Environ. Health.* 1990;45:74-79.
11. Haley NJ, Colosimo SG, Axelrad CM, Harris R, Sepkovic DW. Biochemical validation of self-reported exposure to environmental tobacco smoke. *Environ. Res.* 1989;49:127-135.
12. Bos RD, Theuws JLG, Henderson PT. Excretion of mutagens in human urine after passive smoking. *Cancer Lett.* 1983; 19:85-90.
13. Environmental Protection Agency. Airborne asbestos health assessment update. Prepared by Environmental Criteria and Assessment Office, Research Triangle Park, NC. EPA 600/8-84/003F. 1986
14. Hirayama T. Non-smoking wives of heavy smokers have a higher risk of lung cancer: a study from Japan. *Br. Med. J.* 1981; 282:183-185.
15. Tricholopoulos D, Kalandidi A, Sparros L, MacMahon B. Lung cancer and passive smoking. *Int. J. Cancer.* 1981; 27:1-4.
16. Garfinkel L. Time trends in lung cancer mortality among nonsmokers and a note on passive smoking. *J. Natl. Cancer.* 1981; 6:1066-1066.
17. Hill AB. The environment and disease: association or causation? *Proc. R. Soc. Med.* 1965; 58:295-300.

18. U.S. Department of Health and Human Services. *The Health Consequences of Involuntary Smoking. A Report of the Surgeon General.* DHHS Pub. No. (PHS) 87-8398. U.S. Department of Health and Human Services, Public Health Service, Office of the Assistant Secretary for Health, Office of Smoking and Health, 1986.
19. Brownson RC, Alavanja MCR, Hock ET, Loy TS. Passive smoking and lung cancer. *Am. J. Public Health.* 1992; 82:1525-1530.
20. Stockwell HG, Goldman AL, Lyman GH, et al. Environmental tobacco smoke and lung cancer risk in nonsmoking women. *J. Natl. Cancer Inst.* 1992; 84:1417-1422.
21. Fontham ETH, Correa P, Reynolds P, Wu-Williams A, Buffler PA, Greenberg RS, Chen VW, Alterman T, Boyd P, Austin DF, Liff J. Environmental tobacco smoke and lung cancer in nonsmoking women; a multicenter study. *JAMA.* 1994; 271:1752-1759.
22. Wu-Williams AH, Dai XD, Blot W, et al. Lung cancer among women in northeast China. *Br. J. Cancer.* 1990; 62:982-987.
23. Louis TA, Fineberg HV, Mosteller F. Findings for public health from meta-analysis. *Annu. Rev. Public Health.* 1985; 6:1-20.
24. MacMahaon B, Hutchinson GB. Prenatal x-ray and childhood cancer: A review. *Acta Unio Int. Cancer.* 1964; 2:1171-74.
25. Greenland S. Quantitative methods in the review of epidemiologic literature. *Epidemiol. Rev.* 1987; 9:1-30.
26. Dickersin K, Berlin JA. Meta-analysis: State-of-the-science. *Epidemiol. Rev.* 1992; 14:154-176.
27. LeMarch L, Wilkins LR, Hankin JH, Haley NJ. Dietary patterns of female nonsmokers with and without exposure to environmental tobacco smoke. *Cancer Causes Control.* 1991; 2:11-16.
28. Fontham ETH, Correa P, Wu-Wiiliams A, et al. Lung cancer in non-smoking women: A multi-center case control study. *Cancer Epidemiol.: Biomarkers Prevention.* 1991; 1:35-43.
29. Liu Z, He X, Chapman RS. Smoking and other risk factors for lung cancer in Xuanwei, China. *Int. J. Epidemiol.* 1991; 20:26-31.
30. Chan WC, Fung SC. Lung cancer in non-smokers in Hong Kong. In: Grundmann, E., ed. Cancer campaign, Vol. 6, Cancer Epidemiology. Stuttgart: Gustav Fischer Verlag, 1982, pp. 199-202.
31. Kabat GC, Wynder EL. Lung cancer in nonsmokers. *Cancer* 1984; 53:1214-1221.
32. Buffler PA, Pickle LW, Mason TJ, Contant C. The causes of lung cancer in Texas. In: Mizell, M., Correa, P., eds. Lung Cancer: Causes and Prevention. New York: Verlag Chemie International, 1984, pp. 83-99.
33. Janerich DT, Thompson WD, Varela LR, et al. Lung cancer and exposure to tobacco smoke in the household. *N. Engl. J. Med.* 1990; 323:632-636.
34. Lee PN, Chamberlain J, Alderson MR. Relationship of passive smoking to risk of lung cancer and other smoking-associated diseases. *Br. J. Cancer.* 1986; 54:97-105.
35. Sobue T, Suzuki R, Nakayama N, Inubuse C, Matsuda M, Doi O, Mori T, Furuse K, Fukuoka M, Yasumitsu T, Kuwabara O, Ichigaya M, Kurata M, Nakahara K, Endo S, Hattori S. Passive smoking among nonsmoking women and the relationship between indoor air pollution and lung cancer incidence — results of a multicenter controlled study. *Gan No Rinsho* 1990; 36(3):329-333.
36. Shimizu H, Morishita M, Mizuno K, Masuda T, Ogura Y, Santo M, Nishimura M, Kunishima K, Karasawa K, Nishiwaki K, Yamamoto M, Hisamichi S, Tominaga S. A case-control study of lung cancer in nonsmoking women. *Tohoku J. Exp. Med.* 1988; 154:389-397.

37. Gao Y, Blot WJ, Zheng W, Ershow AG, Hsu CW, Levin LI, Zhang R, Fraumeni JF. Lung cancer among Chinese women. *Int. J. Cancer.* 1987; 40:604-609.

38. Svensson C, Pershagen G, Klominek J. Smoking and passive smoking in relation to lung cancer in women. *Acta Oncol.* 1989; 28:623-629.

39. Pershagen G, Hrubec Z, Svensson C. Passive smoking and lung cancer in Swedish women. *Am. J. Epidemiol.* 1987; 125(1):17-24.

40. Garfinkel L, Auerbach O, Joubert L. Involuntary smoking and lung cancer: a case-control study. *J. Natl. Cancer Inst.* 1985; 75:463-469.

41. Wu AH, Henderson BE, Pike MD, Yu MC. Smoking and other risk factors for lung cancer in women. *J. Natl. Cancer Inst.* 1985; 74(4):747-751.

42. Akiba S, Kato H, Blot WJ. Passive smoking and lung cancer among Japanese women. *Cancer Res.* 1986; 46:4804-4807.

43. Brownson RD, Reif JS, Keefe TJ, Ferguson SW, Pritzl JA. Risk factors for adenocarcinoma of the lung. *Am. J. Epidemiol.* 1987; 125:25-34.

44. Hirayama T. Cancer mortality in nonsmoking women with smoking husbands based on a large-scale cohort study in Japan. *Prev. Med.* 1984; 13:680-690.

45. Koo LC, Ho JH, Saw D, Ho CY. Measurements of passive smoking and estimates of lung cancer risk among non-smoking Chinese females. *Int. J. Cancer.* 1987; 39:162-169.

46. Kalandidi A, Katsouyanni K, Voropoulou N, et al. Passive smoking and diet in the etiology of lung cancer among non-smokers. *Cancer Causes Control.* 1990; 1:15-21.

47. Lam TH, Kung ITM, Wong CM, Lam WK, Kleevens JWL, Saw D, Hsu C, Seneviratne S, Lam SY, Lo KK, Chan WC. Smoking, passive smoking and histological types in lung cancer in Hong Kong Chinese women. *Br. J. Cancer.* 1987; 6:673:678.

48. Lam WK. A clinical and epidemiological study of carcinoma of lung in Hong Kong [doctoral thesis]. Hong Kong: University of Hong Kong, 1985.

49. Butler TL. The relationship of passive smoking to various health outcomes among Seventh-Day Adventists in California [dissertation], Los Angeles: University of California at Los Angeles, 1988.

50. Correa P, Fontham E, Pickle L, Lin Y, Haenszel W. (1983) Passive smoking and lung cancer. *Lancet.* 1983; 2:595-597.

51. Tricholopoulos D, Kalandidi A, Sparros L. Lung cancer and passive smoking: conclusion of Greek study. [Letter] *Lancet.* 1983; 667-668.

52. Geng G, Liang Z-H, Zhang GL. On the relationship between smoking and female lung cancer. Smoking and Health, Elsevier Science Publishers, 1988; pp. 483-486.

53. Hole DJ, Gillis CR, Chopra C, Hawthorne VM. Passive smoking and cardiorespiratory health in a general population in the west of Scotland. *Br. Med. J.* 1989; 299:423-427.

54. Humble CG, Samet JM, Pathak DR. Marriage to a smoker and lung cancer risk. *Am. J. Public Health.* 1987; 77:598-602.

55. Inoue R, Hirayama T. Passive smoking and lung cancer in women. In: Smoking and Heath. Elsevier Science Publishers, 1988; pp. 283-285.

56. Reif JS, Dunn K, Ogilvie GK, Harris CK. Passive smoking and canine lung cancer risk. *Am. J. Epidemiol.* 1992; 135:234-239.

CHAPTER 20

Radon and Lung Cancer Revisited

Jonathan M. Samet

INTRODUCTION

Indoor radon was already widely recognized as a significant public health problem by 1984, the year of the first Oak Ridge National Laboratory Life Sciences Symposium on indoor air. Substantial international research had been in progress since the 1970s, and the finding of homes with extremely high radon levels had catapulted radon into prominence in the U.S. The early 1980s also marked the beginning of the Radon Program of the U.S. Environmental Protection Agency (EPA). The 1984 symposium included four presentations on indoor radon: the first addressed concentrations (Nero, 1985); the second, dosimetric and risk models (Harley, 1985); the third, epidemiology and risk assessment (Ellett and Nelson, 1985); and the fourth, European radon surveys (Steinhausler, 1985). A presentation by Lowder (Lowder, 1985) summarized the state of knowledge (Table 1).

Ten years later, our understanding of indoor radon and lung cancer has advanced significantly. Yet Lowder's 1984 conclusions still largely reflect perceptions of the current state of knowledge, with the exception of the accumulation of exposure information from a number of large, representative surveys. Uncertainties persist in the risk assessment for indoor radon and lung cancer, and these uncertainties contribute to maintained debate on policies for radon control (Abelson, 1991; Cole, 1993).

This presentation, 10 years after the 1984 symposium, reviews the evolution of the scientific evidence during that period, touching on the expansion of the epidemiologic literature on lung cancer in miners, the increasing epidemiologic investigation of radon and lung cancer in the general population, and new risk assessment approaches (Table 2). Comprehensive reviews and meeting proceedings have been published which provide a more complete picture of the evidence

Table 1   Knowledge of Indoor Radon and Lung Cancer

- We know how to measure population exposure and have techniques for carrying out surveys; information on the distributions of indoor exposure is emerging.
- Soil gas is the main source of indoor radon and there is general understanding of the role of ventilation. Criteria are needed for screening for high risk areas.
- Exposure increases lung cancer risk but there will always be uncertainty in the unit risk.
- A risk estimate can be obtained from studies of uranium miners but there are uncertainties in the exposure estimates for the miners and in translating the risks to the general population.
- Risk estimates are variable among the approaches used by different agencies by about a factor of 6.
- The problem of indoor radon and lung cancer should be controllable.

Summarized by Lowder, W. M. (1985) in *Indoor Air and Human Health,* Lewis Publishers, Inc., Chelsea, MI, 39-41.

Table 2   Some Key Events Related to Indoor Radon and Lung Cancer, 1984–1994

1984:	Publication of NCRP Report No. 78
1987:	Publication of ICRP Report No. 50
1988:	Publication of the BEIR IV Report
1989:	First DOE/CEC workshop on case-control studies of indoor radon and lung cancer
1991:	Publication of the NRC radon dose panel report
1991:	Hanford Symposium: Indoor Radon and Lung Cancer: Myth or Reality?
1991:	Second DOE/CEC Workshop
1993:	Publication of ICRP Report No. 65
1994:	Start of BEIR VI

on radon and lung cancer risk (National Research Council and Committee on the Biological Effects of Ionizing Radiation, 1988; Nazaroff and Nero, 1988; Samet, 1989; Cross, 1992a,b).

EPIDEMIOLOGIC STUDIES OF MINERS

As of 1984, the findings of only three large cohort or prospective studies of lung cancer in underground miners with estimated exposures to radon had been reported: the U.S. Public Health Service study of Colorado Plateau uranium miners (Lundin et al., 1971); the study of Czech uranium miners (Sevc et al., 1976); and the study of Swedish underground metal miners in Malmberget (Radford and Renard St. Clair, 1984). On the basis of the findings of these and several other studies as well as the historically documented excess lung cancer in the miners of Schneeberg and Joachimsthal, radon was generally considered to be a cause of lung cancer (Holaday, 1969; Lundin et al., 1971; National Research Council, 1980). Further support for the carcinogenicity of radon was found in the results of exposures of animals to radon and radon progeny and in the emerging understanding of mechanisms of injury by alpha particles.

From 1984 through the early 1990s, the findings of seven additional cohort studies of underground miners all incorporating estimates of radon progeny exposure were published, and the information on exposures to radon progeny of the Newfoundland fluorspar miners was updated (Table 3). These reports confirmed the excess risk of lung cancer observed in the previous studies and expanded the data base for developing models of the relationship between radon progeny exposure and lung cancer risk. The more recent cohorts include three groups of Canadian uranium miners (Muller et al., 1984; Howe et al., 1986, 1987); Chinese tin miners (Xuan et al., 1993), notable for the large population of exposed miners and the number first exposed as children; New Mexico uranium miners (Samet et al., 1991a); Australian uranium miners from the Radium Hill mine (Woodward et al., 1991); and French uranium miners (Tirmarche et al., 1993).

Table 3   Excess Relative Risk[a] of Lung Cancer per WLM (ERR/WLM) by Study Cohort and for the Pooled Data

Study	Cases	(ERR/WLM)%	95% CI
China	908	0.16	0.1-0.2
Czechoslovakia	661	0.34	0.2-0.6
Colorado	294	0.42	0.3-0.7
Ontario	291	0.89	0.5-1.5
Newfoundland	118	0.76	0.4-1.3
Sweden	79	0.95	0.1-4.1
New Mexico	69	1.72	0.6-6.7
Beaverlodge	65	2.21	0.9-5.6
Port Radium	57	0.19	0.1-0.6
Radium Hill	54	5.06	1.0-12.2
France	45	0.36	0.0-1.3
Combined	2701	0.49[b]	0.2-1.0[c]

[a] Background lung cancer rates adjusted for age (all studies), other mine exposure (China, Colorado, Ontario, New Mexico, France), an indicator of Rn exposure (Beaverlodge) and ethnicity (New Mexico). Colorado data restricted to exposures under 3200 WLM.
[b] p value for test of homogeneity of ERR/WLM across studies, $p < 0.001$. Joint 95% CI based on random effects model.

Adapted from Lubin et al., 1994.

While the human carcinogenicity of radon progeny had been well established by 1984, the new studies added to the literature on radon progeny and lung cancer, contributing estimates of the exposure-relationship between radon progeny exposure and lung cancer risk, and new evidence on key uncertainties regarding the lung cancer risk associated with radon. Based on the greater quantitative risks in Beaverlodge compared with Port Radium uranium miners, Howe and colleagues (Howe et al., 1987) suggested that the difference could represent an effect of exposure rate, an inverse dose-rate effect. The average exposure rate had been substantially lower for the Beaverlodge miners. Further followup of the Czechoslovakian miners also provided evidence of an inverse dose-rate effect (Tomasek et al., 1994) and information on cancers other than lung cancer (Tomasek et al., 1993). In this cohort, lung cancer risk varied with the exposure rate, stratified by the authors at 10 Working Levels (WL). Among men whose exposure rates never exceeded 10 WL, the risk increased linearly with time-weighted cumulative exposure, was higher in younger men, and declined with lengthening interval since exposure.

The study of Chinese tin miners included a group that had started working underground as children; age at first exposure did not modify the subsequent risk of lung cancer (Xuan et al., 1993). The study of New Mexico uranium miners included information on smoking for most miners (Samet et al., 1991a). The data indicated a multiplicative interaction between smoking and radon progeny exposure, although the combined effect of the two agents was not described with precision because of the limited number of lung cancer deaths (N = 68). A case-control study conducted within the New Mexico cohort found that the presence of chest radiograph abnormalities indicative of silicosis was not associated with lung cancer risk (Samet et al., 1994). The studies of Port Radium, Radium Hill, and French uranium miners provided new data for levels of exposure lower than in some of the earlier cohorts.

EPIDEMIOLOGIC STUDIES OF RADON AND LUNG CANCER IN THE GENERAL POPULATION

Overview

In 1984, a few small case-control studies of indoor radon and lung cancer had been reported along with the results of several ecologic studies (Axelson et al., 1979; National Research Council and Committee on the Biological Effects of Ionizing Radiation, 1988). These studies provided an indication of increased lung cancer risk associated with higher levels of indoor radon exposure, but did not offer risk coefficients suitable for the purpose of risk assessment. Since 1984, a number of large case-control studies have been designed and implemented and the findings of several of these studies have been reported. Additional ecologic studies have also been conducted. The case-control studies have been implemented to address the risk of lung cancer directly, and thereby remove the uncertainties inherent in extrapolating risks from underground miners to the general population. The ecologic studies, which used readily available data on lung cancer mortality, were conducted to determine whether patterns of lung cancer occurrence were consistent with the hypothesis that indoor radon causes lung cancer.

Ecologic Studies

In ecologic studies, the units of analysis are groups, usually defined by geography, rather than individuals. Methodological problems inherently limit the usefulness of ecologic studies (Piantadosi et al., 1988; Greenland and Morgenstern, 1989; Greenland, 1992; Stidley and Samet, 1994) and these studies are generally considered to be appropriate for developing new hypotheses regarding exposure-disease associations rather than for quantitative risk estimation.

Stidley and Samet (1994) provide a detailed review of the results and methods of 15 ecologic studies of lung cancer and indoor radon exposure published through 1992. A few additional studies were subsequently reported (Neuberger et al., 1992, 1994). Exposure estimates used in these studies included both surrogate measures such as the geologic characteristics of an area, and estimates based on current measurements of indoor radon levels from samples of homes in an area.

Results of the 15 studies considered by Stidley and Samet varied, with seven reporting positive associations, six no association, and two negative associations. Of the seven positive studies, only one included adjustment for smoking, and six were based on surrogate measures for indoor radon exposure. Stidley and Samet considered all of the studies to have low power for detecting effects under reasonable alternative hypotheses. The two studies showing negative associations both used regression methods to assess the relationship between the index of radon and lung cancer risk; both used exposure estimates based on samples of measurements and included adjustment for smoking. A recent study that compared lung cancer rates in high and low exposure areas also found a negative association, but did not include adjustment for smoking (Neuberger et al., 1992).

In addition to the problems inherent in any ecologic study, the ecologic studies of indoor radon and lung cancer are potentially subject to bias from errors in the exposure indices and confounding. Using data for geographic regions of the size for which vital statistics are available, e.g., counties, may narrow the range of exposures considered and thereby reduce statistical power. Additionally, the surrogate measures used in some studies, such as geographic features, inherently misclassify exposures; even indices based on measurements introduce error because they are based on sample data (Stidley and Samet, 1994). Estimates of current concentration are also likely to represent past exposures with substantial error.

Ecologic data on radon and lung cancer are subject to confounding if the geographic regions chosen reflect differences in other risk factors, e.g., cigarette smoking. Confounding can be addressed in ecologic studies through stratification and regression, but complete control of confounding cannot be achieved (Greenland and Morgenstern, 1989). It is often not possible to control adequately for potential confoundens either because adequate data are not available, or because, with grouped data, adjustment for confounders cannot be correctly carried out unless the risk is a linear function of both indoor radon levels and other variables for which adjustment is needed. In evaluating lung cancer risks, smoking is the factor that is of greatest concern, and data are generally not available for finely defined geographic regions. Stidley and Samet (Stidley and Samet, 1994) show that even a small negative correlation of radon levels and smoking could induce a negative correlation of radon levels and lung cancer risks.

Ecologic studies offer little as a tool for evaluating the lung cancer risk associated with indoor radon. This study design has been used primarily as a basis for developing hypotheses to be tested using other study designs. Radon is already an established carcinogen and the major uncertainty is the magnitude of risk at residential levels. The methodological shortcomings of ecologic studies led Stidley and Samet to conclude that ecologic studies of residential radon have been uninformative regarding the association of lung cancer risk with exposure to indoor radon.

Case-Control Studies

In the case-control design, exposures of persons with the disease of interest are compared with those of appropriate controls not having the disease of interest. The case-control design was endorsed by the 1989 International Workshop on Residential Radon Epidemiology as the method of choice for the direct assessment of risks from residential radon exposure (Samet et al., 1991b), this same workshop recommended against further ecologic studies. In a 1992 review, Neuberger (Neuberger, 1992) listed 20 studies including a total of over 12,000 lung cancer cases and 19,000 controls.

In the case-control studies addressing residential radon, exposures to indoor radon of lung cancer cases are compared with those of controls. Information on cigarette smoking and other potentially important co-factors can be obtained by interview with either the case or with a surrogate respondent for persons who have died. In some of the early studies, exposures were indirectly estimated based on type of residential construction or on residence location (National Research Council and Committee on the Biological Effects of Ionizing Radiation, 1988). In the more recent studies, exposures to indoor radon have been estimated by making longer term measurements of radon concentration in the current and previous residences of the cases and controls (U.S. Department of Energy, 1991; Neuberger, 1992). Several of the studies have been restricted to nonsmokers in order to estimate the lung cancer risk in this group as precisely as possible.

Since 1984, results have been reported for several case-control studies having relatively large sample sizes and incorporating measurements of radon concentrations. A study in New Jersey women, which included 480 cases and 442 controls, showed a statistically significant association of lung cancer with radon exposure, although the results depended strongly on a small number of cases in the highest exposure category (Schoenberg et al., 1990). A study in Stockholm County, Sweden, which included 210 cases and 400 controls, showed evidence of a positive but not statistically significant association (Pershagen et al., 1992), while a study in China (Blot et al., 1990) with 308 cases and 356 controls found no evidence of an association. Pooled analyses of data from these three studies were carried out by Lubin et al. (1994). In spite of the seemingly different findings of the two "positive" and the one "negative" studies, this analysis showed that the findings of the three studies were not statistically heterogeneous. With pooling of the data from the three studies, the resulting combined estimate of risk was so imprecise that it covered a range extending from no effect to the estimate derived from the pooled analysis of four studies of underground miners by the Biological Effects of Ionizing Radiation (BEIR) IV Committee (National Research Council and Committee on the Biological Effects of Ionizing Radiation, 1988).

In 1994, the results of the largest case-control study completed to date were reported. In this nationwide study in Sweden, Pershagen and colleagues (1994) studied 586 female and 774 male cases of lung cancer and 1380 female and 1467 male controls. Information was obtained on smoking, and an attempt was made to measure radon for a three-month interval in all homes occupied during approximately the last 35 years. Lung cancer risk was found to increase with increasing cumulative exposure risk with a dose-response relationship compatible with that observed in the studies of miners. The combined effect of smoking and radon exposure exceeded the expectation based on additivity.

The findings of the case-control studies reported to date indicate the need for pooling of the data from the individual studies as recommended by participants in the international workshops of investigators for these studies held by the U.S. Department of Energy and the Commission of European Communities (U.S. Department of Energy, 1991). Pooling is needed to obtain as much precision as possible in estimating the risk of indoor radon and to assess modification of the risk of radon by such factors as smoking, sex, and age at exposure.

The need for pooling was predicted on the basis of sample size calculations published by Lubin, Samet, and Weinberg (1990). The sample size calculations considered the consequences of measurement error and residential mobility, which together tend to reduce the variability of population exposure. Lubin and colleagues addressed the sample sizes required for three distinct purposes: (1) to detect association of lung cancer with residential radon exposure assuming that the risk is the same as predicted from the underground miner data; (2) to detect an effect in the general population that is 50% lower than predicted by the underground miner data; and (3) to detect significant departure from additive interaction of smoking and radon exposure if the true interaction is assumed to be multiplicative. Sample size calculations were performed for hypothetical case-control studies directed at these questions; a log-normal distribution of exposure was assumed, comparable to that described by Nero et al. (1986) and the studies were assumed to have two controls for each case.

Table 4 Effect of Measurement Error, f^a, and Mobility Pattern on Sample Size Required to Reject No Trend with Rn Exposure, $\beta_0{}^a = 0$, when the True Trend is $\beta_1 = 0.015$. Study Based on a Control-To-Case Ratio of 2.

	Mobility pattern					
	60 y		3 × 20 y		6 × 10 y	
f	Cases	Power[b]	Cases	Power[b]	Cases	Power[b]
0.00	251	0.90	529	0.66	938	0.46
0.30	288	0.86	656	0.58	1,303	0.37
0.50	365	0.79	916	0.48	2,050	0.28
1.00	973	0.48	2,987	0.23	8,002	0.14
1.50	4,186	0.22	13,934	0.12	39,456	0.09
2.00	29,542	0.12	100,308	0.08	287,644	0.07

[a] f is a proportional index of error on a logarithmic scale and β is the excess relative risk coefficient in units of % per WLM of exposure.
[b] Power relative to a study with 251 cases, 502 controls, and no error in exposure ($f = 0$).

Adapted from Lubin et al., 1990.

A need for large sample sizes was projected under plausible scenarios of exposure error and population mobility; for each of the three hypotheses of interest, the majority of sample size estimates projected a need for thousands of cases and controls. Table 4 provides example calculations of sample size requirements to test the null hypothesis that residential radon does not cause lung cancer if the true relative risk is properly predicted by the risk model of the BEIR IV Committee. The calculations demonstrate that the combined impact of error in exposure assessment and population mobility on sample size requirements.

Case-control studies are further subject to misclassification of outcome and incomplete control of other relevant factors, including cigarette smoking, which may confound or modify the association between indoor radon and lung cancer. Because lung cancer is rapidly fatal, interviews with surrogate respondents may be necessary to obtain information on cigarette smoking, occupation, and other risk factors for lung cancer. Information from surrogate sources is subject to error, and complete control of confounding may not be possible in case-control studies of lung cancer and radon. Taken together, the sample size requirements of the case-control studies and the other inherent methodologic limitations led Lubin and colleagues to recommend pooling of data from the studies in progress.

RISK ASSESSMENT

Overview

By the time of the 1984 symposium, only a few risk assessment approaches for indoor radon and lung cancer had been developed (see National Research Council and Committee on the Biological Effects of Ionizing Radiation, 1988 for a review). These risk models were primarily based on risk coefficients derived from one or a few of the then extant studies of underground miners. Thomas and McNeill (1982) published a particularly notable report for the Atomic Energy Control Board of Canada. In work that presaged subsequent approaches, Thomas and McNeill (1982) applied a relative risk model to summary data in published reports on the Czechoslovakian, Ontario, and Colorado Plateau uranium miners, the Newfoundland fluorspar miners, and the Swedish metal miners. Their final model was linear in cumulative exposure, except at the highest values, and estimated the increment in excess relative risk as 2.28 per 100 working level months (WLM).

Risk Models

In 1984, the National Council on Radiation Protection and Measurements (NCRP, 1984) published Report 78 which estimated risk using an additive risk model and a risk coefficient based on the then available data from miners. The model assumed that the effect of exposure declined over time on the basis of postulated repair or death of injured cells. Additional models published during the 1980s that have received widespread application include the relative risk models in Report 50 of International Commission on Radiological Protection

(1987) and in the BEIR IV Report (National Research Council and Committee on the Biological Effects of Ionizing Radiation, 1988).

The ICRP model assumes a constant relative risk and a threefold greater risk for exposures received before 20 years of age. The latter assumption was based on considerations of biologic susceptibility and the findings in the study of the atomic bomb survivors. The risk coefficient was drawn from studies of underground miners.

The model developed by the BEIR IV Committee (National Research Council and Committee on the Biological Effects of Ionizing Radiation, 1988) and published in 1988 has been used as the basis for the risk assessment of indoor radon conducted by the EPA. Data from four cohorts, Malmberget iron miners, Colorado Plateau uranium miners, Beaverlodge, Canada uranium miners, and Ontario, Canada uranium miners were analyzed by the committee to develop a risk model. These four groups comprised the only studies having data on radon progeny exposures of individual participants to which the committee could gain access. The data analysis was accomplished with relative risk regression modeling that facilitated a broad assessment of the temporal pattern of risk and of the effect of confounding and modifying factors of interest.

The committee recommended the following relative-risk, termed the time-since-exposure model for $r(a)$, the age-specific lung-cancer mortality rate:

$$r(a) = r_o(a)\left[1 + 0.025\ \gamma(a)\left(W_1 + 0.5\,W_2\right)\right],$$

where $r_o(a)$ is the age-specific background lung-cancer mortality rate; $\gamma(a)$ is 1.2 when age a is less than 55 years, 1.0 when a is 55 to 64 years, and 0.4 when a is 65 years or more; W_1 is WLM received between 5 and 15 years before age a; and W_2 is WLM incurred 15 years or more before age a. This model inherently assumes that the effects of other factors determining lung cancer risk, e.g., cigarette smoking, are multiplicative with those of exposure to radon progeny.

The BEIR IV model may soon be supplemented by a model based on a pooled analysis of data from 11 cohorts of radon-exposed underground miners (Table 3). With the publication of additional studies and the opportunity to work with the team investigating the Czechoslovakian uranium miners, the U.S. National Cancer Institute analyzed data from 11 studies of underground miners (Lubin et al., 1994). The cohorts included uranium, tin, iron, and fluorspar miners; various methods were used to estimate exposure to radon progeny, six studies had some data on cigarette smoking, and only a few studies had information on exposures other than radon progeny. The pooled data set included over 2700 lung cancer deaths among 68,000 miners followed for nearly 1.2 million person-years of observation. A full description of the analysis has been published as a National Career Institute Monograph (Lubin et al., 1994).

The data were analyzed with Poisson regression method comparable to those used by the BEIR IV Committee. Most analyses were based on a linear excess relative risk (ERR) model:

$$RR = 1 + \beta w$$

where RR is relative risk, β is a parameter measuring the unit increase in ERR per unit increase in w, and w is cumulative exposure to radon progeny in WLM. As in the BEIR IV analysis, cumulative exposure was also divided into the amounts received during windows defined by time since exposure.

As in the BEIR IV analysis, ERR was linearly related to cumulative exposure to radon progeny. The ERR/WLM varied significantly with other factors; it decreased with attained age, time since exposure, and time after cessation of exposure, but was not affected significantly by age at first exposure. Over a wide range of total cumulative exposures to radon progeny, lung cancer risk was increased as exposure rate declined. This finding of an exposure rate effect in the pooled analysis confirms the pattern reported from the Colorado Plateau study (Hornung and Meinhardt, 1987), and supports the prior hypothesis of an inverse dose-rate effect (Darby and Doll, 1990). The inverse exposure rate has potentially significant implications for risk estimation at typical indoor levels using the miner studies.

Information on tobacco use was available from six cohorts. The combined data were consistent with a relationship between additive and multiplicative for the joint effect of smoking and exposure to radon progeny. Over 50,000 person-years, including 64 lung cancer deaths, were accrued by miners who were identified as never smokers. In this group, there was a linear exposure-response trend which was about threefold greater than observed in smokers.

The monograph described two sets of preferred models; the sets differ in being either categorical or contiguous in the parameterization of the variables. Each includes one model (TSE/AGE/WL-cat model) incorporating radon progeny exposure during time-since exposure windows and variables for effect modification by attained age and exposure rate and one model (TSE/AGE/DUR-cat model) similarly incorporating exposure in the same windows and variables for attained age and duration of exposure. The two categorical models are:

TSE/AGE/WL-cat model:

$$RR = 1 + \beta \times (w_{5-14} + \theta_2\, w_{15-24} + \theta_3\, w_{25+}) \times \phi_{age} \times \gamma WL$$

where $\beta = 0.0611$, $\theta_2 = 0.81$, $\theta_3 = 0.40$,

$$
\begin{aligned}
\phi_{age} = \ & 1.00 \text{ for age} < 55 \\
& 0.65 \text{ for } 55 \leq \text{age} < 65 \\
& 0.38 \text{ for } 65 \leq \text{age} < 75 \\
& 0.22 \text{ for } 75 \leq \text{age}
\end{aligned}
$$

$$
\begin{aligned}
\gamma WL = \ & 1.00 \text{ for WL} < 0.5 \\
& 0.51 \text{ for } 0.5 \leq \text{WL} < 1.0 \\
& 0.32 \text{ for } 1.0 \leq \text{WL} < 3.0 \\
& 0.27 \text{ for } 3.0 \leq \text{WL} < 5.0 \\
& 0.13 \text{ for } 5.0 \leq \text{WL} < 15.0 \\
& 0.10 \text{ for } 15.0 \leq \text{WL}
\end{aligned}
$$

TSU/AGE/DUR-cat model:

$$RR = 1 + \beta \times (w_{5-14} + \theta_2 w_{15-24} + \theta_3 w_{25+}) \times \phi_{age} \times \gamma DUR$$

where $\beta = 0.0039$, $\theta_2 = 0.76$, $\theta_3 = 0.31$,

$$\phi_{age} = \begin{array}{l} 1.00 \text{ for age} < 55 \\ 0.57 \text{ for } 55 \le \text{age} < 65 \\ 0.34 \text{ for } 65 \le \text{age} < 75 \\ 0.28 \text{ for } 75 \le \text{age} \\ 1.00 \text{ for DUR} < 5 \text{ y} \\ 3.17 \text{ for } 5 \le \text{DUR} < 15 \text{ y} \end{array}$$

$$\gamma DUR = \begin{array}{l} 5.27 \text{ for } 15 \le \text{DUR} < 24 \text{ y} \\ 9.08 \text{ for } 25 \le \text{DUR} < 35 \text{ y} \\ 13.6 \text{ for } 35 \le \text{DUR} \end{array}$$

The models are similar to the BEIR IV model with the additional terms for either rate of exposure or duration of exposure. Additionally, the larger data set available for this analysis made it possible to estimate the effect of exposure in four time windows rather than three as in the BEIR IV model.

In applying this model, uncertainty arises as to the most biologically appropriate extrapolation of the exposure rate effect to typical levels of indoor exposure. Uncertainty related to exposure rate was an implicit limitation of the application of the BEIR IV model, not an explicit limitation as in the new model based on the 11 studies.

Risk Assessments

In 1984, few quantitative risk assessments of the burden of lung cancer attributable to radon had been made, although radon was considered by many to be a significant public health problem. While the BEIR IV Report did not offer quantitative estimates of the numbers of lung cancer cases attributable to radon, it offered a general approach for doing so. In addition to extending the risk model to the lung cancer experience of the general population, it is also necessary to adjust for differences in exposure-dose relationships for the circumstances of exposure in mines and in homes. A wide array of physical and biological factors are relevant to assessing exposure-dose relations (National Research Council and Committee on the Biological Effects of Ionizing Radiation, 1988; National Research Council and Panel on Dosimetric Assumptions Affecting the Application of Radon Risk Estimates, 1991). The BEIR IV Committee concluded that exposure yielded equivalent doses of alpha energy to target respiratory cells for exposures received in homes and in mines; a follow-up National Research Council Committee charged with reviewing the dosimetry of radon progeny concluded that exposures indoors yielded slightly lower doses than in mines (National

Research Council and Panel on Dosimetric Assumptions Affecting the Application of Radon Risk Estimates, 1991).

Using the BEIR IV model and assuming equivalent exposure-dose relationships in homes and mines, Lubin and Boice (1994) calculated the proportion of lung cancer deaths attributable to radon exposure of residents of single-family homes in the U.S. They assumed an exposure distribution comparable to that found by Nero et al. (1986). The BEIR IV risk model predicted that 14% of lung cancer deaths or approximately 13,300 deaths annually were attributable to radon. The EPA estimated 13,600 lung cancer deaths annually, using a modification of the BEIR IV model and adjustment for a slightly lower dose in homes (U.S. Environmental Protection Agency, 1992). The range for this estimate extended from 7000 to 30,000 deaths.

The new model based on the 11 cohorts of underground miners has also been used to estimate the numbers of lung cancer deaths attributable to radon (Lubin et al., 1994). The assumptions with regard to dosimetry and exposure were comparable to those made by the EPA in its 1992 report (U.S. Environmental Protection Agency, 1992). The new model predicted 14,400 lung cancer deaths for 1993 among residents of single-family homes in the U.S. This calculation assumed a higher relative risk values for never smokers than for smokers as found in the pooled analysis.

THE FUTURE

The 10 years since the first Oak Ridge National Laboratory Life Sciences Symposium on Indoor Air Pollution have brought substantial advances in our understanding of radon and lung cancer. Among these advances are the evolution of the epidemiologic evidence from studies of underground miners, improved lung dosimetry of radon progeny, the implementation and completion of epidemiologic studies in the general population, and continued deepening of understanding of mechanisms of alpha carcinogenesis. The pooled analysis of the underground miner studies offers new risk models based on data at exposures only one or two orders of magnitude greater than those received by most members of the population. Key scientific uncertainties remain unaddressed, however; principal among these are the extrapolation from the generally higher doses and dose rates of the miners to the lower doses and dose rates of the general population and the combined effect of smoking and radon progeny.

Can these and other uncertainties be addressed through further research? During the next 10 years, needed data should become available from two lines of investigation: the ongoing epidemiologic studies in the general population and laboratory-based studies of carcinogenesis. The epidemiologic studies may provide an informative assessment of risk, particularly if pooling is accomplished, and the laboratory studies should provide guidance on the biological appropriateness of extrapolating studies of miners to the general population.

The advances during the last 10 years have been considered as warranting a new study of radon and lung cancer by the National Research Council. In 1994, the BEIR VI Committee was formed to review the new evidence and to develop new risk models for radon and lung cancer. The committee's report, projected to be published in 1997, will likely provide a new risk model and reduce uncertainties in risk estimates. Undoubtedly, however, radon will remain a controversial carcinogen at the end of the century, as it was at the century's start.

REFERENCES

Abelson, P. H. (1991). "Mineral dusts and radon in uranium mines (editorial)." *Science* **254**: 777.

Axelson, O., C. Edling, et al. (1979). "Lung cancer and residency — a case-referent study on the possible impact of exposure to radon and its daughters in dwellings." *Scand. J. Work Environ. Health* **5**: 10-15.

Blot, W. J., Z.-Y. Xu, et al. (1990). "Indoor radon and lung cancer in China." *J. Natl. Cancer Inst.* **82**: 1025-1030.

Cole, L. A. (1993). *Elements of Risk: The Politics of Radon.* Washington, D.C., AAAS Press.

Cross, F. T., Ed. (1992a). *Indoor Radon and Lung Cancer: Reality or Myth? Part 1.* Columbus, Battelle Press.

Cross, F. T., Ed. (1992b). *Indoor Radon and Lung Cancer: Reality or Myth? Part 2.* Columbus, Battelle Press.

Darby, S. C. and R. Doll (1990). "Radiation and exposure rate." *Nature* **344** (April 26, 1990): 824.

Ellett, W. H. and N. S. Nelson (1985). Epidemiology and risk assessment: testing models for radon-induced lung cancer. *Indoor Air and Human Health.* Chelsea, MI, Lewis Publishers, Inc. 79-107.

Greenland, S. (1992). "Divergent biases in ecologic and individual-level studies." *Stat. Med.* **11**: 1209-1223.

Greenland, S. and H. Morgenstern (1989). "Ecological bias, confounding, and effect modification." *Int. J. Epidemiol.* **18**: 269-274.

Harley, N. H. (1985). Comparing radon daughter dosimetric and risk models. *Indoor Air and Human Health.* Chelsea, MI, Lewis Publishers, Inc.

Holaday, D. A. (1969). "History of the exposure of miners to radon." *Health Phys.* **16**: 547-52.

Hornung, R. W. and T. J. Meinhardt (1987). "Quantitative risk assessment of lung cancer in U.S. uranium miners." *Health Phys.* **52**: 417-430.

Howe, G. R., R. C. Nair, et al. (1986). "Lung cancer mortality (1950-1980) in relation to radon daughter exposure in a cohort of workers at the Eldorado Beaverlodge uranium mine." *J. Natl. Cancer Inst.* **77**: 357-362.

Howe, G. R., R. C. Naire, et al. (1987). "Lung cancer mortality (1950-1980) in relation to radon daughter exposure in a cohort of workers at the Eldorado Port Radium uranium mine: possible modification of risk by exposure rate." *J. Natl. Cancer Inst.* **79**: 1255-1260.

International Commission on Radiological Protection (1987). *Lung cancer risk from indoor exposures to radon daughters. Publication 50.* Oxford, Pergamon Press.

Lowder, W. M. (1985). Part one: overview. *Indoor Air and Human Health.* Chelsea, MI, Lewis Publishers, Inc. 39-41.

Lubin, J. H., J. D. Boice, et al. (1994). *Radon and Lung Cancer Risk: A Joint Analysis of 11 Underground Miner Studies.* Washington, D.C., U.S. Department of Health and Human Services.

Lubin, J. H., Z. Liang, et al. (1994). Radon exposures in residences and lung cancer among women: combined analysis of three studies. *Cancer Causes Control* 5:114-128.

Lubin, J. H., J. M. Samet, et al. (1990). "Design issues in epidemiologic studies of indoor exposure to radon and risk of lung cancer." *Health Phys.* **59**: 807-817.

Lundin, F. D., Jr., J. K. Wagoner, et al. (1971). *Radon daughter exposure and respiratory cancer, quantitative and temporal aspects. NIOSH-NIEHS Joint. Monograph No. 1.* Springfield, VA, National Technical Information Services.

Muller, J., W. C. Wheeler, et al. (1984). *Study of mortality of Ontario miners.* International Conference on Occupational Radiation Safety in Mining, Toronto, Ontario, Canada.

National Council on Radiological Protection and Measurements (NCRP) (1984). *Evaluation of Occupational and Environmental Exposures to Radon and Radon Daughters in the United States.* Bethesda, MD, NCRP.

National Research Council (1980). *The Effects on Populations of Exposure to Low Levels of Ionizing Radiation: 1980.* Washington, D.C., National Academy Press.

National Research Council and Committee on the Biological Effects of Ionizing Radiation (1988). *Health Risks of Radon and Other Internally Deposited Alpha-Emitters: BEIR IV.* Washington, D.C., National Academy Press.

National Research Council and Panel on Dosimetric Assumptions Affecting the Application of Radon Risk Estimates (1991). *Comparative Dosimetry of Radon in Mines and Homes. Companion to BEIR IV Report.* Washington, D.C., National Academy Press.

Nazaroff, W. W. and A. V. Nero Jr. (1988). *Radon and Its Decay Products in Indoor Air.* New York, John Wiley & Sons, Inc.

Nero, A. V., Jr. (1985). Indoor concentrations of radon-222 and its daughters: sources, range, and environmental influences. *Indoor Air and Human Health.* Chelsea, MI, Lewis Publishers, Inc. 43-67.

Nero, A. V., M. B. Schwehr, et al. (1986). "Distribution of airborne radon-222 concentrations in U.S. homes." *Science* **234**: 992-997.

Neuberger, J. S. (1992). "Residential radon exposure and lung cancer: an overview of ongoing studies." *Health Phys.* **63**: 503-509.

Neuberger, J. S., F. Frost, et al. (1992). "Residential radon exposure and lung cancer: evidence of an inverse association in Washington State." *J. Environ. Health* **55**: 23-25.

Neuberger, J. S., C. F. Lynch, et al. (1994). "Residential radon exposure and lung cancer: evidence of an urban factor in Iowa." *Health Phys.* **66**(3): 263-69.

Pershagen, G., Z. H. Liang, et al. (1992). "Residential radon exposure and lung cancer in women." *Health Phys.* **63**: 179-186.

Piantadosi, S., D. Byar, et al. (1988). "The ecological fallacy." *Am. J. Epidemiol.* **127**: 893-904.

Radford, E. P. and K. G. Renard St. Clair (1984). "Lung cancer in Swedish iron ore miners exposed to low doses of radon daughters." *N. Engl. J. Med.* **310**: 1485-1494.

Samet, J. M. (1989). "Radon and lung cancer." *J. Natl. Cancer Inst.* **81**: 745-757.

Samet, J. M., D. R. Pathak, et al. (1994) "Silicosis and lung cancer risk in underground uranium miners." *Health Phys.* **66**: 450-453.

Samet, J. M., D. R. Pathak, et al. (1991a). "Lung cancer mortality and exposure to radon decay products in a cohort of New Mexico underground uranium miners." *Health Phys.* **61**: 745-752.

Samet, J. M., J. Stolwijk, et al. (1991b). "Summary: international workshops on residential radon epidemiology." *Health Phys.* **60**: 223-227.

Schoenberg, J. B., J. B. Klotz, et al. (1990). "Case-control study of residential radon and lung cancer among New Jersey women." *Cancer Res.* **50**: 6520-6524.

Sevc, J., E. Kunz, et al. (1976). "Lung cancer mortality in uranium miners and long-term exposure to radon daughter products." *Health Phys.* **30**: 433-437.

Steinhausler, F. (1985). European radon surveys and risk assessment. *Indoor Air and Human Health.* Chelsea, MI, Lewis Publishers, Inc. 109-29.

Stidley, C. A. and J. M. Samet (1994). "A review of ecological studies of lung cancer and indoor radon." *Health Phys.* **65**: 234-251.

Stidley, C. A. and J. M. Samet (1994). "Assessment of ecologic regression in the study of lung cancer and indoor radon." *Am. J. Epidemiol.* **139**(3): 312-22.

Tirmarche, M., A. Raphalen, et al. (1993). "Mortality of a cohort of French uranium miners exposed to relatively low radon concentrations." *Br. J. Cancer* **67**: 1090-1097.

Thomas, D. C., K. G. McNeill, "Risk estimates for the health effects of alpha radiation." Info-0081. Ottawa, Canada. Atlantic Energy Control Board, 1982.

Tomasek, L., S. C. Darby, "Patterns of lung cancer mortality among uranium miners in West Bohemia with varying rates of exposure to radon and its progeny." *Radiation Res.* **137**: 251-261.

Tomasek, L., S. C. Darby, et al. (1993). "Radon exposure and cancers other than lung cancer among uranium miners in West Bohemia." *Lancet* **341**: 919-923.

U.S. Department of Energy, C. O. E. C. (1991). *Report on the Second International Workshop on Residential Radon.* Springfield, VA, National Technical Information Service, U.S. Department of Commerce.

U.S. Environmental Protection Agency (1992). *Technical support document for the 1992 citizen's guide to radon. EPA/400-R-92-011.* Washington, D.C., U.S. EPA.

Woodward, A., D. Roder, et al. (1991). "Radon daughter exposures at the Radium Hill uranium mine and lung cancer rates among former workers, 1952-87." *Cancer Causes Control* **2**: 213-220.

Xuan, X.Z., Lubin, J.H., et al. (1993). "A cohort study in southern China of workers exposed to radon and radon decay products." *Health Phys.* **64**: 120-131.

CHAPTER 21

Asbestos Exposures of Building Occupants and Workers

Rashid A. Shaikh

ABSTRACT

Thousands of buildings in the U.S. contain asbestos-containing materials (ACM). There has been a concern that the occupants and workers in such buildings may be exposed to airborne asbestos and therefore be at risk for asbestos-induced diseases, particularly mesothelioma and lung cancer. In 1991, a panel of experts assembled by the Health Effects Institute-Asbestos Research (HEI-AR) reviewed the data available on asbestos levels in buildings. Based on information on 1377 samples collected in 198 buildings, the panel calculated an average airborne asbestos level of 0.00027 f/mL (fibers longer than 5 μm, analyzed by transmission electron microscope using the direct method). Although the samples were not collected in a statistically representative group of buildings, and there were other limitations to the data, based on all the available information, the HEI-AR panel concluded that the added lifetime risk of cancer for general occupants appeared to be relatively low.

Unlike general building occupants, janitorial, custodial, maintenance, and renovation workers may experience peak exposure episodes in the course of their work because of disturbance or damage to ACM, dust or debris, which may release relatively high, transient concentrations of fibers. Operations and maintenance (O&M) programs for the management of ACM and for control of asbestos exposures are being widely implemented in buildings, but little information is available on the effectiveness of O&M strategies. We have analyzed data on airborne asbestos concentrations in 394 samples (191 area and 203 personal) collected during 106 jobs that were part of an O&M program at a hospital that contained a variety of ACM. Analysis of the data showed that the average concentrations, as determined

1-56670-144-9/96/$0.00+$.50

by phase contrast microscopy, for personal and area samples were 0.11 and 0.02 f/mL, respectively. When the data were used to calculate 8-h time-weighted average (TWA) concentrations for personal samples, 95% of the values were below 0.1 f/mL and 99% were below 0.2 f/mL. The results of our study are consistent with the few other reports in the literature on airborne levels in buildings where O&M programs had been implemented.

INTRODUCTION

The term asbestos is applied to several naturally occurring silicate minerals which are extremely fibrous in nature. Asbestos fibers are good insulators, have high tensile strength, and are flexible. These qualities have made asbestos minerals commercially useful and, during the last century, asbestos has been mined, processed, and used in thousands of products. Of the various forms of asbestos, chrysotile has been used most widely in the U.S.; amosite and crocidolite has also been used but to a lesser degree.

ACMs have been used in buildings as surface applied finishes and for thermal insulation as well as in equipment used in buildings (Table 1). ACM may contain the mineral in an unbound form (as in fire retardant applications) or in a bound state as in floor tiles. The U.S. Environmental Protection Agency (EPA) has estimated, based on a national sample consisting of 231 buildings, that roughly 733,000 buildings, or approximately 20% of the 3.6 million total population of U.S. public and commercial buildings, had some type of friable ACM (sprayed or trowelled-on surfacing, thermal system insulation, and ceiling tile).[1]

Asbestos is a well-recognized source of malignant and nonmalignant disease among workers in the asbestos mining, processing, manufacturing, and installation industries. The inhalation of airborne asbestos dust in an occupational setting can produce asbestosis, a fibrotic lung disease, as well as cancer of the lungs, and mesothelioma of the pleura and peritoneum; asbestos may also pose a risk of cancer at other, distant sites. In view of the widespread use of ACM in buildings, there has been concern regarding the potential for exposure of workers and occupants to airborne asbestos in the indoor environment. Experience with one building at Yale University during the 1970s showed that friable ACM in poor condition could produce high exposure levels.[2]

During the 1980s, concern about indoor asbestos largely focused on schools after apparently elevated concentrations were found in some schools. Adding to the concern about schools was the possibility of heightened risk from exposure during childhood, either because of inherent susceptibility or the lengthening of the interval over which disease could develop. In response to these concerns, Congress enacted the Asbestos Hazard Emergency Response Act (AHERA) in 1986. AHERA mandated the EPA to develop and implement a comprehensive regulatory framework for inspection, management, planning, and operations and maintenance activities as well as appropriate abatement actions to control ACM in the nation's schools.

Table 1   Asbestos-Containing Products Found in Buildings by Physical State

Unbound asbestos in inorganic mixtures	Bound asbestos composites[a]	Asbestos textiles

Surface treatments[b]
  Fireproofing (steel structures)
  Acoustical applications
  Decorative surfaces
  Equipment lagging
  Moisture barrier
 	
Dry applications
  Feathered (generally friable)
 	
Wet applications
  Tamped (generally nonfriable)
  Untamped
	Thermal systems insulation
	Insulating products: boiler covers,
	  cements, pipe lagging
	Vinyl tiles and floor coverings
	Ceiling tiles
	Ceiling and wall boards
	Cement products
	Papers (including pipe covering)
	Acoustic plasters
	Spackling, patching, taping
	  compounds
		Packings (valves
		  and flanges)
		Plumbing cords and
		  ropes
		Electrical wire
		  insulation

[a] All generally nonfriable.
[b] Note that "surface treatments" in the U.S. also include asbestos bound in cements and plasters. The use of the word "unbound" for these applications is European in origin.

From the HEI-AR Panel Report.[3] With permission. This classification is modified from a scheme used in the International Program on Chemical Safety, Environment Health Criteria *53:* Asbestos and Other Natural Mineral Fibres. United Nations Environment Program, International Labor Office, World Health Organization, Geneva (1986).

In 1988, the U.S. Congress mandated that the Health Effects Institute, a nonprofit, private organization in Cambridge, MA, address several issues related to asbestos exposures in public and commercial buildings. In particular, the Congress mandated that the institute conduct research to determine airborne asbestos concentrations prevalent in buildings, characterize peak exposure episodes and their significance, and evaluate the effectiveness of asbestos management and abatement strategies in a scientifically meaningful manner. In response, the HEI-AR was formed. Initially, HEI-AR convened a panel which published a comprehensive report on asbestos in public and commercial buildings.[3,4] This paper describes some of the findings summarized by the HEI-AR Panel; it also summarizes other data that have been analyzed by the institute subsequent to publication of the report.

GENERAL BUILDING OCCUPANT EXPOSURE

During the 1970s and 1980s, concern mounted about the potential for asbestos exposure of general building occupants. Potential mechanisms of exposure for general building occupants included the "peaks" generated by activities of various trade groups, contamination of heating, ventilation and air-conditioning systems, fiber releases from impact and abrasion of accessible asbestos, and resuspension of asbestos-contaminated surface dust.

The HEI-AR Literature Review Panel compiled all published and much unpublished data on buildings that had been samples for ambient asbestos levels.

The panel found information on 1377 samples that were collected in 198 buildings; these samples had been analyzed using the direct method by the transmission electron microscope (Table 2). The average airborne asbestos level for fibers longer than 5 μm, which are considered most relevant to disease risk, was 0.00027 f/mL. Asbestos levels tended to be somewhat higher in schools, presumably reflecting the higher level of activity in schools in comparison with public and commercial buildings. The results from nearly 300 buildings that were sampled for litigation purposes were similar to those summarized above.

The HEI-AR Report acknowledged that the data had not been collected in buildings chosen within a statistical framework and therefore may not be representative of all U.S. buildings. In addition to problems related to the analytical protocols and procedures used by the different laboratories that analyzed the samples, other sources of uncertainty and variability included factors such as sampling location within the building, extent of ACM damage, level of activity in the building, and whether and what measures to control asbestos releases had been undertaken.

Within the constraints of the reservations mentioned above, the panel estimated risk for general building occupants based on linear extrapolation from effects observed in workers with heavy occupational exposure (Table 3). For industrial workers who were exposed for 20 years at a level of 10 f/mL, lifetime increase in cancer risk was estimated to be about 1 in 5. When this dose-response relationship was extrapolated to the airborne asbestos concentrations reported in public and commercial buildings, the predicted lifetime risk for 20 years of exposure during working hours was estimated to be about 4 per million. The panel stated that the relative impact of the public health consequences of asbestos exposure could be judged by comparing the risk from asbestos exposure to the much greater risks of lung cancer that have been projected to occur as a result of lifetime exposure to indoor radon and environmental tobacco smoke.

Little information is available regarding the long-term ambient levels in buildings that have asbestos in well-maintained condition. In one office building in the Southeastern U.S. that contained sprayed on ACM in return air plenum and thermal pipe insulation, air was sampled on nine occasions (328 samples) from 1985 to 1988; an O&M program (see below) was operational during the entire period. The average for all fibers longer than 5 μm determined using the transmission electron microscope was 0.0004 f/mL (Table 4).[4]

EXPOSURE OF CUSTODIAL AND MAINTENANCE WORKERS

In contrast to general building occupants, building custodial and maintenance workers, during the course of their routine occupational activities, may disturb or damage ACM or resuspend contaminated dust, and thus generate brief, but relatively high, exposure episodes. Two early studies demonstrated that in the absence of control practices, asbestos exposures of building workers can be high

Table 2 Distribution of Building Average Airborne Asbestos Concentrations by Building Type

| Building type | No. of buildings | 10th | | | 90th | | |
		Minimum	Percentile	Median	Mean	Percentile	Maximum
School	48	0	0	0	0.00051	0.0016	0.0080
Residence	96	0	0	0	0.00019	0.0005	0.0025
Public and commercial	54	0	0	0	0.00020	0.0004	0.0065
All buildings	198	0	0	0	0.00027	0.0007	0.0080

Note: Fibers longer than 5 μm; samples analyzed using the direct method by the transmission electron microscope.

From the HEI-AR Panel Report.[3] With permission.

Table 3   Estimated Lifetime Cancer Risks for Different Scenarios of Exposure to Airborne Asbestos Fibers[a]

Conditions	Premature cancer deaths (lifetime risks) per million exposed persons
Lifetime, continuous outdoor exposure	
• 0.00001 f/mL from birth (rural)	4
• 0.0001 f/mL from birth (high urban)	40
Exposure in a school containing ACM, from age 5 to 18 years (180 days/year, 5 h/day)	
• 0.0005 f/mL (average)[b]	6
• 0.005 f/mL (high)[b]	60
Exposure in a public building containing ACM, age 25 to 45 years (240 days/year, 8 h/day)	
• 0.0002 f/mL (average)[b]	4
• 0.002 f/mL (high)[b]	40
Occupational exposure, from age 25 to 45	
• 0.1 f/mL (current occupational levels)[c]	2,000
• 10 f/mL (historical industrial exposures)	200,000

[a] This table represents the combined risk (average for males and females) estimated for lung cancer and mesothelioma for building occupants exposed to airborne asbestos fibers under the circumstances specified. These estimates should be interpreted with caution because of the reservations concerning the reliability of the estimates of average levels and of the risk assessment models; see Chapter 8 of the HEI-AR Report.

[b] The "average" levels for the sampled schools and buildings represent the means of building averages for the buildings reviewed. The "high" levels for schools and public buildings, shown as 10 times the average, are approximately equal to the average airborne levels of asbestos recorded in approximately 5% of schools and buildings with asbestos-containing materials (ACM). If the single highest sample value were excluded from calculation of the average indoor asbestos concentration in public and commercial buildings, the average value is reduced from 0.00020 to 0.00008 f/mL, and the lifetime risk is approximately halved.

[c] The concentration shown (0.1 f/mL) represents the permissible exposure limit (PEL) proposed by the OSHA. Actual worker exposure, expected to be lower, will depend on a variety of factors including work practices and use and efficiency of respiratory protective equipment.

From the HEI-AR Panel Report.[3] With permission.

and can reach the exposures experienced by industrial workers.[2,5] However, little other information was available to the Literature Review Panel on past or current exposures of custodial and maintenance workers. Using a risk estimate based on data from asbestos industry workers and an exposure level of 0.1 f/mL, the permissible exposure level proposed by OSHA, the panel estimated 2000 premature cancer deaths per million exposed workers for 20 years of exposure (Table 3). Therefore, the panel emphasized that appropriate controls should be used to minimize asbestos emissions and exposures, and the potential risk of exposure of custodial and maintenance workers should be the primary determinant of any remedial action.

During the last decade, it has been widely recognized that the asbestos removal is generally not a good option for management of ACM in buildings and that in-place

Table 4   Average Airborne Asbestos Concentrations in an Office Building with an Operations and Maintenance Program

Sampling period	Number of samples	Mean fiber concentration	
		Fibers ≤5 μm f/mL	Fibers of all sizes s/L
8/85	35	—[a]	9.70
10/86	94	0.00001	0.85
12/86	30	0.00003	0.47
3/87	31	ND[b]	0.32
6/87	31	ND	0.71
9/87	23	ND	0.22
12/87	30	ND	0.37
4/88	29	0.00003	0.48
11/88	25	0.00034	4.96
Total, Mean	328	0.00004	1.89

[a] Data not available.
[b] ND = Not detected.

From Supplement to the HEI-AR Panel Report.[4] With permission. Data provided to HEI-AR by McCrone Environmental Services, Inc., Norcross, GA. Samples analyzed using the direct method by transmission electron microscopy. Data analyzed by HEI-AR.

management is the preferable option.[6] O&M programs for in-place management of ACM and for control of asbestos exposures are being widely implemented in buildings. An O&M program may be viewed as an administrative framework that prescribes the application of specific work procedures for activities that may disturb or damage ACM, dust, or debris; the work procedures include engineering controls, worker protection, and clean up and disposal procedures.[3] However, despite the widespread use, as well as the availability of documents that provide practical guidelines,[7] little information is available on the effectiveness of O&M strategies for controlling asbestos exposures.

Until recently, relatively little information was available to assess the effectiveness of O&M programs. We have recently analyzed data for airborne asbestos fiber concentrations for 394 samples (191 area and 203 personal) collected during 106 jobs that were part of an O&M program at a hospital.[4,8] These data were provided by H⁺GCL, Inc., a Boston based firm that was responsible for development and management of the program over an 18-month period during the late 1980s. All samples were analyzed using phase contrast microscopy. Analyses of the data show that the average airborne concentration for personal samples was 0.11 f/mL and that for the area samples was 0.02 f/mL (Table 5). When the data were used to calculate 8-h TWA concentrations for personal samples, 95% were below 0.1 f/mL and 99% were below 0.2 f/mL; 0.1 and 0.2 f/mL are the current and previous permissible exposure limits established by OSHA.[9]

Table 5   Airborne Fiber Levels in an Asbestos Operations and Maintenance Program at a Hospital Facility[a]

Sample type	Number of samples f/mL	Arithmetic average	Percentile	Concentration f/mL
Personal	203	0.11	50th	0.06
			90th	0.24
			95th	0.42
			Max	0.84
Area	190	0.02	50th	0.01
			90th	0.03
			95th	0.05
			Max	0.42
Background[b]	45	0.01		

[a] These samples were collected over an 18-month period in several buildings at a hospital facility in northeastern U.S. Samples analyzed using the phase contrast microscope (PCM). Note that all concentrations are for fibers longer than 5 μm and that the PCM does not allow a discrimination between asbestos and other fibers. For further details, see Shaikh et al. (1994).[8]

[b] These samples were collected before the O&M program began.

The results of our analysis of airborne fiber levels in the hospital O&M program are consistent, within the errors typically associated with such measurements, with other recent studies in which worker exposures during asbestos O&M programs have been reported. For example, Perkins et al.[10] analyzed data on asbestos levels in the vicinity of glovebags during asbestos removal; the average of 430 samples taken at the glovebag was 0.037 f/mL, whereas the average of 386 samples collected 15 to 25 feet from the glovebag was 0.028 f/mL. Corn et al.[17] have analyzed approximately 500 area and personal air samples obtained in five buildings. Although arithmetic averages are not given, the information provided (the 8-h TWA and average work time) can be used to calculate the average fiber levels during electrical/plumbing activities, cable pulling, and HVAC work; these averages were 0.0075, 0.0059, and 0.0074 f/mL, respectively. In Washington during a variety of O&M activities 916 area samples obtained in a large office building had an average value of 0.0059 f/mL.[11] Kaselaan and D'Angelo[12] summarized air monitoring data for 178 samples obtained in 5 commercial buildings; the averages in the buildings ranged from 0.011 to 0.073 f/mL. Price et al.[13] have compiled and analyzed data submitted to OSHA by a number of organizations; the mean of 1227 samples was 0.045 f/mL (however, note that Price and colleagues have reported volume-weighted averages).

CONCLUSION

Exposure of the general public to asbestos in buildings has been an area of concern during the last 20 years. Interpretation of the available data on ambient levels of asbestos in buildings is difficult because the samples were not collected in a statistically representative group of buildings, and also because of other limitations. However, the available evidence suggests that concentrations of indoor airborne asbestos in buildings in which ACM is well maintained are very low, and the added lifetime risk of cancer for general building occupants from such exposures is very small. Therefore, even though ACM in some buildings may represent a higher potential asbestos exposure hazard, there does not appear to be sufficient risk to justify arbitrary removal of intact ACM in well-maintained buildings.

The strategies for asbestos management have matured substantially during the past 20 years. It is now generally accepted that removal of ACM is often not the best option, and the thrust of asbestos control strategies has shifted to in-place management, generally with an O&M program. In an O&M program, appropriate work practices, which have evolved during the last 10 to 12 years, play the dominant role in keeping worker exposure low. Indeed, in earlier studies where stringent measures (in comparison with contemporary practices) were not taken,[2,5] or in recent studies where ACM disturbance has been simulated without precautions,[14] considerably higher levels of asbestos fiber exposure in the building environment have been demonstrated.

The only data to evaluate the effectiveness of O&M programs are from a small number of recently reported studies, all are based on data obtained from a few, non-randomly selected buildings. (Some of these and other data were presented at a workshop on asbestos O&M programs organized by HEI-AR in March 1993.[15]) The available data appear to suggest that the use of appropriate work practices, probably aided by being a part of an overall asbestos O&M program, can meet the goal of minimizing worker exposure and avoiding environmental contamination. Indeed, it appears that worker exposures can be generally kept below the OSHA mandated permissible exposure limit (PEL) under O&M programs.

If the above data are taken to support the usefulness of O&M programs, they also underscore the need for continued, strict vigilance in minimizing asbestos exposures and managing ACM in place. It is possible that the biggest challenge to managing asbestos in buildings is related to implementation of O&M procedures and adherence to applicable regulations. Roger Morse has speculated that only some 10% of U.S. buildings with ACM have O&M programs in place.[16] If true, this suggests that there may be substantial numbers of buildings where uncontrolled exposure is taking place and where workers and general building occupants are being exposed to unnecessary and preventable risk from asbestos.

ACKNOWLEDGMENT

The HEI-AR is an independent, nonprofit organization, which is funded jointly and equally by the EPA and a number of private parties (including asbestos manufacturers, real estate groups, and insurance companies) that have an interest in asbestos in public and commercial buildings. This work was funded with partial support by the EPA under Assistance Agreement X-815497 granted to HEI-AR; however, this work has not been subjected to the agency's peer and administrative review and therefore may not necessarily reflect the views of the agency and no official endorsement should be inferred. Similarly, this work has not been reviewed by the private parties that support HEI-AR and therefore may not reflect the views or policies of these parties and no endorsement should be inferred.

REFERENCES

1. Environmental Protection Agency (EPA): EPA Study of Asbestos Containing Materials in Buildings: A Report to Congress. U.S. Environmental Protection Agency, Washington, D.C. (1988).
2. Sawyer, R.N.: Asbestos Exposure in a Yale Building. *Environ. Res.* 13:146-169 (1977).
3. Health Effects Institute-Asbestos Research (HEI-AR): Asbestos in Public and Commercial Buildings: A Literature Review and Synthesis of Current Knowledge. Literature Review Panel, Health Effects Institute-Asbestos Research, Cambridge, MA (1991).
4. Health Effects Institute-Asbestos Research (HEI-AR): Asbestos in Public and Commercial Buildings: Supplementary Analyses of Selected Data Previously Considered by the Literature Review Panel. Health Effects Institute-Asbestos Research, Cambridge, MA (1992).
5. Paik, N.W.; Walcott, R.J.; Brogan, P.A.: Worker Exposure to Asbestos During Removal of Sprayed Material and Renovation Activity in Buildings Containing Sprayed Material. *Am. Ind. Hyg. Assoc. J.* 44(6):428-432 (1983).
6. United State Environmental Protection Agency (EPA): Managing Asbestos in Place: A Building Owner's Guide to Operations and Maintenance Programs for Asbestos-Containing Materials. EPA Pub. No. 20T-2003. EPA, Washington, D.C. (1990).
7. National Institute of Building Sciences (NIBS): The NIBS Guidance Manual: Asbestos Operations and Maintenance. 1992, Washington, D.C. NIBS.
8. Shaikh, R.A.; Satterfield, M.H.; Kinney, P.: Airborne Fiber Levels in a Hospital Operations and Maintenance Program. *Appl. Occup. Environ. Hyg.* 9(11):811-824 (1994).
9. Occupational Safety and Health Administration, Department of Labor (OSHA): Occupational Exposure to Asbestos; Final Rule. Federal Register 59(153) 40963-41162, August 10, 1994.
10. Perkins, J.L.; Rose, V.E.; Cleveland, M.S.: Analyses of PCM Asbestos Air Monitoring Results for a Major Abatement Project. *Appl. Occup. Environ. Hyg.* 7(1):27-32 (1992).
11. Kinney, P.; Satterfield, M.H.; Shaikh, R.A.: Airborne Fiber Levels During Asbestos Operations and Maintenance Work in a Large Office Building. *Appl. Occup. Environ. Hyg.* 9(11): 825-835 (1994).
12. Kaselaan & D'Angelo Associates, Inc.: Worker Exposure to Asbestos Resulting from Small-Scale Short Duration Operations: An Historical Sample. Prepared for Real Estate's Environmental Action League. Kaselaan & D'Angelo Associates, Inc., Haddon Heights, NJ (1991).
13. Price, B.; Crump, K.S.; Baird, III, E.C.: Airborne Asbestos Levels in Buildings: Maintenance Worker and Occupant Exposures. *J. Exp. Anal. Environ. Epidemiol.* 2(3):357-374 (1992).

14. Keyes, D.L.; Chesson, J.; Ewing, W.M.; et al.: Exposure to Airborne Asbestos Associated with Simulated Cable Installation Above a Suspended Ceiling. *Am. Ind. Hyg. Assoc. J.* 52(11):479-484 (1990).

15. Health Effects Institute-Asbestos Research (HEI-AR): Proceedings of Workshop on Asbestos Operations and Maintenance in Buildings. Lippmann, M.; Samet, J.M.; Shaikh, R. *Guest Editors. Appl. Occup. Environ. Hyg.* 9, 11 (1994).

16. Morse, R.: In: Summary, Synthesis and Outlook. Proceedings of Workshop on Asbestos Operations and Maintenance in Buildings. *Appl. Occup. Environ. Hyg.* 9(11):918-923 (1994).

17. Corn, M.; McArthur, B.; Dellarco, M.: Asbestos Exposures of Building Maintenance Personnel (Abstract). Presented at the HEI-AR Workshop on Asbestos Operations and Maintenance in Buildings; March 8-9, 1993. Health Effects Institute-Asbestos Research, Cambridge, MA.

Index

Index

355

107   90% of time indoors ← see 197 for a different statement.

127-8 Summary of Part II -

151 respiration rate   10,000 - 20,000 Liters/day

171 little information on bioaerosols

179   Good section on Endotoxins ← 315

223-4 Overview of neurotoxicity (Part III)

291   Risk assessment - 4 questions ⟶ 315 !!

293 The low dose problem. → 315 also pg. 295

297 A, B₁, B₂ Carcinogen Classification explained - 315.
309 - Chap 14 much good epidemiology stuff - 315
321 Classic paper on causality ← get a copy of this !